S0-AKY-071

Steric Exclusion Liquid Chromatography Of Polymers

CHROMATOGRAPHIC SCIENCE

A Series of Monographs

Editor: JACK CAZES
Fairfield, Connecticut

Volume 1: Dynamics of Chromatography (out of print)
J. Calvin Giddings

Volume 2: Gas Chromatographic Analysis of Drugs and Pesticides
Benjamin J. Gudzinowicz

Volume 3: Principles of Adsorption Chromatography: The Separation of Nonionic Organic Compounds (out of print)
Lloyd R. Snyder

Volume 4: Multicomponent Chromatography: Theory of Interference (out of print)
Friedrich Helfferich and Gerhard Klein

Volume 5: Quantitative Analysis by Gas Chromatography
Joseph Novák

Volume 6: High-Speed Liquid Chromatography
Peter M. Rajcsanyi and Elisabeth Rajcsanyi

Volume 7: Fundamentals of Integrated GC-MS (in three parts)
Benjamin J. Gudzinowicz, Michael J. Gudzinowicz, and Horace F. Martin

Volume 8: Liquid Chromatography of Polymers and Related Materials
Jack Cazes

Volume 9: GLC and HPLC Determination of Therapeutic Agents (in three parts)
Part 1 edited by Kiyoshi Tsuji and Walter Morozowich
Part 2 and 3 edited by Kiyoshi Tsuji

Volume 10: Biological/Biomedical Applications of Liquid Chromatography
Edited by Gerald L. Hawk

Volume 11: Chromatography in Petroleum Analysis
Edited by Klaus H. Altgelt and T. H. Gouw

Volume 12: Biological/Biomedical Applications of Liquid Chromatography II
Edited by Gerald L. Hawk

Volume 13: Liquid Chromatography of Polymers and Related Materials II
Edited by Jack Cazes and Xavier Delamare

Volume 14: Introduction to Analytical Gas Chromatography: History, Principles, and Practice
John A. Perry

Volume 15: Applications of Glass Capillary Gas Chromatography
Edited by Walter G. Jennings

Volume 16: Steroid Analysis by HPLC: Recent Applications
Edited by Marie P. Kautsky

Volume 17: Thin-Layer Chromatography: Techniques and Applications
Bernard Fried and Joseph Sherma

Volume 18: Biological/Biomedical Applications of Liquid Chromatography III
Edited by Gerald L. Hawk

Volume 19: Liquid Chromatography of Polymers and Related Materials III
Edited by Jack Cazes

Volume 20: Biological/Biomedical Applications of Liquid Chromatography IV
Edited by Gerald L. Hawk

Volume 21: Chromatographic Separation and Extraction with Foamed Plastics and Rubbers
G. J. Moody and J. D. R. Thomas

Volume 22: Analytical Pyrolysis: A Comprehensive Guide
William J. Irwin

Volume 23: Liquid Chromatography Detectors
Edited by Thomas M. Vickrey

Volume 24: High-Performance Liquid Chromatography in Forensic Chemistry
Edited by Ira S. Lurie and John D. Wittwer, Jr.

Volume 25: Steric Exclusion Liquid Chromatography of Polymers
Edited by Josef Janča

Other Volumes in Preparation

Chem
Sep/ae

Steric Exclusion Liquid Chromatography Of Polymers

edited by

Josef Janča

Institute of Analytical Chemistry
Czechoslovak Academy of Sciences
Brno, Czechoslovakia

MARCEL DEKKER, INC. New York and Basel

Library of Congress Cataloging in Publication Data
Main entry under title:

Steric exclusion liquid chromatography of polymers.

(Chromatographic science ; v. 25)
Bibliography: p.
Includes index.
1. Polymers and polymerization--Analysis. 2. Gel permeation chromatography. I. Janča, Josef, [date]. II. Series.
QD381.S7 1984 547.7 83-20938
ISBN 0-8247-7065-X

MARCEL DEKKER, INC.
270 Madison Avenue, New York, New York 10016

Current printing (last digit):
10 9 8 7 6 5 4 3 2 1

PRINTED IN THE UNITED STATES OF AMERICA

FOREWORD

Natural polymers have been an integral part of human existence since people first appeared on earth. Our physical well being is due to the polymers in our bodies; we are clothed, sheltered, and fed with polymers; we use polymers to record our activities.

But, although organic chemistry of small molecules became understood more than a century ago, polymers were not at the forefront of organic chemical research until the early part of the twentieth century. It was only then that the physical methods needed to characterize macromolecular substances started to become available. For the most part, these methods were initially confined to examination of bulk, unfractionated polymers and did not provide any insight into the nature of a polymer's molecular weight distribution (MWD), a key property needed to understand completely the behavior of polymers. Although some crude fractionation methods did become available, they did not lend themselves to routine application, because they were experimentally difficult and time-consuming and they yielded results that were less than adequate for a thorough understanding of a polymer's MWD.

For routine, day-to-day needs, scientists relied heavily on guessing the MWD by measuring and comparing two or more bulk properties, such as light-scattering, viscosity, and osmotic pressure.

But, as synthetic polymers became increasingly important and a synthetic polymer industry developed, particularly during and after World War II, the need to know more about a polymer's MWD became progressively acute; only through a detailed knowledge of the MWD could polymer properties be understood and could materials be tailored for specialized applications.

The "need-to-know" syndrome often serves as the catalyst or driving force for scientific and technological development. Scientific innovation and creativity provide the means. Gel permeation chromatography (GPC), often referred to as steric exclusion chromatography (SEC) today, developed from a need to know the MWDs and other properties of polymers and through the innovation and creativity of many workers throughout the world.

Porath and Flodin [1] demonstrated steric exclusion of proteins in columns packed with cross-linked dextran particles. Although this was not the first such report, it did conclusively demonstrate the significance and potential utility of SEC. Moore's pioneering work [2] with cross-linked styrene-divinylbenzene copolymers led to the development and introduction of a commercial gel permeation chromatograph in the mid-1960s. For the first time, we could routinely generate data within two or three hours that allowed us to *see* the MWD for a polymer instead of having to *guess* it.

Development of polymers and processes for producing them moved ahead by leaps and bounds. We were able to make polymers having desired properties rather than "take what comes along and hope for the best!" If a particular batch didn't come up to our expectations, we could *see*, via its MWD, what went wrong and we were better able to correct the situation for production of a new batch of polymer.

But, GPC has come a long way during the past twenty years. The mechanism of GPC has been put on a firmer footing, calibration techniques have been refined, and MWD-property relationships have been established. Methods for refining and using GPC data are now routine and GPC has become the method of choice for determining MWDs of polymers.

All of this has been brought about because of the catalytic need to know and through the innovative efforts of many scientists. Albert Einstein once said, "If I have been able to look into the future, it is because I stood on the shoulders of giants." Continuing this tradition of scientific development through the insights of one's predecessors, Dr. Janča, has succeeded in bringing together several of the most innovative giants in this field. Each has contributed significantly. In this monograph, we are able to *look* over their shoulders to see where future advances will lie.

Jack Cazes

REFERENCES

1. J. Porath and P. Flodin, Nature, *183*, 1657, 1959.
2. J. C. Moore, Gordon Conference on Separations, New London, NH, Aug., 1962.

PREFACE

Steric exclusion liquid chromatography is a separation method whose historically significant publications date from the late 1950s and early 1960s. During that time experimental observations that separation of substances of different molecular size occurs in chromatographic columns packed with porous gels began to proliferate rapidly. The further development of this separation method and the successive recognition and discovery of its fundamental principles have also been reflected in the numerous names by which the method has been denoted up to the present time. Gel filtration and gel filtration chromatography are among the oldest. Later terms reflect an effort to express the principal mechanism by which the separation is effected. Gel exclusion chromatography, restricted diffusion chromatography, size separation chromatography, and gel permeation chromatography are examples. Finally, the most recent terms focus on the working procedure. This group includes the names "gel chromatography" and "molecular sieve chromatography."

Each of these terms has certain advantages; however, each also suffers from numerous drawbacks. Since the fundamental mechanism of this chromatographic method is very complicated and hence cannot be completely rendered into one term, the choice of the most suitable one is, to a certain extent, arbitrary.

Steric exclusion liquid chromatography—in abbreviated form, steric exclusion chromatography (SEC)—classifies this method of liquid chromatography based on the type of interactions governing the separation; it is the term that has been used most frequently in the recent literature and is the one used in this book. Particular chapters deal with the most important fundamental aspects of SEC, starting with theoretical principles and going on to discuss calibration methods, correction of axial dispersion, evaluation and interpretation of data, prospects and limitations of SEC, possible interactions between the substance under separation, solvent and column packing, and major applications. A prime goal was to correlate fundamental separation mechanisms and secondary interactions in SEC with possibilities of studying and analyzing structure and behavior of macromolecules in solution and, vice versa, to point out how the properties of macromolecules in solution influence (or are associated with) separation processes. An effort was made to present a critical review of the existing state of the above aspects of SEC in polymer analysis and to attract attention to important problems that have not been solved adequately yet, and thus to contribute to the understanding of the comple: character of SEC. It was not our endeavor to provide a detailed description of experimental technique or practical instructions for carrying out experiments. We believe that existing books treat this aspect of the subject sufficiently and present complete information both on techniques of SEC in particular and on liquid chromatography in a wider view.

In polymer analysis, SEC is understood as a mere separation method from the results of which any quantitative conclusions can be drawn by using some sort of calibration. Universal calibration, which is closely associated with the question of effective dimensions of macromolecules in solution and with the model of separation mechanism, is here the most important problem. The resulting chromatogram includes information on molecular weight distribution of the polymer under study and also on zone broadening. This fact must be taken into consideration and corrected for when experimental data are evaluated quantitatively. The mechanism of pure steric exclusion is sometimes complicated by additional interactions. In some instances they can be used actively for the separation itself or can provide important additional information on the character and structure of the polymer. An understanding of the influence of experimental variables on SEC results permits an explanation of its separation mechanisms in further detail and, as a consequence of this, also a more profound knowledge of physicochemical parameters of the analyzed polymers. Real possibilities of SEC for determining molecular weight distribution, structure, solution properties, chemical composition, and additional characteristics not only of polymers but also of column packings were demonstrated by typical examples of analyses of the most important groups of synthetic polymers. Finally, attention was directed to possible expansion of applicability of SEC and to coupling this separation method with other methods and instruments to make it possible to analyze the architecture of polymers in more detail.

By far, not all fundamental problems of SEC have unambiguously been solved as yet. Even in experimental work some controversial findings exist. It was our intention not to evade expressing and describing conflicting opinions. We consider this kind of scientific freedom as a basic principle of challenging and creative inspiration.

The present book is intended primarily for researchers in the field of macromolecular chemistry, physics, and chromatography who have a deeper interest in the study of polymer characteristics by means of chromatography and have already had certain minimal knowledge of and experience in chromatography.

This monograph has been brought into existence through the idea of Dr. Maurits Dekker and the permanent encouragement of Dr. Jack Cazes. Professor Claude Quivoron has significantly contributed by his suggestions as to the final content of the book. Critical comments of Dr. Josef Novák concerning the first chapter are gratefully acknowledged. The assistance of Dr. Milan Rinchenbach has contributed to the smooth organizational and technical work in preparing the manuscript.

Josef Janča

CONTRIBUTORS

JOHN V. DAWKINS* Institute of Materials Science, University of Connecticut, Storrs, Connecticut

ARCHIE E. HAMIELEC Department of Chemical Engineering, Institute for Polymer Production Technology, McMaster University, Hamilton, Ontario, Canada

JOSEF JANČA Institute of Analytical Chemistry, Czechoslovak Academy of Sciences, Brno, Czechoslovakia

SADAO MORI Department of Industrial Chemistry, Faculty of Engineering, Mie University, Tsu, Mie, Japan

SVATOPLUK POKORNÝ Laboratory of Instrumental Analytical Methods, Institute of Macromolecular Chemistry, Czechoslovak Academy of Sciences, Prague, Czechoslovakia

CLAUDE QUIVORON Laboratory of Macromolecular Physicochemistry, École Supérieure de Physique et de Chimie Industrielle, Université Pierre et Marie Curie, Paris France

BENGT STENLUND Laboratory of Polymer Technology, University of Åbo Akademi, Turku/Åbo, Finland

CARL-JOHAN WIKMAN Laboratory of Polymer Technology, University of Åbo Akademi, Turku/Åbo, Finland

*Permanent address: Department of Chemistry, Loughborough University of Technology, Loughborough, Leicestershire, England

CONTENTS

Steric Exclusion Liquid Chromatography Of Polymers

— 1 —

PRINCIPLES OF STERIC EXCLUSION LIQUID CHROMATOGRAPHY

JOSEF JANČA / *Institute of Analytical Chemistry, Czechoslovak Academy of Sciences, Brno, Czechoslovakia*

1.1. INTRODUCTION

The first successful separation of a synthetic polymer by the method of steric exclusion liquid chromatography (SEC) appears to be that described in Vaughan's paper [1]. He succeeded in separating low molecular weight polystyrene in benzene on a weakly cross-linked polystyrene gel. In 1964, Moore [2] prepared polystyrene gels cross linked with divinylbenzene, on which he separated high molecular weight polystyrene while employing tetrahydrofuran as a solvent. In the subsequent years, a very quick development of this method, as to both the elucidation of its principles and the extension of its applications to solve various problems of macromolecular chemistry, could be observed.

Although the knowledge about SEC as well as its practical applications have reached a high level, there are still many theoretical as well as practical problems that have not, for various reasons, been solved satisfactorily. In many cases, especially with regard to theoretical models, there are many contradictory concepts. There are in the literature a number of reviews [3-32] and monographs [33-37] (the enumeration of which is not at all complete) in which various concepts of SEC separation are evaluated. These concepts always reflect certain evolutionary stages of the knowledge and understanding of the fundamental physicochemical processes that bring about separation in SEC and/or accompany such separation.

From the initial concepts of SEC separation, based on very simplified models which, however, sometimes differ from each other rather substantially in emphasizing a certain, precisely specified physicochemical process to explain separation, a much more complex view of SEC has now been arrived at. Separation and the accompanying phenomena are looked upon as a group of participating processes that can, to varying extents, share in the resultant effects. From this point of view we shall approach the description of the basic mechanisms and models of separation in SEC. Owing to the relatively complex nature of the basic separation phenomena in SEC, it has not yet been possible to integrate the participating mechanisms and the complexity of the concept of SEC into a consistent theoretical model. Although the main goal of this chapter is to provide an introduction to the problems of SEC as a separation method, simplifying concepts and imperative judgments are avoided. This will probably place more demand on creating one's own ideas and judgments, but should stimulate the endeavor after further progress in this theoretical area.

1.2. SEPARATION MECHANISMS

1.2.1. Retention

The mechanism of retention in SEC has not yet been made unequivocally clear in all details [9,26]. However, an understanding of it is very important if we are to make full use of all the possibilities that this separation method offers for the characterization and/or analysis of polymers. In a simplified view, separation by SEC can be looked upon as a specific type of distribution of the (solute) separated molecules between the solvent outside the porous particles in the column (the mobile phase) and the solvent filling the pores (the stationary phase), the cause of which is the steric exclusion of the molecules according to the relation of their sizes to that of the pores. The total volume of a packed chromatographic column, V_T, is given by the sum [38] of the total volume of the pores, V_p, the volume occupied by the matrix of the gel, V_m, and the interstitial volume, V_0, between the gel particles (the term *gel* is used for the sake of simplicity; e.g., porous glass is not gel but its functioning in separation does not differ in principle from porous silica or cross-linked polystyrene gel):

$$V_T = V_p + V_m + V_0 \qquad (1.1)$$

The retention volume V_R of a monodisperse substance with a certain molecular weight, which, to a first approximation, is the volume of the solvent (eluant) that passed through the column from the moment of sample (solute) introduction to the moment when the solute left the column in its maximum concentration, lies within V_0 and $(V_T - V_m)$. Alternatively, retention can be expressed in terms of time instead of volume, and then we speak of retention time. The largest molecules cannot enter the pores of the porous packing and pass only through the interstitial volume, so that their retention volume practically equals the volume V_0. Very small molecules (e.g., the molecules of the eluant itself) can penetrate all, or almost all, of the pores of the porous packing, and their retention volume equals the sum

$(V_0 + V_p)$. The other molecules are, according to their size, excluded from part of the pores while accommodated by another part, so that their retention volumes lie within the foregoing limiting values. Hence it follows that the retention volume V_R of a monodisperse substance can be written as

$$V_R = V_0 + K_{SEC}V_p \tag{1.2}$$

where K_{SEC} is a formal analog of the distribution coefficient. Equation (1.2) was derived by Martin and Synge [39] for partition chromatography. However, this equation does not say anything about the mechanism by which the solute is distributed between the phases. As long as the solute molecules are excluded totally from the pores, $K_{SEC} = 0$. In those cases when they can partially and/or completely permeate the pores of the gel and no other interaction with the gel surface occurs in addition to steric exclusion, $0 \leqq K_{SEC} \leqq 1$. If the distribution coefficient is larger than 1, it is certain that other interactions (e.g., adsorption) also occur that enhance retention. Although such interactions are secondary from the point of view of SEC, they can either improve or impair the resulting separation. They will be dealt with later, especially in Chapter 4. Here we are interested primarily in an exact physicochemical interpretation of the distribution coefficient K_{SEC} at the dominating mechanism of steric exclusion and in the dependence of K_{SEC} on the molecular weight and/or dimensions of the solute molecules.

1.2.2. Equilibrium Geometrical Models

In the early history of the SEC method, a number of theoretical papers explained the mechanism of separation by simple steric exclusion [38], that is, in the way we described it to a first approximation at the beginning of Section 1.2. This was also associated with the simplest geometrical models used to describe the process of separation. Porath [40] considered the porous gel structure to be a system of conical holes that spherical molecules could penetrate. By virtue of a simple geometrical concept, the distribution coefficient for such a system was described by the equation

$$K_{SEC} = k\left(1 - \frac{2R}{A}\right)^3 \tag{1.3}$$

where A is the maximum diameter of the cone, R an effective radius of the separated macromolecule, and k a proportionality constant. A similar model for a somewhat more complicated system of cylindrical, conical, and crevicelike pores was elaborated independently by Squire [41]. In this case, the distribution coefficient was described by a relationship similar to equation (1.3), where the proportionality constant k also involved the distribution of pores of different shapes.

A very interesting model was published by Laurent and Killander [42], who employed Ogston's [43] theory of space accessible to spheres of radius R within a gel structure constituted by randomly situated cylindrical rods of radius r_0. The resultant equation for the distribution coefficient has the form

$$K_{ave} = \exp\,[-\pi L(R + r_0)^2] \tag{1.4}$$

where L is the concentration of the rods, expressed in centimeters per cubic centimeter of the gel. For pragmatic reasons, the distribution coefficient was defined in a somewhat different form as

$$K_{ave} = \frac{V_R - V_0}{V_T - V_0} = K_{SEC}\,\frac{V_p}{V_p + V_m} \tag{1.5}$$

The Laurent and Killander model [42] based on Ogston's theory [43] is very close to the physical reality represented by the three-dimensional network of a polymeric gel swollen in a solvent, provided that the spatial density of the network is relatively low and that it is not necessary to take into account corrections consequent to the fact that two different parts of the rods cannot simultaneously occupy a common space. In such a structure, separation of macromolecules takes place according to what part of the free space is accessible to them. In principle, this model makes it possible to determine from chromatographic data the structure of the three-dimensional network. This procedure, called reverse SEC, is dealt with in detail in Chapter 5.

De Vries et al. [44] also assumed that the mechanism by which SEC separation takes place was steric exclusion and supposed the separation of macromolecules of different sizes to be due to pore size distribution in gels and/or porous silica or porous glass particles. They calculated the retention volume by the equation

$$V_R = V_0 + \int_R^{\infty} \phi(r)\, dr \qquad (1.6)$$

where $\phi(r)\, dr$ is the total volume of pores whose radii lie within r and r + dr, and R is an equivalent radius of the macromolecular solute. However, their experimental results, that is, the correlation of the molecular weights and V_R values of different polystyrene standards (called the calibration curve), did not comply with the course of the pore size distribution curves determined by mercury porosimetry for silica gels with different mean pore sizes.

If the distribution coefficient for an individual pore [45], depending on the ratio of the pore size to the size of the macromolecule, is taken into consideration, the equilibrium concentration can be expressed by the relationship

$$K(R,r)_{SEC} = \frac{C_p}{C_0} \qquad (1.7)$$

where the concentrations C_p and C_0 refer to the space inside the pores and the interstitial volume, respectively. According to the classical definition,

$$V_R = V_0 + V_p \frac{C_p}{C_0} \qquad (1.8)$$

and therefore

$$V_R = V_0 + \int_R^{r_{max}} K(R,r)_{SEC}\phi(r)\, dr \qquad (1.9)$$

Hence equation (1.9) expresses the fact that the retention volume of a separated macromolecule is determined by both the accessibility of a part of the volume of the individual pores and the size distribution of the entire system of pores in the gel. However, the true value of C_p remains problematic. In the definition equations (1.7) and/or (1.8), C_p means an average value. In the interstitial volume surrounding the gel particle the average value of C_0 has its defined meaning, as the concentration in such a delimited space can justifiably be considered as being practically homogeneous. However, this condition may not apply inside the pore; moreover, within the constrained space a macromolecule can be supposed not to behave as it would in a free space; that is, its actual activity there may differ from the activity it has in the interstitial volume. This problem will be discussed later.

Although different gels display almost identical courses regarding the dependence of V_R on the molecular weight, porosimetric measurements reveal varying cumulative distributions of the inner pores. This means that there is no simple function correlating the volume and/or the size of the separated macromolecules and the size and distribution of pores, as expressed by equation (1.9). According to Berek et al. [46], the shape of pores, which can be inferred from the ratio of the area and volume of the inner pores, is also essential.

Kubin [47] has solved this problem theoretically. He assumed a complex structure of the pore as represented in Figure 1.1. Because of random fluctuations in the cross section of a real pore, the probability that a macromolecule with a diameter smaller than the mean diameter of the pore will penetrate to a certain depth in the gel particle increases as the effective diameter of the macromolecule decreases. A solution is possible provided that there is a certain accessible surface layer of given thickness and an unaccessible core of the spherical gel particle. To describe the dynamic processes, Kubin employed Lapidus and Amundson's [48] partial differential mass-balance equations. By means of Laplace-Carson's transformation he obtained an analytical solution for the statistical moments of an elution curve, and for the distribution coefficient K_{SEC} he obtained the relationship

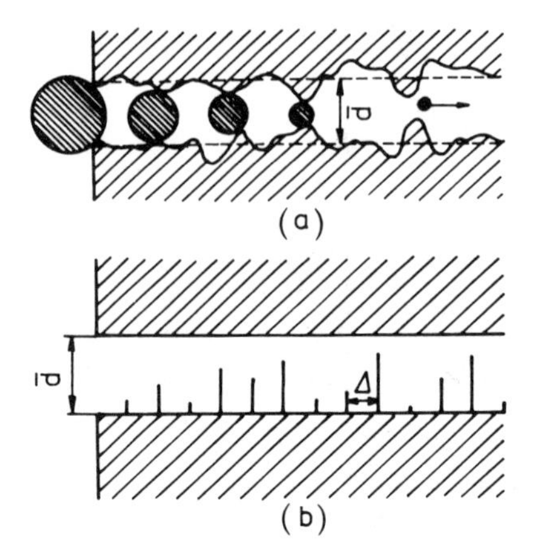

Figure 1.1. Penetration of macromolecules of different size into the pore of a complicated shape: (a) cross section of the real pore; (b) schematic idealization of the pore. (From Ref. 47.)

$$K_{SEC} = \frac{C_p}{C_0}(1 - \rho^3) \tag{1.10}$$

where ρ is the ratio of the radius of the impermeable core to that of the gel particle. The second term in parentheses on the right-hand side of equation (1.10) can have the meaning of a contribution of restricted diffusion, which will be dealt with below. Hence the distribution coefficient K_{SEC} in equation (1.10) is expressed as the product of a contribution of steric exclusion [or the equilibrium distribution coefficient (C_p/C_0)] and a contribution of restricted diffusion. The formal decomposition of K_{SEC} into the two contributions above, expressed by

$$K_{SEC} = K_{rd}K_{se} \tag{1.11}$$

has already been suggested by Yau [49] to substantiate differences between distribution coefficients measured under equilibrium stationary conditions and dynamically by a chromatographic experiment.

As will be shown later, the influence of restricted diffusion is not considered an important contribution governing the retention. Kubin [47] himself mentioned that $(1 - \rho^3)$ has the meaning of restricted diffusion and thus limited the importance of his own theoretical model. But when attributing to this expression (without mathematical formulation as well as Kubin did) the meaning of total exclusion of the solute from the inner core, this difficulty is overcome.

Van Kreveld and Van den Hoed [50] correlated K_{SEC} values obtained experimentally with those calculated theoretically and arrived at a very good agreement. They proceeded from the experimental finding, ascertained by electron microscopy, that particles of porous silica consisted of agglomerates of microparticles of spherical shape. For a description of the relationship between the number of randomly spaced uniform spheres, the volume of the individual spheres, and the volume fraction of the void space (created inside the microparticle agglomerates), they employed the equation derived by Haller [51]. Further, they supposed that distribution equilibrium was established instantaneously under the chromatographic conditions and arrived at the relationship

$$\ln \phi'(R) = \left(\frac{r + R}{r}\right)^3 \ln \phi \tag{1.12}$$

where ϕ is the volume fraction of the pores in a gel constituted by spherical microparticles of diameter r and $\phi'(R)$ is the volume fraction of the space accessible to macromolecules of radius R. It also holds that

$$K_{SEC} = \frac{\phi'(R)}{\phi} \tag{1.13}$$

Although this model has been derived considering a real structure of gel particles, as found by electron microscopy, it belongs to the category of steric exclusion geometrical models.

Later, these authors studied the mass-transfer phenomena in SEC [52]. Carrying out very precise experimental measurements, they obtained extraordinarily good agreement with a theoretical model similar to Kubin's model described above [47].

1.2.3. Restricted Diffusion Models

All the models of separation by SEC described above have been based on the assumption that all the processes involved take place at diffusional equilibrium--in other words, that the time necessary to transport the separated macromolecules between the phases is much shorter than the time the chromatographic zone spends at a certain place in the column. A mechanism of restricted diffusion has been suggested by Ackers [53]. According to this model, migration is restricted for those macromolecules that can penetrate the pores but are not small enough to move there as freely as they do in the interstitial space between the gel particles. Hence their effective diffusion coefficients inside the pores are smaller than in the interstitial volume between the gel particles. By virtue of this model, Ackers [53] explained the differences in K_{SEC} values measured under static and dynamic conditions on gels with large pores (i.e., in a region of high molecular weights). For a quantitative description of the dependence of K_{SEC} on the dimensions of the separated macromolecules and the pores, he employed Renkin's [54] equation:

$$K_{SEC} = \left(1 - \frac{R}{r}\right)^2 \left[1 - 2.104\frac{R}{r} + 2.09\left(\frac{R}{r}\right)^3 - 0.95\left(\frac{R}{r}\right)^5\right] \tag{1.14}$$

A similar model of retention in SEC, based exclusively on restricted diffusion, has been elaborated by Yau and Malone [55].

When comparing equation (1.14) with equation (1.3) or (1.4), and/or with (1.9), (1.10), and (1.12), it can be seen that the mutual dimensions of the porous gel structure and the separated macromolecules occur in all cases, regardless of whether the model of separation is described by steric exclusion or by restricted diffusion. That is, equation (1.14) has been derived with the assumption that the ratio of the diffusion coefficient of the macromolecule inside the pore to that in the interstitial space, D_p/D, is a function of the ratio R/r (see, e.g., Figure 1.2). The formal relationships that result are therefore very similar, although the philosophies of the physical concepts of various models are quite different.

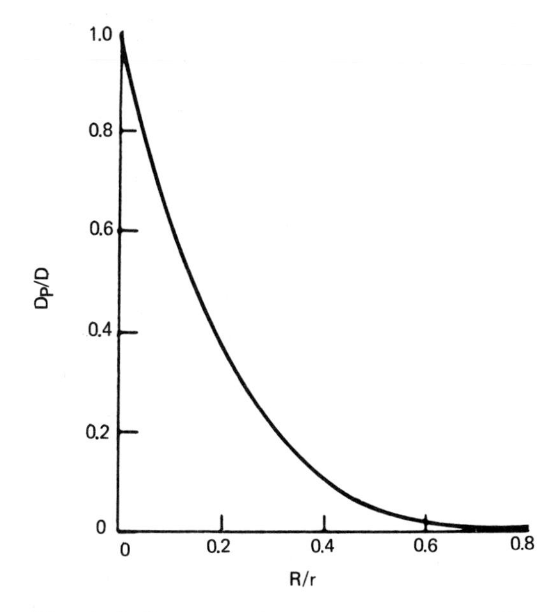

Figure 1.2. Dependence on the ratio of macromolecule radius R to pore radius r of the ratio of diffusion coefficient D_p in a pore to D in a free solution of a spherical macromolecule. (Reprinted with permission from Biochim. Biophys. Acta, *59*, 137, 1962.)

This consideration will, of course, apply to other models of SEC separation as well, as discussed below.

The model of restricted diffusion is at variance with the fact that the effect of the mobile-phase flow rate on K_{SEC} values is insignificant within wide limits. Despite Ackers' [53] attempts at an explanation of this contradictory fact by the existence of a stagnant solvent layer on the surface of gel particles that diminishes as the flow rate increases and thus makes it possible for the separated macromolecules to diffuse more easily into the pores, this model is not generally accepted now, especially because the differences in K_{SEC} values measured statically and dynamically can be explained in other ways, as will be shown below. Moreover, the total volume of the stagnant layer is very small compared with V_p and therefore cannot bring about an appreciable compensation of such changes in V_R upon flow rate changes. A practical insensitivity of V_R to changes in temperature is also considered to be evidence of the invalidity of the theoretical model based on restricted diffusion. However, the possibility that the model of the mechanism of restricted diffusion suggested by Ackers [53] may apply in part in SEC separation [56] is not out of the question. Smith and Kollmansberger [57] observed a dependence of the retention volume on the inverse of the diffusion coefficient with low molecular weight solutes. This finding apparently supports, to a certain extent, the hypothesis regarding the validity of the mechanism of restricted diffusion. Similar observations have been made by Andrews [58] in connection with the study of proteins.

Yau et al. [36] have recently rejected quite resolutely the effect of restricted diffusion on retention in SEC. As will be discussed in Sections 1.3.1 and 1.3.2, restricted diffusion in pores undoubtedly affects the dispersion of the chromatographic zone. However, restricted diffusion may partially influence retention, as the latter was found experimentally [59,60] to depend on the mobile-phase flow rate.

1.2.4. Hydrodynamic Models

A very interesting model of separation by flow was described by Di Marzio and Guttman [61-64] and, later, by Verhoff and Sylvester [65]. According to this model, the porous structure of gel particles can be approximated as a system of cylindrical capillaries of different diameters. The macromolecular coil is approximated as a spherical particle moving down the capillary by the action of flow. As there is a parabolic flow velocity profile in the capillary under the conditions of laminar Newtonian flow, and because an isolated spherical macromolecule, being subject to Brownian migration, can get no nearer the wall than to a distance determined by its radius, this macromolecule is moved down the capillary at a velocity larger than the mean flow velocity of the fluid (see Figure 1.3). The larger the particle in a capillary of a given diameter, the larger will be the velocity of its relative movement. With a cylindrical capillary, the retention volume is given by the relationship

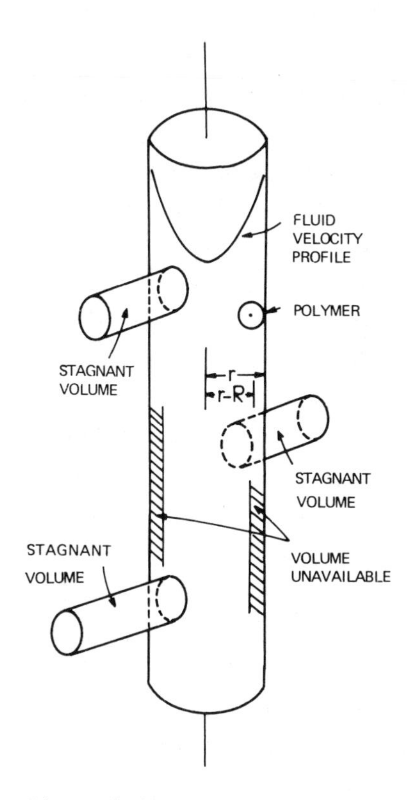

Figure 1.3. Representation of flow down a capillary. The fluid velocity profile causes polymer separation by virtue of the unavailable volume. (From Ref. 62.)

$$V_R = \frac{\pi \ell r^2}{2[1 - (1 - R/r)^2/2 - \gamma'(R/r)^2]} \tag{1.15}$$

The factor γ' is a function of the density distribution in the macromolecule (for a homogeneous spherical particle, $\gamma' = 2/3$) and ℓ is the length of the capillary. In the limit of low flow velocities it holds for the distribution coefficient

$$K_{SEC} = \left(1 - \frac{R}{r}\right)^2 \qquad R \leqslant r \tag{1.16}$$

and

$$K_{SEC} = 0 \qquad R \geqslant r \tag{1.17}$$

Obviously, the authors [61-64] have also taken into account the pore size distribution, and in that case they described the retention volume by equation (1.9). This model is of particular interest because of the premise that separation due to flow can occur both in the porous structure of gel particles (i.e., in the volume V_p) and, to a certain extent, in the interstitial volume V_0. Evidence that this is indeed so under certain conditions was provided by Small [66], who succeeded in separating particles of a polystyrene latex on packings with large pore diameters as well as on nonporous column packings. He called this method *hydrodynamic chromatography*. Mori et al. [67] have demonstrated the separation by flow using soluble high molecular weight polystyrenes on nonporous column packings.

There are several facts that speak against a general acceptance of this SEC separation mechanism. First, this model assumes that the percolating fluid can pass through the system of pores (i.e., the volume V_p) in gel particles, which, in most cases, especially with polymeric gels, will not be true. It can be supposed that only the largest pores, particularly those situated near the surface of gel particles, may comply with this condition, but the other pores would hardly do that. Further, with regard to the Hagen-Poiseuille equation,

$$q = \frac{\pi r^4 \Delta P}{8\eta \ell} \tag{1.18}$$

the flow rate q through a capillary of length ℓ at a given pressure drop ΔP and the viscosity η of the fluid is proportional to the fourth power of the radius of the capillary. Therefore, small and medium-size pores (about 10-100 nm), those which are operative in SEC separation, have practically no flow-through capability. With gel particle sizes of about 10-100 μm, the ratio of the mean effective diameter of the capillaries in the interstitial volume to that of the pores is about 10-1000. Only the largest pores, those 250-1000 nm in diameter, can exhibit appreciable flow-through at the usual chromatographic conditions. Hence it follows that only with molecular weights in the region of about $M > 10^6$ is it possible to expect a more significant contribution of the mechanism of hydrodynamic separation by flow. Thus, although this model successfully explains separation in hydrodynamic chromatography, particularly that of particles, it is only of limited importance in the SEC of polymers.

1.2.5. Stochastic Models

The stochastic model of separation was developed by Carmichael [68-70] with the use of Giddings and Eyring's [71] and McQuarrie's [72] theories. The retention volume or retention time is a function of the probability with which the macromolecule will penetrate the pore of the gel. The probability that a macromolecule will pass from the interstitial volume into a pore is given by the relationship

$$\text{Prob (mobile state} \longrightarrow \text{trapped state)} = \Lambda_1 \, \Delta t + o(\Delta t) \qquad (1.19)$$

and the probability of the reverse process is given by

$$\text{Prob (trapped state} \longrightarrow \text{mobile state)} = \Lambda_2 \, \Delta t + o(\Delta t) \qquad (1.20)$$

where Λ_1 is the rate constant of penetration multiplied by the concentration of the pores ($\Lambda_1 = \Lambda_1' C$) a macromolecule of given dimensions can permeate, Λ_2 is the rate constant of elution of the macromolecule out of the pore into the interstitial volume, and $o(\Delta t)$ is the probability of occurrence of two or more steps within a time interval $(t, t + \Delta t)$. Equations (1.19) and (1.20) result in a Poisson distribution $P(t)$ of the time a macromolecule spends in a pore and in the interstitial volume

$$P(t) = [\exp\,(\Lambda_2 t - \Lambda_1 t_0)] \left(\frac{\Lambda_1 \Lambda_2 t_0}{t}\right)^{0.5} I_1 (4\Lambda_1 \Lambda_2 t t_0)^{0.5} \qquad (1.21)$$

where I_1 is a modified Bessel function of the first order and t_0 is the retention time of totally excluded macromolecules (i.e., those that cannot enter any pores). The retention time t_R is then given by the equation

$$t_R = \frac{t_0 \Lambda_1}{\Lambda_2} + t_0 \qquad (1.22)$$

This model has been developed with the assumption that all the pores in the gel have equal dimensions. The parameter Λ_1 is proportional to the probability P_e with which the macromolecule will be captured in the pore. It is further supposed for P_e that

$$P_e = \text{Prob } (R < r) \tag{1.23}$$

where R is again the effective size of the macromolecule. Under the given conditions, P_e is a function of the molecular weight distribution of the separated polymer, and Λ_2 is independent of the size of the macromolecule. A comparison of this theoretical model with published experimental data has rendered good agreement.

In his later work [73,74], Carmichael also took into consideration the pore size distribution in the gel. He supposed at first a set of discrete values $\Lambda_{11}, \Lambda_{12}, \ldots, \Lambda_{1n}$ and $\Lambda_{21}, \Lambda_{22}, \ldots, \Lambda_{2n}$ to exist for a set of n different pore sizes and modified equation (1.22) as follows:

$$t_R = \frac{t_0 \sum_{i=1}^{n} a_i \Lambda_{1i}}{\Lambda_{2i}} \tag{1.24}$$

A Gaussian pore size distribution G(r) was assumed and therefore

$$\sum_{i=1}^{n} a_i = 1 \qquad a_i = G(r) \tag{1.25}$$

provided that G(r) is given by the equation

$$G(r) = \frac{1}{\sqrt{2\pi}\sigma} \exp\left[\frac{-(r_i - \bar{r})^2}{2\sigma^2}\right] \tag{1.26}$$

where σ is the standard deviation of the Gaussian distribution and $\bar{r}$ is the mean radius of the given set of pores. Among other findings, the theoretical considerations and calculations resulted in the conclusion that the width of the pore size distribution in the gel distinctly affected the course of the dependence of the retention time on the dimensions of the macromolecules [e.g., the unperturbed root-mean-square end-to-end distance $(\langle r_0^2 \rangle)^{0.5}$], but that the course of this calibration dependence was not very sensitive to changes in σ (see Figure 1.4). Later, Kubin [75] arrived at a similar conclusion

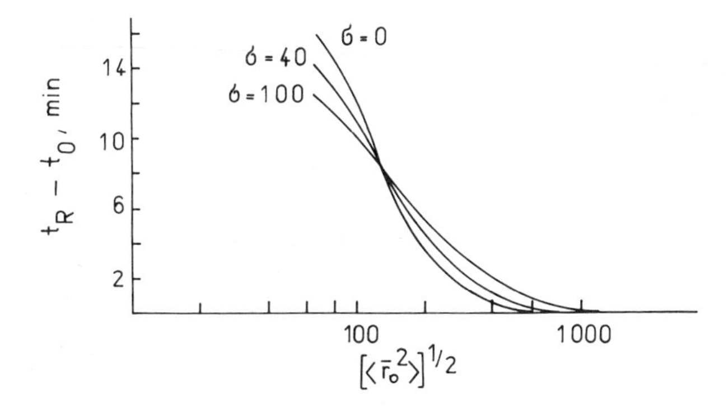

Figure 1.4. Effect on retention time of the variance of a Gaussian pore size distribution with a given mean radius. (From Ref. 73.)

by virtue of model calculations. Hence there follows the important conclusion that great caution must be used when evaluating the results of inverse SEC used to study the distribution of pores in a gel. Compared to the case above, much better agreement of experimental results with the theoretical model, even with nonflexible biopolymers [74], was obtained by Carmichael [73]. Yau et al. [36], who have evaluated this model rather critically, consider the analogy between the rate of adsorption (for which the theory was elaborated originally [71, 72]) and the parameter of macromolecular dimensions to be arbitrary and confusing.

1.2.6. Thermodynamic Models

As is generally true of separations based on phase equilibria, in SEC solute transport usually takes place at conditions near equilibrium. This means that equilibrium concentration conditions represent a good approximation of the actual concentration distribution of the solute in the system. This applies for a change in the Gibbs free energy at a constant temperature T and pressure P,

$$dG = dH - T\,dS \qquad (1.27)$$

where H and S are enthalpy and entropy, respectively. Separation processes take place in open systems which render an exchange of mass possible and for which it holds that

$$dG = \left(\frac{\partial G}{\partial n_i}\right)_{T,P,n_{j\neq i}} dn_i \tag{1.28}$$

Equation (1.28) describes an infinitesimal change in the Gibbs free energy of the system, brought about by a flux of the mass of component i, dn_i being the number of moles of component i entering the system. The chemical potential of component i is defined as

$$\mu_i = \left(\frac{\partial G}{\partial n_i}\right)_{T,P,n_{j\neq i}} \tag{1.29}$$

The chemical potential of component i is related to the activity of this component, a_i, in a given system by the known relationship

$$\mu_i = \mu_i^* + RT \ln a_i \tag{1.30}$$

where R is the molar ideal-gas constant and μ_i^* is the chemical potential of component i in a standard state. When defining the activity of component i (at the temperature and pressure of the system) in both phases as $a_i = \gamma_i^* C_i$ (where γ_i^* and C_i are a rational activity coefficient and the concentration of component i, respectively), normalizing the activity coefficient so that $\gamma_i^* = 1$ as $C_i \rightarrow 0$, and provided that the solute concentrations in both phases range within a region of the validity of Henry's law, it is possible to write, in place of equation (1.30),

$$\mu_i = \mu_i^\circ + RT \ln C_i \tag{1.31}$$

where μ_i° is the chemical potential of component i in a hypothetical standard state of infinite dilution at the temperature and pressure of the system. At equilibrium, dG = 0, and it is possible to write

$$\mu_i^m = \mu_i^s \tag{1.32}$$

On substituting from equation (1.31) into equation (1.32), we obtain

$$\frac{C_{si}}{C_{mi}} = \exp\left(\frac{\Delta\mu_i^\circ}{RT}\right) \tag{1.33}$$

The ratio of equilibrium concentrations in the left-hand side of equation (1.33) is identical to the coefficient defined by equation (1.7), which can be expressed as

$$K_{SEC} = \exp\left(\frac{-\Delta\mu_i^\circ}{RT}\right) \tag{1.34}$$

By a formal decomposition and rearrangement of equation (1.34), we obtain

$$K_{SEC} = \exp\left(\frac{-\Delta H^\circ}{RT}\right) \exp\left(\frac{\Delta S^\circ}{R}\right) \tag{1.35}$$

The first quantitative thermodynamic interpretation of the dependence of a general distribution coefficient on the solute molecular weight was given by Brönsted [76], who derived the equation

$$K = \exp\ (kM) \tag{1.36}$$

where k is a constant characterizing, among other things, the chemical properties of the solute and the phases, and M is the molecular weight of the solute. A rigorous application of this concept to SEC leads to the conclusion that the distribution coefficient will depend on both the chemical character of the solute molecules and that of the solvent, and on the matrix constituting the porous structure of gel particles. Marsden [77] attributed solute-gel interactions (i.e., either adsorption or incompatibility) to the enthalphic term. Dawkins and Hemming [78] consider the enthalpic term of the product on the right-hand side of equation (1.35) as a distribution coefficient (K_p) the value of which is unity provided that steric exclusion is the only type of interaction in separation (i.e., $\Delta H^\circ = 0$). If adsorption and/or partition of the solute takes place, $K_p > 1$ (ΔH° is negative), and if there is incompatibility between the solute and gel, the K_p values range between zero and unity (i.e., ΔH° is positive). According to this concept, Dawkins and Hemming supposed the entropic term on the right-hand side of equation (1.35) to be the distribution coefficient K_D, which represents pure steric exclusion. It acquires values between zero and unity as discussed above, and in accord with experimental findings, is independent of temperature. Hence the distribution coefficients K_P and K_D can be specified as

$$K_P = \exp\left(\frac{-\Delta H^\circ}{RT}\right) \tag{1.37}$$

$$K_D = \exp\left(\frac{\Delta S^\circ}{R}\right) \tag{1.38}$$

If $K_P \neq 1$, nonexclusion interactions operate in separation; these interactions are discussed in detail in Chapter 4. Here we shall deal only with cases in which $K_P = 1$ and therefore $K_D = K_{SEC}$--cases in which separation is based on entropic effects only, in other words, in which solely steric exclusion phenomena apply. A slight dependence of V_R on temperature, sometimes found experimentally, can be explained as being due to the dependence of the effective dimension of the separated macromolecules or of the porous network on temperature; otherwise, it indicates enthalphic solute-pore interactions.

In a series of papers, Casassa [79-84] and Giddings et al. [85, 86] developed very exact theoretical models which agreed well with the foregoing thermodynamic considerations, while respecting the diversity in the shape of both the pores and the separated macromolecules.

The starting point of Casassa's theory is random flight statistics [79], describing the probability $P_n(x)$ of finding at a point x an nth step by a macromolecular chain the beginning of which is in the cavity center. For a mathematical description he used the differential equation

$$\frac{\partial P_n(x)}{\partial n} = \frac{b^2}{6} \nabla^2 P_n(x) \tag{1.39}$$

where b^2 is the mean-square step length. By a rather complicated mathematical procedure he arrived at expressions of K_{SEC} for different simple geometrical pore shapes, as functions of the pore dimensions r and the mean-square radius of unconfined linear chain $(R^2)_{lin}$, the latter being given by

$$(R^2)_{lin} = \frac{Nb^2}{6} \tag{1.40}$$

where N is the total number of segments in the chain of the macromolecule. For spherical, cylindrical, and slab-shaped pores, respectively, he obtained the following equations for the distribution coefficient K_{SEC}:

$$K_{SEC} = \frac{6}{\pi^2} \sum_{m=1}^{\infty} \frac{1}{m^2} \exp\left[-m^2\pi^2\left(\frac{R}{r}\right)^2\right] \qquad (1.41)$$

$$K_{SEC} = 4 \sum_{m=1}^{\infty} \frac{1}{\beta_m^2} \exp\left[-\beta_m^2 \left(\frac{R}{r}\right)^2\right] \qquad (1.42)$$

$$K_{SEC} = \frac{8}{\pi^2} \sum_{m=1}^{\infty} \frac{1}{(2m+1)^2} \exp\left[-\frac{(2m+1)^2\pi^2}{4}\left(\frac{R}{r}\right)^2\right] \qquad (1.43)$$

β_m being the roots for $I_{1,0}(\beta) = 0$, where $I_{1,0}$ is a Bessel function of the first kind and of zero order. Hence the distribution of a macromolecular solute between the phases is governed by the loss of conformational entropy [79,86]. For the limiting case when the macromolecule penetrating a pore is much smaller than the dimensions of the pore (i.e., when K_{SEC} approaches unity), it was possible to formulate a generalized equation for K_{SEC} [80]:

$$K_{SEC} = 1 - 2\Lambda\psi(f)f^{1/2}\left(\frac{nfb^2}{6r^2}\right)^{1/2} + O\,\frac{nb^2}{6r^2} \qquad (1.44)$$

where Λ has a value of 1, 2, and 3 for a slab, cylinder, and sphere, respectively. There holds for a linear macromolecule

$$\psi(1) = \psi(2) = \pi^{1/2} \qquad (1.45)$$

and for a branched macromolecule with f chains

$$\psi(f) = \left(\frac{f}{\pi}\right)^{1/2} \int_0^{\infty} \operatorname{erf}\left(t^{1/2}\right)^{f-1} e^{-t}\, dt \qquad (1.46)$$

where t is the time and therefore

$$N = nf \qquad (1.47)$$

The relationships derived for the distribution coefficients apply to equilibrium conditions. Such equilibrium conditions can be implemented experimentally by mixing dry porous material with a solution of the polymer and measuring the increase in the concentration of the polymer in the supernatant. This will be dealt with in detail in Section 1.2.7. There remains a question of how far a real chromatographic experiment approaches the equilibrium conditions. Casassa and Tagami [80] have discussed this problem with reference to the stochastic model of Carmichael [68]. The process of trapping the solute macromolecules in the pores and releasing them back to the interstitial volume is repeated many times in the chromatographic column during elution. Under such conditions, the retention volume can be described by the equation

$$V_R = V_0 + K_{SEC}V_p\left(1 - \frac{3}{2\bar{N}}\right) \tag{1.48}$$

Thus equation (1.48) characterizes quasiequilibrium in the column, as the expression in parentheses on the right-hand side actually specifies the bias between a static and a dynamic experiment. The larger the number of trappings per molecule, $\bar{N} = \Lambda_1 t_0$, the smaller the difference between data calculated using equations (1.48) and (1.2). Obviously, the smaller the flow rate of the solvent through the column, the closer the process of separation will approach a state of equilibrium.

Equation (1.40) and the subsequent relationships have been derived under the assumption that the linear polymer is dissolved in a thermodynamically poor solvent (i.e., under theta conditions). It holds at such conditions that $(R^2)_{lin}$ is related directly to the hydrodynamic volume of the macromolecule, defined by equation (2.25). This implies that Casassa's model also theoretically substantiates the validity of the foregoing theory and justifies, at least under theta conditions, the use of the empirically determined [87] universal calibration parameter $[\eta]M$ ($[\eta]$ is the intrinsic viscosity of the polymer in a given solvent at a given temperature, and M is its molecular weight), which is discussed in detail in Chapter 2. In accordance with Casassa [82], the applicability of this universal calibration parameter can be extended to involve good solvents to a fairly good approximation, at the expense of a loss of elegance.

However, this approximation is possible only as long as macromolecules of similar type [84] (e.g., flexible chains, linear as well as with a limited degree of branching, and/or rigid macromolecules, etc.), within a limited range of molecular weights, are correlated. In a good solvent, a flexible macromolecule inside a pore is smaller than when it is in the interstitial space, and the expansion of the macromolecule on its transition from the pore to the interstitial volume manifests itself by a slight enthalpy change. Regarding equations (1.37) and (1.38) as well as the preceding discussion, K_p will apparently be less than 1 in this case. As the dimensions of the macromolecule inside the pore are smaller than those outside, the respective K_D value will be somewhat larger than that corresponding to the dimensions in free space. This brings about a certain compensation. Of course, the facts above somewhat modify the stringent limitation of the mechanism of pure steric exclusion only to cases where $\Delta H^\circ = 0$. With low-degree cross-linked polymer gels (e.g., cross-linked polysaccharide Sephadex) the latter model of separation may not be quite correct [80]. It is possible that some models of restricted diffusion or of differential transport into the pores and out of them are more suitable for the description of SEC separation in such cases. However, a fundamental problem sets in when actual determination of V_0 and V_p is required. Totally excluded macromolecules are, for the same steric reasons, also excluded from a part of V_0 around the surface of the column packing particles up to a distance given by the effective radius of the macromolecules. Considerations similar to those involved in the determination of V_p by SEC of small but finite-size molecules also apply here.

Giddings et al. [85] employed the statistical mechanics of both rigid and flexible macromolecules to express K_{SEC} for porous networks with uniform pore shapes and for statistical systems of pores. For rigid spherical macromolecules and for slab-shaped, spherical, and rectangular pores, respectively, they obtained the relations

$$K_{SEC} = 1 - \frac{2R}{r} \tag{1.49}$$

$$K_{SEC} = \left(1 - \frac{2R}{3r}\right)^3 \tag{1.50}$$

$$K_{SEC} = \left[1 - \frac{2R}{r(1 + P_r)}\right]\left[1 - \frac{2RP_r}{r(1 + P_r)}\right] \tag{1.51}$$

where P_r is the ratio of the lengths of the longer and shorter sides of a rectangular pore. The rotation of spherical rigid macromolecules need not be considered. For rigid macromolecules with rotational symmetry (e.g., rod-shaped ones), Giddings et al. [85] have determined that it is the mean projection length of solute macromolecules along various axes, $\bar{L}$, that constitutes the effective dimension (i.e., the quantity governing separation). They derived explicit expressions of $\bar{L}$ for different macromolecular geometries. This dimension is essentially identical to the effective dimension according to Casassa's theory. In fact, there is a distribution of pore sizes, so Giddings et al. approximate the porous structure by a system of surfaces with random location and orientation in space. They have chosen this model with regard to the fact that steric exclusion is essentially a surface-overlap phenomenon. Such a porous structure is shown schematically in Figure 1.5. The resultant expressions of K_{SEC} for a random-plane pore/spherical solute system and a random-plane pore/general rigid solute system, respectively, are

$$K_{SEC} = \exp\left(\frac{-2R}{r}\right) \tag{1.52}$$

$$K_{SEC} = \exp\left(\frac{-\bar{L}}{r}\right) \tag{1.53}$$

The theoretical dependence of K_{SEC} on $\bar{L}/r$ for different types of porous networks is shown in Figure 1.6.

Casassa [81] reviewed critically different separation mechanisms in SEC and pointed out the fact, already mentioned in Section 1.2.3, that despite different physical concepts of separation processes, the individual formal dependences of K_{SEC} on the dimensions and shapes of the macromolecules and pores are very similar. On the basis of a critical analysis he prefers the equilibrium models to those of restricted diffusion and hydrodynamic separation by flow. Carmichael

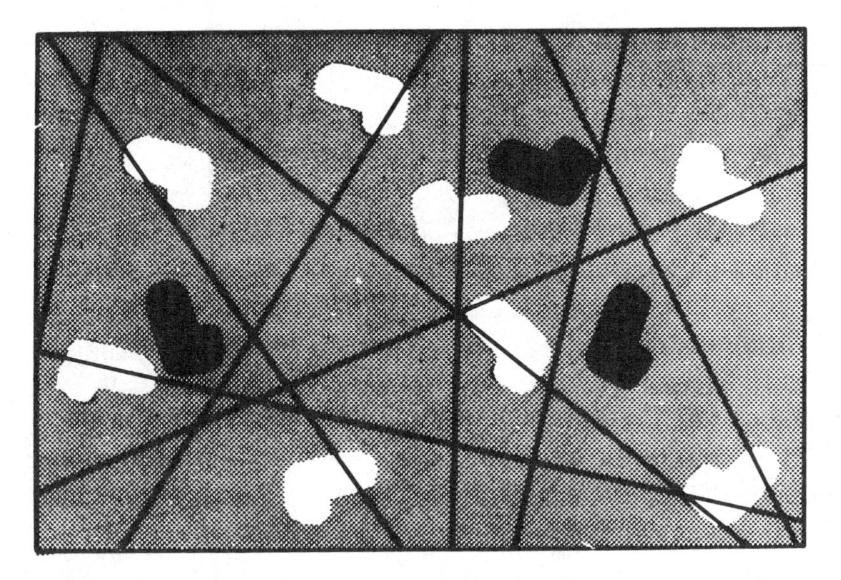

Figure 1.5. Steric exclusion of macromolecules (unshaded bodies cut by random planes creating a pore network) and permeation of macromolecules (shaded bodies between the random planes). (From Ref. 85.)

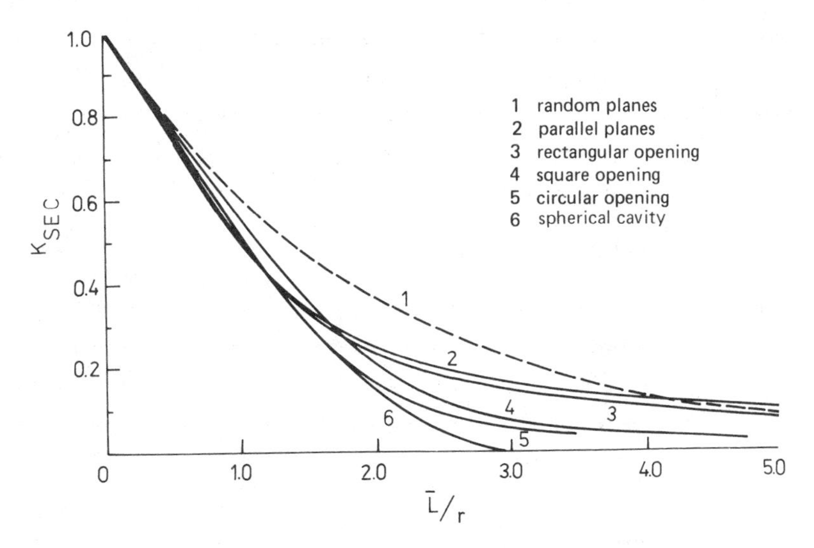

Figure 1.6. Dependence on the dimensionless ratio $\bar{L}/r$ of the distribution coefficient K_{SEC} for thin rods of length $\bar{L}$. (From Ref. 85.)

[70] compared his stochastic model with Casassa's thermodynamic models [79], declared a formal agreement to exist in a number of definitions and approximations, and considered the stochastic model, elaborated on the basis of earlier theories [71,72], to be a more generalized form of the thermodynamic model.

Pouchly [88] has generalized the equilibrium theory of Casassa and Giddings, taking into account solute-gel surface interactions (i.e., adsorption). His model logically leads to relationships such as equation (4.7), by means of which other authors later interpreted the secondary mechanisms in SEC, cases in which pure steric exclusion was not the only factor operative in separation.

1.2.7. Equilibrium at Stationary Conditions

The stationary experiment, consisting of mixing dry porous material with a solution of polymer solute and measuring the solute concentration changes in the supernatant after equilibration, provides equilibrium distribution coefficients that can be correlated with the corresponding K_{SEC} values measured chromatographically. Such a correlation makes it possible to decide whether equilibrium models constitute good approximations of actual separation processes in SEC. Yau et al. [89] and Yau [49] carried out measurements using porous silica and a polymeric cross-linked gel (Styragel; styrene cross linked with divinylbenzene) under both stationary and chromatographic conditions. For a stationary system at equilibrium,

$$C_i V_i' = C_0 (V_i' - V_p) + C_p V_p \qquad (1.54)$$

where C and V' are the solute concentration and the volume of the solution, and the subscripts i and 0 refer to the initial state before mixing the components of the system and the final state after equilibrium is attained, respectively. The substitution from equation (1.7) and rearrangement results in

$$1 - \frac{C_i}{C_0} = \frac{V_p}{V_i'} [1 - K(R,r)_{SEC}] \qquad (1.55)$$

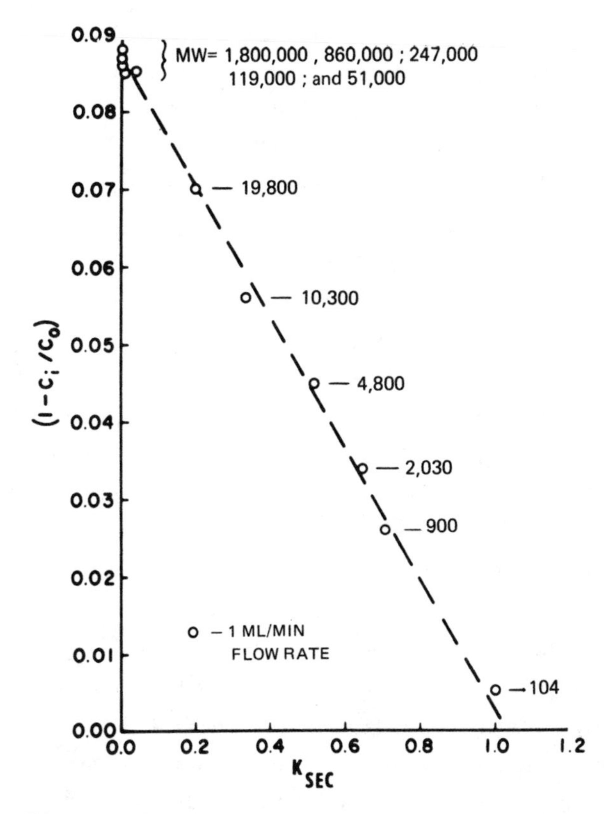

Figure 1.7. Static mixing data versus distribution coefficient K_{SEC} for various polystyrene standards in chloroform. (From Ref. 89.)

Yau et al. correlated data calculated by the expression on the left-hand side of equation (1.55) with K_{SEC} values measured chromatographically for polystyrene standards over a wide range of molecular weights. The results are illustrated in Figure 1.7. The very good linear correlation shown in Figure 1.7 constitutes a supporting argument for the equilibrium models of SEC separation. The authors have stated, in accord with Grubisic and Benoit [90], that the absolute concentration of the polymer in the solution has no effect on the solute distribution at stationary conditions. As the effective volume of flexible macromolecules in a good solvent depends on their concentration (see Chapters 2 and 4), this finding is at variance with the linear correlation represented in Figure 1.7. This problem has been studied more thoroughly by Janča et al. [91], who have proved [91] the existence of an evident concentration dependence of the distribution coefficient measured at stationary conditions. The results obtained cannot yet render an unequivocal corroboration of the validity of certain models of SEC separation [91]. Doubts arise especially due to the fact that it is impossible to eliminate various other effects that operate together with the merely entropic ones, as already stated by Casassa [80,81]. Hitherto unpublished results by Janča and Pokorný [92] have indicated a strong adsorption of polystyrene on porous silica in a stationary experiment using a theta solvent, although a corresponding dynamic experiment has not shown any remarkable adsorption. Undoubtedly, further investigation in this area is needed.

With all the models suggested so far, only interactions of the macromolecule with the porous network were considered; interactions of the individual macromolecules with each other were supposed to be zero. Such conditions may exist in a limiting case when the concentration of the macromolecules in the solution approaches zero. As long as the concentration is finite, as it is in real cases, this factor has to be taken into account. A more detailed discussion of the concentration effects in SEC separation is presented in Chapters 2 and 4. Here we quote only the results of the work by Anderson and Brannon [93], who have theoretically investigated the effect of solute concentration on solute-solute and solute-pore interactions, although the model they elaborate has the character of a complementary mechanism in SEC. If a rigid spherical particle of radius R is inside a rigid pore of radius r, the interaction of the former with the wall of the pore can be considered as equivalent to the interaction between two identical particles. In view of particle-particle interactions, the volume that is simultaneously excluded by the pore wall and the particle is considered as the excluded volume (see Figure 1.8). This excluded volume is that by which the space accessible to particle-particle interactions is reduced in the area surrounding the basic

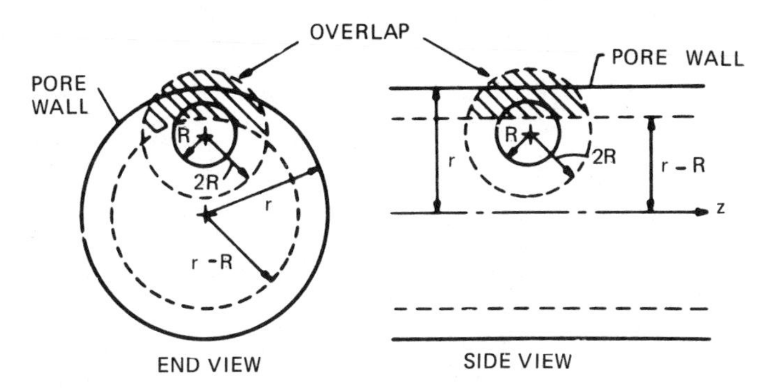

Figure 1.8. Overlap of excluded volumes. The volume excluded by the base macromolecule is a sphere of radius 2R, while exclusion due to the pore wall is the region outside the cylinder of radius r - R. The crosshatched region represents the volume from which a test macromolecule is excluded simultaneously by the base macromolecule and the pore wall. (From Ref. 93.)

particle. It means that a decrease in spherical symmetry surrounding a particle inside the pore leads to a reduction of interparticle interactions. The result of this is that particle-particle interactions in the free space outside the pore bring about a shift of the distribution coefficient toward an increase of local concentration inside the pore, which constitutes a reversed effect with respect to the exclusion of a single particle due to particle-pore wall interactions. If electrorepulsive forces between the particles are combined with purely steric interactions, the resultant effect is even more marked. Anderson and Brannon [93] have presented an exact mathematical formulation of this phenomenon, derived on the basis of statistical mechanics. The effect of solute concentration on K_{SEC} can be described by the virial expansion

$$K_{SEC} = K_{SEC(C=0)}(1 + a_1 C_\infty + a_2 C_\infty^2 + \cdots) \qquad (1.56)$$

where C_∞ is the bulk solute concentration in the interstitial volume. A theoretically calculated dependence involving the use of the first virial coefficient of equation (1.56) is shown in Figure 1.9, in which the higher coefficients of the virial expansion were neglected.

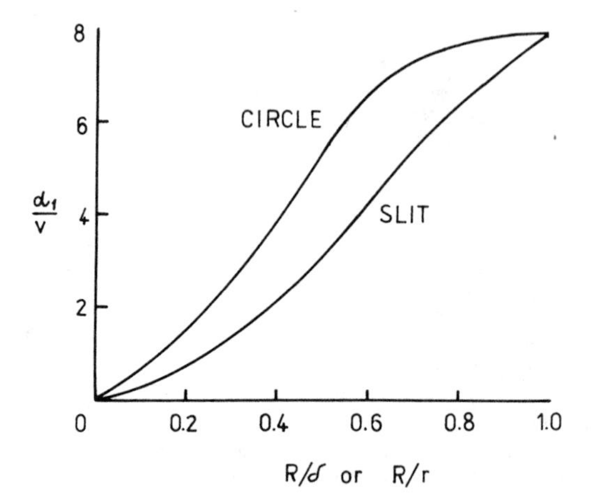

Figure 1.9. First virial coefficient of the distribution coefficient for hard sphere-hard wall interactions of spherical macromolecules of radius R; r is the radius of the circular pore, and δ is one-half the width of the slit pore. (From Ref. 93.)

In addition to the models of SEC separation discussed in the preceding sections, several other papers have been published which constitute primarily either modifications of earlier models or approaches that differ from those models only in details. They do not contribute significantly to the chronologically formed consistent conception of the retention mechanism.

1.3. ZONE DISPERSION

As shown in Section 1.2, SEC differs rather substantially from other types of chromatography by its unique mechanism of separation of solutes. On the other hand, the dynamic processes and mechanisms by which the natural tendency of the system to disperse solute zones developed on their migration down the column manifests itself are, for the most part, nearly identical in SEC and in other types of chromatography. It is due to these processes that even if an infinitesimally short pulse of a monodisperse (in view of the effective dimensions deciding about separation) solute is introduced at the column inlet in SEC, so that no separation takes place (which does not mean that this monodisperse solute is unretained), this zone is subject to broadening upon migration down the column. This dispersive process has been given many names, including band broadening, zone broadening, zone dispersion, axial dispersion, and longitudinal spreading. The term *zone dispersion* is preferred here, as we shall deal with the fundamental processes that lead to the formation of a solute-concentration distribution in the zone. This concentration distribution differs from the initial distribution of the sample that exists at the moment of injection, and it is forces of a dispersive character that constitute the cause or essence of the processes necessitating these changes.

In cases involving phenomenological aspects of these processes, such as the width of an elution curve or the shape and width of the chromatogram and/or their correction, we feel that it is not necessary to adhere strictly to a single term. Rather, it is most suitable to choose the term whose meaning is most descriptive of the given situation.

The problems of zone dispersion are described in detail in the specialized literature (see, e.g., [94], so we shall confine the discussion in this section to an explanation of the basic principles that must be known to comprehend SEC in its entirety and interpret experimental results correctly. This will be essential particularly in relation to the necessary corrections for axial dispersion, which are dealt with in Chapter 3. The problems that are specific for zone dispersion with macromolecular solutes under the conditions of SEC will be dealt with below in greater detail.

1.3.1. Plate Height Concept

It follows from the general theory of chromatography [94] that the shape of the elution curve (chromatogram) of a monodisperse solute can in most cases be well approximated by a Gaussian distribution. This chromatogram shape is the result of the action of random dispersive processes that take place in the chromatographic column during elution and bring about solute mixing. If all the quantities are expressed consistently (e.g., in terms of volume), the V_R corresponding to the position of peak maximum is equal to the first statistical moment, μ_1', of the Gaussian elution curve: that is,

$$V_{R_{max}} = \mu_1' = \frac{\int_0^{+\infty} F(V)V \, dV}{\int_0^{+\infty} F(V) \, dV} \tag{1.57}$$

where a Gaussian chromatogram F(V) is defined by the equation

$$F(V) = \frac{1}{\sqrt{2\pi\sigma^2}} \exp\left[\frac{(-1/2)(V - V_{R_{max}})^2}{\sigma^2}\right] \tag{1.58}$$

The width of the elution curve can be described by the standard deviation σ, the second power of which is called the variance, σ^2 = var (V), and equals the second central moment, μ_2, of the Gaussian curve,

$$\mu_2 = \frac{\int_0^{+\infty} F(V)(V - \mu_1')^2 \, dV}{\int_0^{+\infty} F(V) \, dV} \tag{1.59}$$

For a quantitative evaluation of the column efficiency, the height equivalent to a theoretical plate (HETP, or simply H) is used, defined by

$$H = L \frac{\sigma^2}{V_R^2} \tag{1.60}$$

The physical meaning of H follows from the random-walk theory of chromatography [94], according to which the standard deviation σ, expressed in units of length, is determined by the relationship

$$\sigma = l\sqrt{n} = \sqrt{HL} \tag{1.61}$$

where l and n are the length and number of random steps, respectively, of the individual statistical processes taking place in a column of length L. Under the terms of the random-walk theory, we can explain any movement of a solute molecule between regions (streamlines) of different velocities of the mobile phase flow within the column.

There are several partial processes that contribute to the final width of the elution curve, and with regard to their overall effect, it can be assumed that the respective second central moments (or variances) are additive: that is,

$$\sigma^2 = \sum_{i=1}^{n} \sigma_i^2 \quad (1.62)$$

In addition to H, it is possible to employ a dimensionless quantity, the number of plates N_p, to describe the efficiency of the separation system:

$$N_p = \frac{L}{H} = \frac{\mu_1'^2}{\mu_2} \quad (1.63)$$

A consistent model of chromatographic zone dispersion in SEC, based on the classical random-walk theory, was first elaborated by Giddings and Mallik [95]. They described the quantity H by the equation

$$H = \frac{B}{v} + \frac{C}{v} + \Sigma \frac{1}{(1/A_i) + (1/C_{mi}v)} \quad (1.64)$$

where v is the linear velocity of the mobile phase in the column. The first, second, and third terms on the right-hand side of equation (1.64) describe, respectively, the effect of longitudinal molecular diffusion, nonequilibrium effects in the stationary phase, and the effects of streaming and nonequilibrium in the mobile phase. The diffusion coefficient of the macromolecular solute in the stationary phase (in pores), D_p, should equal that in the mobile phase, D. However, obstruction, combined with permeation, reduces the diffusion coefficient D_p by an obstructive factor $\gamma_s = D_p/D$. Hence the quantity B in equation (1.64) is given by the relationship

$$B = 2\gamma_o D + \frac{2\gamma_s D(1 - R_s)}{R_s} \quad (1.65)$$

where γ_o is an obstructive factor of diffusion in the interstitial space, and R_s is the ratio of the solute zone velocity to the mean velocity of the mobile phase. For spherical column packing particles of diameter d_p,

$$C = \frac{1}{30} R_s(1 - R_s) \frac{d_p^2}{\gamma_s D} \quad (1.66)$$

the parameters A_i and C_{mi} being defined by the relationships

$$A_i = 2\Lambda_i d_p \tag{1.67}$$

$$C_{mi} = \frac{\omega_i d_p^2}{D} \tag{1.68}$$

The quantity A_i describes the effect of eddy diffusion and manifests itself appreciably when the contribution of nonequilibrium in the mobile phase is large, that is, at high mobile-phase flow velocities, when $1/C_{mi}v \ll 1/A_i$. At very low flow velocities $1/C_{mi}v \gg 1/A_i$ and the contribution of A_i becomes relatively insignificant. The geometrical parameters Λ_i and ω_i are defined [94] as

$$\Lambda_i = \frac{\omega_\beta^2 \omega_\Lambda}{2} \tag{1.69}$$

$$\omega_i = \frac{\omega_\alpha^2 \omega_\beta^2}{2} \tag{1.70}$$

where

$$\omega_\beta = \frac{\Delta v}{\bar{v}} \tag{1.71}$$

ω_β being the ratio of the difference between the extreme and mean velocities to the mean velocity,

$$\omega_\alpha = \frac{1}{d_p} \tag{1.72}$$

and finally,

$$\omega_\Lambda = \frac{L}{nd_p} \tag{1.73}$$

Giddings [94] distinguishes between five different kinds of A_i and C_{mi} (and, consequently, Λ_i and ω_i) according to the distances over which the effects of flow velocity differences are considered [see equation (1.71)]. These effects have been specified as transchannel, transparticle, short-range interchannel, long-range interchannel, and transcolumn. A demonstration of these effects is shown

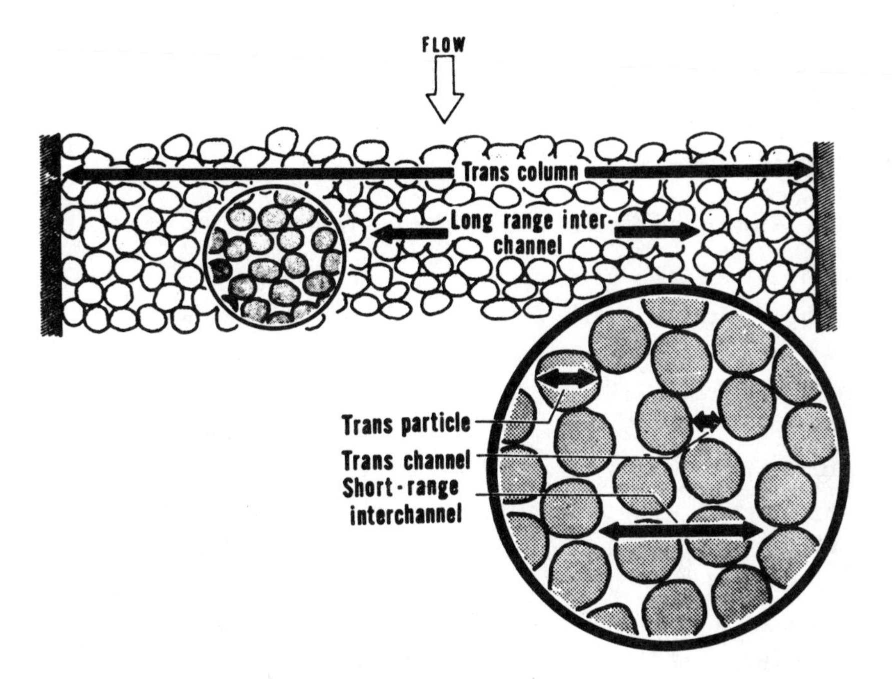

Figure 1.10. Location and distance covered by the various exchange processes between velocity extremes in the mobile phase. (From Ref. 94.)

in Figure 1.10. Giddings [94] has presented discrete values for the individual Λ_i and ω_i parameters and has specified the latter as having a strictly structural character, that is, a property that depends on the structure of the column packing, not on the nature of the mobile phase. After introducing the reduced plate height h and reduced velocity v_r,

$$h = \frac{H}{d_p} \tag{1.74}$$

$$v_r = \frac{d_p v}{D} \tag{1.75}$$

and carrying out some simplifications (e.g., $D_p/D = 2/3$) that follow from experience, Giddings and Mallik [95] obtained

$$h = \frac{4}{3R_s}\frac{1}{v_r} + \frac{1}{20} R_s(1 - R_s)v_r + \Sigma \frac{1}{1/(2\Lambda_i) + 1/(\omega_i v_r)} \tag{1.76}$$

The condition that $h \leqq 10$ at $v_r \doteq 0.2$-10 can be considered a necessary criterion regarding the column efficiency for use in SEC. Complementary information can be found in original works [94,95] and in reviews [9,96-98].

1.3.2. Experimental Verification of the Plate Height Concept

The general validity of the plate height concept has been verified by several authors. Smith and Kollmansberger [99] and Heitz and Coupek [100,101] found optimum conditions to exist at $v_r \doteq 2$ together with the existence of a single function of h versus v_r for different dimensions of column packing particles. This is at variance with the findings of De Vries et al. [44] and LePage et al. [45], who have stated larger h values for smaller particles at the same v_r, even when employing reduced quantities. This fact can be explained as being due to a lower homogeneity of settling with smaller particles of the packing, which is closely associated with the technique used to pack the column. When working at flow rates of up to 35 ml/min, Little et al. [102-104] found the effect of the flow velocity on the zone width to be smaller than that of the theoretical model. Billmeyer et al. [105] have described the solute transport processes taking place in the chromatographic bed by the equations

$$\frac{\partial C_m}{\partial t} + v\frac{\partial C_m}{\partial x} = D\frac{\partial^2 C_m}{\partial x^2} + \frac{k}{\phi_m}\left(C_s - \frac{C_m}{K_{SEC}}\right) \tag{1.77}$$

$$\frac{\partial C_s}{\partial t} = \frac{k}{\phi_s}\left(\frac{C_m}{K_{SEC}} - C_s\right) \tag{1.78}$$

for the mobile and stationary phases, C, k, and ϕ being the solute concentration in the mobile (m) or stationary (s) phase, solute mass-transfer coefficient, and volume fraction, respectively. By solving the equations for the chosen boundary conditions through Laplace transformation, they obtained the following modification of the classical van Deemter equation:

$$H = \frac{2D}{v} + \frac{2\phi_s^2 v}{\phi_m K_{SEC}}\left(1 + \frac{\phi_s}{\phi_m K_{SEC}}\right)^2 k \tag{1.79}$$

They also pointed out the necessity of carrying out corrections for zone broadening, due to the finite time involved in charging the sample.

They investigated experimentally different courses for the H versus v function [106]. In all cases their study concerned totally excluded solute macromolecules, for which $H = 2D/v$ in the absence of any mass transfer between the phases. Further, they investigated in great detail zone dispersion as a function of mobile-phase flow rate and other experimental parameters for the sample introduction and detection systems alone, for columns with nonporous particles, and for various kinds of porous packing [107-109]. The conclusions drawn from this complex work corroborated the general validity of Giddings' concept and specified the individual contributions to the resulting zone dispersion [110]. The most important contributions are molecular diffusion, resistance to mass transfer, and the velocity profile. Some other possible effects, such as those relating to viscosity and concentration, have not been included in this theory, which may account for discrepancies between theory and experiment where these occur.

Unless no specific interactions with the solute occur, the chemical nature of the column packing generally has no effect on zone dispersion [100,111-113], and the effect of pore size distribution on efficiency is minor [44,101,112-114]. The effect of particle size on efficiency is described by equations (1.66) to (1.68), and the validity of these theoretical relationships has been proven experimentally [44,45,101,108,115,116]. The particle size distribution plays some role, which obviously is associated with the homogeneity of column packing. For instance, the column efficiency can be improved by compressing the packing [117-119]. The effect of temperature involves changes in the mobile-phase viscosity and, consequently, changes in the solute diffusion coefficients. A very detailed investigation of zone dispersion in all parts of the separation system, including the connecting capillaries, detector, and so on, was carried out by Biesenberger and Ouano [120] and Ouano and Biesenberger [121, 122]. Important practical conclusions have followed from their work, showing, in agreement with theory, that zone dispersion increases with increasing length and diameter of the connecting capillaries and that the effects of solute concentration and flow velocity profile are evident. This theoretical model, and the experimental results, agree both conceptually and actually with the results discussed previously. The effect of solute molecular weight on column efficiency is related to the mobility of the macromolecules in the solution. This mobility decreases with decreasing solute diffusion coefficient, the latter being a function of molecular weight [99, 101,123-125]. An increase in the solute molecular weight leads to an increase in zone asymmetry (i.e., zone tailing) [125,126].

Horvath and Lin [127] have modified original equation (1.64) by employing the mass-transfer coefficient k_e derived by Pfeffer [128]:

$$k_e = \frac{\Omega D v^{1/3}}{d_p} \tag{1.80}$$

where Ω is a function of interparticle porosity, so that instead of the original v, $v^{1/3}$ is used in equation (1.64). This modification is based on the assumption that at the surface of the particles there is a stagnant solvent layer the thickness of which is inversely proportional to $v^{1/3}$ (see Figure 1.11). However, as a result of the existence of this layer and its variable thickness as a function of the mobile-phase flow rate, the value of V_R should (as in the model of restricted diffusion) depend on the flow rate. Horvath used his model, some other more or less empirical equations for h, and equation (1.64) to correlate experimental data. The results are shown in Figure 1.12. A comparison of the theoretical curves and experimental data shows a somewhat better correlation of the experimental data with those obtained using Horvath's equation and the empirical equation. Later the same authors generalized their model to make it applicable to the description of zone dispersion in liquid chromatography, that is, to cases in which sorption occurs [129]. Knox and McLennan [130,131] have introduced, in addition to their contributions to zone broadening discussed above, a contribution of polydispersity, expressed as [132]

Figure 1.11. Cross-sectional area in a packed column. The fluid stream in the middle of the interstitial channels is normal to the cross-sectional plane. The particles are surrounded by quasistagnant fluid. (From Ref. 127.)

$$h_{poly} = \frac{L \ln [\bar{M}_w/\bar{M}_n]}{D_2^2 V_R^2} \tag{1.81}$$

where L, D_2, $\bar{M}_w$, and $\bar{M}_n$ are the column length, slope of the calibration function [defined, e.g., by equation (2.17)], weight-average molecular weight, and number-average molecular weight [defined by equation (5.2)], respectively. This approach was later utilized by Dawkins and Yeadon [132,133] for a precise evaluation of mass-transfer dispersion and polydispersity of samples of polystyrene with very narrow molecular weight distributions. This contribution of polydispersity to the total width of the chromatogram has a certain significance, as this factor is often unjustifiably neglected when evaluating zone dispersion, although its part may be dominant. However, it must be realized that in the strict physical sense, a real separation of

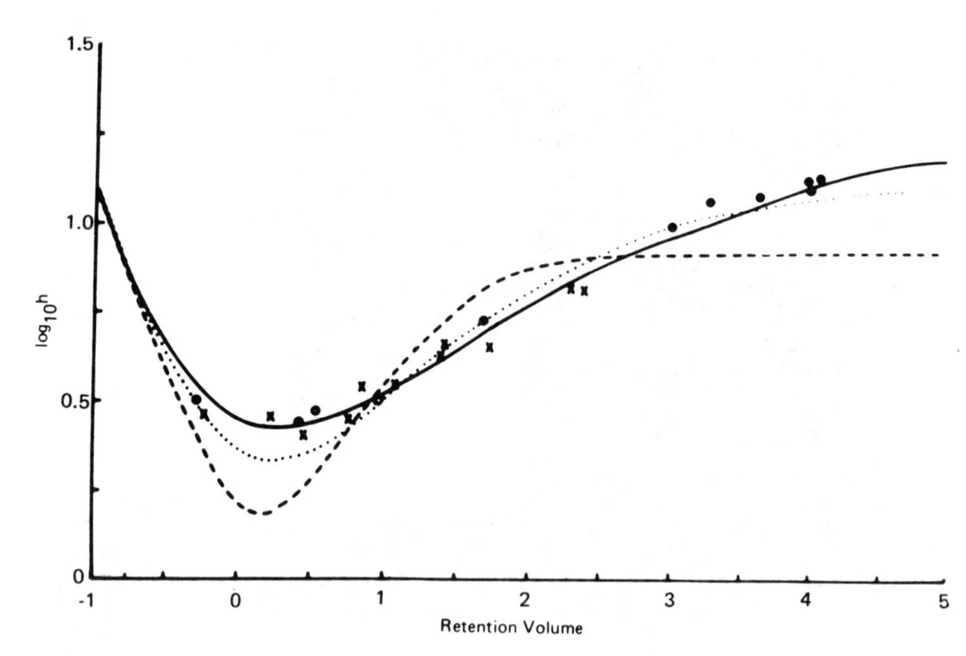

Figure 1.12. Plots of the reduced plate height versus the reduced velocity. The points represent various experimental data. ----, Giddings' equation [equation (1.76)]; ——, Horvath and Lin's equation [127]; ····, Huber's equation (not cited here). (From Ref. 127.)

polydisperse solutes (i.e., differences in retention) appears together with the zone broadening, and that the two processes differ from each other substantially. Whether evaluation of zone dispersion in the original physical sense or correction for zone dispersion is the goal, these two phenomena must not be confused.

Having calculated the statistical moments of an elution curve, the analytical form of which was derived from the solution of diffusion equations of the type (1.77) and (1.78), Hermans [134] declared the first statistical moment of the elution curve (i.e., the V_R value) to be independent of the mobile-phase flow rate, a rise of the second central moment on increasing the flow rate and/or decreasing the solute diffusion coefficient, and finally, a rise of the third central moment in the same sense as with the second one [i.e., a rise of the asymmetry of the elution curve (tailing)]. These conclusions have been corroborated by a computer simulation of the SEC process, consisting of a modeled numerical solution of the diffusion equations [28], and by analog modeling [135].

Giddings et al. [136] have substantiated decisively their theoretical considerations of the effects of the individual factors determined by the basic parameters of equation (1.76) on zone dispersion in SEC. At the same time they demonstrated experimentally some possibilities for determining the extent of the individual contributing processes. They again confirmed SEC to be based on entropy-controlled processes and, consequently, V_R to be practically independent of D, contrary to H. A problem that remains somewhat unclarified is the theoretical formulation of the value of obstructive factor γ_s and/or its correlation with experiment. Contrary to the theoretically substantiated value of $\gamma_s = 2/3$ for small dimensions of solute molecules, γ_s is a function of the ratio of the solute molecule and pore dimensions, provided that these dimensions are mutually commensurate. In this case, values as small as 0.01 by an order of magnitude have been obtained for γ_s [52,131,133,137-139].

Among the contributions to zone dispersion that have not been incorporated in the preceding theoretical considerations is an experimentally demonstrated contribution of convective transport inside pores [140] which manifests itself in the presence of the mechanism of separation by flow. The diffusional and convective transports are mutually coupled in such a case, and therefore

$$H = \left(\frac{1}{H_{flow}} + \frac{1}{H_{diff}} \right)^{-1} \tag{1.82}$$

External exclusion [141], mentioned in Section 1.2.4, may also play some role. There is also a significant contribution from differences in the hydrodynamic behavior of polymer solutions, especially those of very high molecular weight and/or those at higher concentrations. These concentration or viscosity phenomena are discussed in detail in Chapter 4.

1.3.3. Resolution

To evaluate the actual separation capability of chromatographic separation systems, the concept of resolution R was introduced, defined as

$$R_{1,2} = \frac{2(V_{R_2} - V_{R_1})}{w_1 + w_2} \tag{1.83}$$

where $w = 4\sigma$. In the case of SEC separation of two solutes having discrete molecular weight, we speak of practically complete separation if $R_{1,2} = 1$. However, to date, such a situation can be attained only in an oligomer region in SEC.

A concept of selectivity S has been introduced in SEC defined as [142]

$$S = \frac{d \log M}{dV_R} \tag{1.84}$$

The value of S is usually constant over a wide range of SEC separation (see Chapter 2). It is also possible to use the resolution index RI [142],

$$RI = \frac{M_1}{M_2} \exp\left[\frac{w_1 + w_2}{2(V_{R_1} - V_{R_2})}\right] \tag{1.85}$$

and/or the specific resolution R_{sp} [143],

$$R_{sp} = \left(\frac{w \, d \log M}{dV}\right)^{-1} \tag{1.86}$$

To compare the performance of different separation systems, it is expedient that the specific resolution be normalized; that is, quantities R_{sp}^{x} and/or $R_{sp}^{x'}$ [144] are considered,

$$R_{sp}^{x} = \left(\frac{w\sqrt{L} \, d \log M}{dV}\right)^{-1} \tag{1.87}$$

$$R_{sp}^{x'} = \frac{2(V_{R_1} - V_{R_2})}{\left(\dfrac{w_1}{M_{w_1}/M_{n_1}} + \dfrac{w_2}{M_{w_2}/M_{n_2}}\right) \log \dfrac{M_{w_2}}{M_{w_1}}} \tag{1.88}$$

the interrelation between the individual definitions being evident. The objective of increasing resolution is the improvement of separation, perfecting the separative transport while minimizing the dispersive one.

1.3.4. The Shape of the Chromatogram

The position, width, and shape of the elution curve (or chromatogram) provide information on various processes contributing to separation and dispersion, as demonstrated in previous sections, but also on molecular weight distribution of the polymer being analyzed. The latter is the required result of the analysis, and the procedures for evaluation of experimental chromatograms are described in Chapters 2, 3, and 5. Various theoretical models describing dispersion in the chromatographic column predict an almost Gaussian concentration profile for monodisperse solute in the limit of infinite time [94]. The concentration profile represents the distribution of solute concentration along the longitudinal axis of the chromatographic column at a given fixed time. The prediction of a Gaussian concentration profile is valid only for homogeneously packed columns. The elution curve, which is the response of a detector at the end of the separation system to an eluting solute as a function of the time, should be non-Gaussian in such a case. This is caused by the passage of the solute zone through the column end with continuing zone broadening. Whereas a given point of the zone has already been detected, the part of the zone following this point in time still moves in the column and is subjected to additional broadening. In other words, the broadening of the front of the zone is less than that of the back of the zone, because of the shorter time during which the front is present in the column. This effect is usually neglected, and it is supposed that the concentration profile and the elution curve are identical (see Section 1.3.1, and the following text).

The greater the spreading of the concentration profile on its movement along the unit length of the column, the more marked will be the tailing of the elution curve. Hence it is obvious that the concentration profile and the elution curve have a mutual relationship with respect to asymmetry, which is associated with the efficiency.

This effect is general for each type of chromatographic separation. But considering the efficiencies of chromatographic columns when separating low molecular weight solutes, it is often neglected. Because of the low diffusion coefficients of macromolecules in solution and consequently the lower efficiencies by an order of magnitude, it is not always correct to neglect this effect in SEC. Janča [145, 146] derived the relationships between the efficiency of the chromatographic system and the asymmetry of the elution curve based on statistical moments of the elution curve. The asymmetry can also be evaluated by slope analysis at the points of inflection of the elution curve, which is sometimes less complicated than the use of statistical moments for practical evaluations.

When the analysis of the shape of the elution curve is based on statistical moments, the efficiency, characterized by the number of theoretical plates N_p, is given by equation (1.63). By analogy, the parameter-designated asymmetry A is defined [145] as

$$A = \frac{\mu_3}{(\mu_1')^3} \tag{1.89}$$

where μ_1' is given by equation (1.57) and $_3$ is the third central moment of the elution curve:

$$\mu_3 = \frac{\int_0^{+\infty} F(V)\,(V - \mu_1')^3\,dV}{\int_0^{+\infty} F(V)\,dV} \tag{1.90}$$

Skew parameter γ can also be defined as [see also equation (3.25a)]

$$\gamma = \frac{\mu_3}{\mu_2^{1.5}} \tag{1.91}$$

($\gamma \equiv A_3$ in Section 3.3.1), μ_2 being the second central moment [see equation (1.59)].

It has been found by numerical evaluation [145] that for a Gaussian concentration profile, simple approximate relationships exist,

$$A \doteq 3N_p^{-2} \tag{1.92}$$

and

$$\gamma \doteq 2.9N_p^{-0.5} \tag{1.93}$$

relating the asymmetry of the elution curve to the efficiency of the separation system.

For a general dispersion model resulting in a non-Gaussian concentration profile, it has been found [146] that an exact solution exists: that is,

$$A = 4N_p^{-2} \tag{1.94}$$

and

$$\gamma = 4N_p^{-0.5} \tag{1.95}$$

Slope asymmetry SA defined, shown in Figure 1.13, is related to A by an approximative relation for the Gaussian concentration profile:

$$SA = 1.1\sqrt[4]{A} \tag{1.96}$$

and by a more complicated exact equation (see Ref. 146) found analytically for a general dispersion model.

All these parameters--N_p, A, γ, and SA--can be used for the analysis of the shape of the elution curve to find out not only the efficiency of the columns used in SEC but also to evaluate, for example, the homogeneity of the column packing. There were used in practice for comparison of theoretically suggested velocity profiles, formed in capillaries during the flow of polymer solutions, with experimental data [147], thus facilitating the evaluation of transport phenomena of the processes studied.

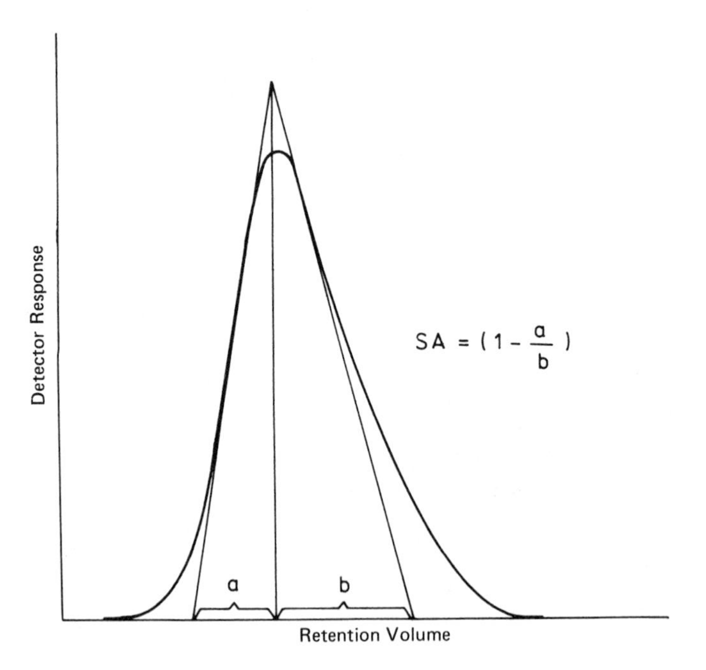

Figure 1.13. Evaluation of the slope asymmetry of the elution curve. (From Ref. 146.)

1.4. POTENTIALITIES OF SEC

As a conclusion to this survey of separation and dispersion mechanisms governing SEC, it is necessary to stress that this chromatographic method is, in view of the attained degree of knowledge, merely a separation method, itself rendering no absolute information on molecular weights and their distribution or on the structure of the polymers studied. From this point of view, the retention volume and the width and shape of an elution curve are characteristic, directly measured quantities for a given sample of the polymer analyzed; all other items of information have to be obtained by means of calibration procedures and/or through combinations of the SEC separation system with a suitable absolute detector. All of these possibilities and related problems are discussed in detail in the following chapters.

REFERENCES

1. Vaughan, M. F., Nature, *188*, 55, 1960.
2. Moore, J. C., J. Polym. Sci., B, *2*, 835, 1964.
3. Anderson, D. M. W., and Stoddart, J. F., Anal. Chim. Acta, *34*, 401, 1966.
4. Cazes, J., J. Chem. Educ., *43*, A567, 1966.
5. Cazes, J., J. Chem. Educ., *43*, A625, 1966.
6. Johnson, J. F., Porter, R. S., and Cantow, M. J. R., J. Macromol. Chem., C, *1*, 393, 1966.
7. Anderson, D. M. W., and Stoddart, J. F., Lab. Pract., *16*, 841, 1967.
8. Flodin, P., Anal. Chim. Acta, *38*, 89, 1967.
9. Altgelt, K. H., Adv. Chromatogr., *7*, 3, 1968.
10. Kranz, D., Kolloid Z. Z. Polym., *227*, 41, 1968.
11. Benoit, H., and Gallot, Z., Column Chromatogr., Int. Symp. Sep. Methods, *5*, 108, 1969.
12. Cazes, J., Appl. Polym. Symp., *10*, 7, 1969.
13. Determann, H., Adv. Chromatogr., *8*, 3, 1969.
14. Laurent, T. C., Obrink, B., Hellsing, K., and Wasteson, A., Prog. Sep. Purif., *2*, 199, 1969.
15. Mindner, K. D., and Berger, R., Plaste Kautsch., *16*, 405, 1969.
16. Mindner, K. D., and Berger, R., Plaste Kautsch., *16*, 882, 1969.
17. Bata, G. L., Hazell, J. E., and Prince, L. A., J. Polym. Sci., C, *30*, 157, 1970.
18. Bly, D. D., Science, *168*, 527, 1970.
19. Cazes, J., J. Chem. Educ., *47*, A461, 1970.
20. Cazes, J., J. Chem. Educ., *47*, A505, 1970.
21. Heitz, W., Angew. Chem., *82*, 675, 1970.
22. Lambert, A., Br. Polym. J., *3*, 13, 1971.
23. Tung, L. H., J. Macromol. Sci., C, *6*, 510, 1971.
24. Smith, W. V., Rubber Chem. Technol., *45*, 667, 1972.
25. Evans, J. M., Polym. Eng. Sci., *13*, 401, 1973.
26. Ouano, A. C., J. Macromol. Sci., C, *9*, 123, 1973.
27. Audebert, R., Analusis, *4*, 399, 1976.
28. Ouano, A. C., Adv. Chromatogr., *15*, 233, 1977.
29. Dawkins, J. V., Dev. Polym. Charact., *1*, 71, 1978.
30. Hatt, B. W., Dev. Chromatogr., *1*, 157, 1978.
31. Dawkins, J. V., Anal. Proc. (Lond.), *18*, 395, 1981.
32. Ouano, A. C., Rubber Chem. Technol., *54*, 535, 1981.
33. Determann, H., *Gel Chromatography*, Springer-Verlag, Berlin, 1967.
34. Altgelt, K. H., and Segal, L., *Gel Permeation Chromatography*, Marcel Dekker, New York, 1971.
35. Kremmer, T., and Boross, L., *Gel Chromatography*, Akadémiai, Budapest, 1979.
36. Yau, W. W., Kirkland, J. J., and Bly, D. D., *Modern Size-Exclusion Liquid Chromatography*, Wiley-Interscience, New York, 1979.

37. Fisher, L., *Gel Filtration Chromatography,* Elsevier North-Holland Biomedical Press, Amsterdam, 1980.
38. Porath, J., and Flodin, P., Nature, *183*, 1657, 1959.
39. Martin, A. J. P., and Synge, R. L. M., Biochem. J., *35*, 1358, 1941.
40. Porath, J., Pure Appl. Chem., *6*, 233, 1963.
41. Squire, P. G., Arch. Biochem. Biophys., *107*, 471, 1964.
42. Laurent, T. C., and Killander, J., J. Chromatogr., *14*, 317, 1964.
43. Ogston, A. G., Trans. Faraday Soc., *54*, 1754, 1958.
44. De Vries, A. J., LePage, M., Beau, R., and Guillemin, C. L., Anal. Chem., *39*, 935, 1967.
45. LePage, M., Beaux, R., and De Vries, A. J., J. Polym. Sci., C, *21*, 119, 1968.
46. Berek, D., Novák, I., Grubisic-Gallot, Z., and Benoit, H., J. Chromatogr., *53*, 55, 1970.
47. Kubin, M., J. Chromatogr., *108*, 1, 1975.
48. Lapidus, L., and Amundson, N. R., J. Phys. Chem., *56*, 984, 1952.
49. Yau, W. W., J. Polym. Sci., A2, *7*, 483, 1969.
50. Van Kreveld, M. E., and Van den Hoed, N., J. Chromatogr., *83*, 111, 1973.
51. Haller, W., J. Chem. Phys., *42*, 686, 1965.
52. Van Kreveld, M. E., and Van den Hoed, N., J. Chromatogr., *149*, 71, 1978.
53. Ackers, G. K., Biochemistry, *3*, 723, 1964.
54. Renkin, E. M., J. Gen. Physiol., *38*, 225, 1955.
55. Yau, W. W., and Malone, C. P., J. Polym. Sci., B, *5*, 663, 1967.
56. Haller, W., Basedow, A. M., and König, B., J. Chromatogr., *132*, 387, 1977.
57. Smith, W. B., and Kollmansberger, A., J. Phys. Chem., *69*, 4157, 1965.
58. Andrews, P., Biochem. J., *96*, 595, 1965.
59. Kolinsky, M., and Janča, J., J. Polym. Sci., Polym. Chem. Ed., *12*, 1181, 1974.
60. Glöckner, G., private communication, 1982.
61. Di Marzio, E. A., and Guttman, C. M., 9th Int. Semin. GPC, Miami Beach, Fla., 1970.
62. Di Marzio, E. A., and Guttman, C. M., J. Polym. Sci., B, *7*, 261, 1969.
63. Di Marzio, E. A., and Guttman, C. M., Macromolecules, *3*, 131, 1970.
64. Guttman, C. M., and Di Marzio, E. A., Macromolecules, *3*, 681, 1970
65. Verhoff, H. F., and Sylvester, N. D., J. Macromol. Sci., A, *4*, 979, 1970.
66. Small, H., J. Colloid Int. Sci., *48*, 147, 1974.
67. Mori, S., Porter, R. S., and Johnson, J. F., Anal. Chem., *46*, 1599, 1974.
68. Carmichael, J. B., J. Polym. Sci., A2, *6*, 517, 1968.
69. Carmichael, J. B., J. Chem. Phys., *49*, 5161, 1968.
70. Carmichael, J. B., Polym. Prepr., *9*, 572, 1968.

71. Giddings, J. C., and Eyring, H., J. Phys. Chem., *59*, 416, 1955.
72. McQuarrie, D. A., J. Chem. Phys., *38*, 437, 1963.
73. Carmichael, J. B., Macromolecules, *1*, 526, 1968.
74. Carmichael, J. B., Biopolymers, *6*, 1497, 1968.
75. Kubin, M., J. Polym. Sci., Polym. Symp., *68*, 209, 1980.
76. Brönsted, J. N., Z. Phys. Chem., Bodenstein-Festband, 257, 1931.
77. Marsden, N. V. B., Ann. N.Y. Acad. Sci., *125*, 428, 1965.
78. Dawkins, J. V., and Hemming, M., Makromol. Chem., *176*, 1795, 1975.
79. Casassa, E. F., J. Polym. Sci., B, *5*, 773, 1967.
80. Casassa, E. F., and Tagami, I., Macromolecules, *2*, 14, 1969.
81. Casassa, E. F., J. Phys. Chem., *75*, 3929, 1971.
82. Casassa, E. F., Sep. Sci., *6*, 305, 1971.
83. Casassa, E. F., J. Polym. Sci., A2, *10*, 381, 1972.
84. Casassa, E. F., Macromolecules, *9*, 182, 1976.
85. Giddings, J. C., Kucera, E., Russell, C. P., and Myers, M. N., J. Phys. Chem., *72*, 4397, 1968.
86. Giddings, J. C., Anal. Chem., *40*, 2143, 1968.
87. Benoit, H., Grubisic, Z., Rempp, P., Decker, D., and Zilliox, J. G., J. Chim., Phys., *63*, 1507, 1966.
88. Pouchly, J., J. Chem. Phys., *52*, 2567, 1970.
89. Yau, W. W., Malone, C. P., and Fleming, S. W., J. Polym. Sci., B, *6*, 803, 1968.
90. Grubisic, Z., and Benoit, H., 7th Int. Semin. GPC, Monaco, 1969.
91. Janča, J., Pokorny, S., Bleha, M., and Chiantore, O., J. Liq. Chromatogr., *3*, 953, 1980.
92. Janča, J., and Pokorny, S., unpublished results, 1981.
93. Anderson, J. L., and Brannon, J. H., J. Polym. Sci., Polym. Phys. Ed., *19*, 405, 1981.
94. Giddings, J. C., *Dynamics of Chromatography*, Marcel Dekker, New York, 1965.
95. Giddings, J. C., and Mallik, K. L., Anal. Chem., *38*, 997, 1966.
96. Kelley, R. N., and Billmeyer, F. W., Sep. Sci., *5*, 291, 1970.
97. Hamielec, A. E., and Friis, N., Adv. Chromatogr., *13*, 41, 1975.
98. De Ligny, C. L., and Hammers, W. E., J. Chromatogr., *141*, 91, 1977.
99. Smith, W. B., and Kollmansberger, A., J. Phys. Chem., *69*, 4157, 1965.
100. Heitz, W., and Čoupek, J., Makromol. Chem., *105*, 280, 1967.
101. Heitz, W., and Čoupek, J., J. Chromatogr., *36*, 290, 1968.
102. Little, J. N., Waters, J. L., Bombaugh, K. J., and Pauplis, W. J., J. Polym. Sci., A2, *7*, 1775, 1969.
103. Little, J. N., Waters, J. L., Bombaugh, K. J., and Pauplis, W. J., Sep. Sci., *5*, 765, 1970.
104. Little, J. N., Waters, J. L., Bombaugh, K. J., and Pauplis, W. J., J. Chromatogr. Sci., *9*, 341, 1971.
105. Billmeyer, F. W., Johnson, G. W., and Kelley, R. N., J. Chromatogr., *34*, 316, 1968.
106. Billmeyer, F. W., and Kelley, R. N., Polym. Prepr., *8*, 1259, 1967.
107. Billmeyer, F. W., and Kelley, R. N., J. Chromatogr., *34*, 322, 1968.

108. Kelley, R. N., and Billmeyer, F. W., Anal. Chem., *41*, 874, 1969.
109. Kelley, R. N., and Billmeyer, F. W., Anal. Chem., *42*, 399, 1970.
110. Kelley, R. N., and Billmeyer, F. W., Sep. Sci., *5*, 291, 1970.
111. Cantow, M. J. R., and Johnson, J. F., J. Appl. Polym. Sci., *11*, 1851, 1967.
112. Meyerhoff, G., Angew. Makromol. Chem., *4-5*, 268, 1968.
113. Cooper, A. R., Cain, J. H., Barrall, E. M. T., and Johnson, J. F., Sep. Sci., *5*, 787, 1970.
114. Cooper, A. R., and Johnson, J. F., J. Appl. Polym. Sci., *15*, 2293, 1971.
115. Čoupek, J., and Heitz, W., Makromol. Chem., *112*, 286, 1968.
116. Maldacker, T. A., and Rogers, L. B., Sep. Sci., *6*, 747, 1971.
117. Peaker, F. W., and Tweedale, C. R., Nature, *216*, 75, 1967.
118. Edwards, V. H., and Helft, J. M., J. Chromatogr., *47*, 490, 1970.
119. Fishman, M. L., and Barford, R. A., J. Chromatogr., *52*, 494, 1970.
120. Biesenberger, J. A., and Ouano, A. C., J. Appl. Polym. Sci., *14*, 471, 1970.
121. Ouano, A. C., and Biesenberger, J. A., J. Appl. Polym. Sci., *14*, 483, 1970.
122. Ouano, A. C., and Biesenberger, J. A., J. Chromatogr., *55*, 145, 1971.
123. Cooper, A. R., Johnson, J. F., and Bruzzone, A. R., Polym. Prepr., *10*, 1455, 1969.
124. Hendrickson, J. G., J. Polym. Sci., A2, *6*, 1903, 1968.
125. Osterhoudt, H. W., and Ray, L. N., J. Polym. Sci., C, *21*, 5, 1968.
126. Osterhoudt, H. W., and Ray, L. N., J. Polym. Sci., A2, *5*, 569, 1967.
127. Horvath, C., and Lin, H. J., J. Chromatogr., *126*, 401, 1976.
128. Pfeffer, R., Ind. Eng. Chem. Fundam., *3*, 380, 1964.
129. Horvath, C., and Lin, H. J., J. Chromatogr., *149*, 43, 1978.
130. Knox, J. H., and McLennan, F., Chromatographia, *10*, 75, 1977.
131. Knox, J. H., and McLennan, F., J. Chromatogr., *185*, 289, 1979.
132. Dawkins, J. V., and Yeadon, G., J. Chromatogr., *188*, 333, 1980.
133. Dawkins, J. V., and Yeadon, G., J. Chromatogr., *206*, 215, 1981.
134. Hermans, J. J., J. Polym. Sci., A2, *6*, 1217, 1968.
135. Laurent, T. C., and Laurent, E. P., J. Chromatogr., *16*, 89, 1964.
136. Giddings, J. C., Bowman, L. M., and Myers, M. N., Macromolecules, *10*, 443, 1977.
137. Klein, J., and Grüneberg, M., Macromolecules, *14*, 1411, 1981.
138. Dawkins, J. V., and Yeadon, G., Polym. Prepr., *21*, 89, 1980.
139. Chiantore, O., and Guaita, M., J. Liq. Chromatogr., *5*, 643, 1982.
140. Grüneberg, M., and Klein, J., Macromolecules, *14*, 1415, 1981.
141. Klein, J., and Grüneberg, M., Macromolecules, *14*, 1419, 1981.

142. Smith, W. V., and Feldman, G. A., J. Polym. Sci., A2, *7*, 169, 1969.
143. Yau, W. W., Kirkland, J. J., Bly, D. D., and Stoklosa, H. J., J. Chromatogr., *125*, 219, 1976.
144. Bly, D. D., J. Polym. Sci., C, *21*, 13, 1968.
145. Janča, J., J. Liq. Chromatogr., *1*, 731, 1978.
146. Janča, J., J. Liq. Chromatogr., *5*, 1605, 1982.
147. Janča, J., J. Liq. Chromatogr., *5*, 1621, 1982.

— 2 —

CALIBRATION OF SEPARATION SYSTEMS

JOHN V. DAWKINS* / *Institute of Materials Science, University of Connecticut, Storrs, Connecticut*

2.1. INTRODUCTION

It is evident in Chapter 1 that polymer fractionation by a steric exclusion mechanism depends on molecular size in solution. For polymers in organic solvents, molecular size fractionation was greatly advanced by Moore's description [1] of the preparation of versatile cross-linked polystyrene gels covering a wide range of

*Permanent address: Department of Chemistry, Loughborough University of Technology, Loughborough, Leicestershire, England.

porosity. After the separation, the concentration by weight of polymer in the eluting solvent was monitored continuously with a differential refractometer [1,2]. The resulting chromatogram is therefore a weight distribution of the polymer as a function of retention volume V. The shape of the chromatogram will be determined by the pore size distribution within the porous packing and the sizes of the polymer molecules in solution. If the theory of the separation mechanism were sufficiently precise and comprehensive, and if the nature of the pore structure and the distribution of pore sizes were determined accurately by experiment, a molecular size distribution could be calculated from an experimental chromatogram for a polymer. Unfortunately, this procedure is inadequate at present.

The polymer scientist is interested in information on molecular weight distribution (MWD), defined as the cumulative weight distribution I(M) or differential weight distribution W(M) as a function of molecular weight M [3]. Molecular size fractionation may be considered an absolute method for the direct determination of the distribution function W(M) provided that the molecular weight of the polymer in the eluting solvent can be measured experimentally. This necessitates a chromatograph equipped with dual detectors for determining both the concentration and molecular weight of eluting polymer. Although several concentration detectors are available, for example, polymer property detectors monitoring refractive index (DRI), ultraviolet absorption (UV), and infrared absorption (IR), only the low-angle laser light-scattering photogoniometer (LALLSP) has been used as an on-line detector to find absolute values of M [4]. Although light scattering gives good scattering intensity for polymers of high molecular weight, there may be little or no detector sensitivity with $M < 10^4$. For a polydisperse polymer, experimental measurement of M for the chromatogram at high V may not be accurate (see Chapter 5). It follows that when average molecular weights are computed from the distribution W(M) derived from data obtained with both concentration and molecular weight detectors, the weight-average molecular weight $\bar{M}_w$ is likely to be more reliable than the number-average molecular weight $\bar{M}_n$, which could be substantially in error.

Many important polymers are polydisperse, having chromatograms covering a wide range of V. It is therefore necessary to consider a steric exclusion separation with a concentration detector as an analytical fractionation method which must be calibrated in order to determine both the molecular size and MWD. The normalized molecular weight distributions I(M) and W(M) are related to the normalized chromatogram by [5-7]

$$W(M) = -\frac{dI(M)}{dM} = -\frac{dI(V)}{dV}\frac{dV}{d \log M}\frac{d \log M}{dM} \tag{2.1}$$

where I(V) is the weight fraction of polymer eluted up to retention volume V. The ordinate of the chromatogram is dI(V)/dV, which is represented simply by F(V). Since d log M/dM is 1/M, the calculation of W(M) from the experimental chromatogram necessitates the determination of the relation

$$\log M = f(V) \tag{2.2}$$

which is invariably dependent on polymer type and structure. In the simplest situation log M is linearly related to the function f(V) and we can write

$$\log M = C_1 - C_2 V \tag{2.3}$$

where C_1 and C_2 are constants. Having calculated W(M) from an experimental chromatogram according to equation (2.1), the polymer scientist may then compute the number-, viscosity-, weight-, and z-average molecular weights, $\overline{M}_n$, $\overline{M}_v$, $\overline{M}_w$, and $\overline{M}_z$, respectively, for the polymer sample.

It follows that this chapter is concerned with methods for establishing the calibration curve defined by equations (2.2) and (2.3). Nilsson and Nilsson [8] considered a classification of seven calibration methods. Here, as in a previous review [9], the methods are placed into three categories. First, the use of reference standards having narrow MWDs is surveyed. Second, calibration methods with polydisperse reference materials are reviewed. Third, it is assumed that the separation is determined by molecular size, so that from a calibration curve in terms of molecular size established experimentally with reference standards, the molecular weight calibration for a polymer may be calculated when the relation between the molecular size of that polymer in solution and molecular weight is known. All these calibration methods require separate experimental study of the reference materials (or standards) and the polymers requiring analysis. The simultaneous separation of the standards and an unknown polymer would generate a complex chromatogram and is an impractical procedure for determining MWD. Because separate experiments are used to establish the calibration curve and to obtain the chromatogram for an unknown polymer, the chromatographic conditions, such as flow rate and composition of solvent, temperature, and concentration and volume of the injected solution, must remain constant. With careful experimentation, the calibration curve is unlikely to change, but it should be checked on a regular basis.

2.2. NARROW MOLECULAR WEIGHT DISTRIBUTION STANDARDS

A calibration curve with reference standards is readily established from chromatograms for each standard by a plot of log (average molecular weight) against peak retention volume V_R. The value of V_R for each standard is calculated from the point of injection to the appearance of the maximum value of F(V) of the chromatogram. The molecular weight of the standard corresponding to V_R is defined as M_{peak}. For reference standards having narrow MWDs, the experimental average molecular weights are considered to be very similar, so we can write

$$M_{peak} \simeq \bar{M}_n \simeq \bar{M}_v \simeq \bar{M}_w \qquad (2.4)$$

The popular and convenient plotting technique is to draw a curve of log M_{peak} versus V_R. For symmetrical narrow chromatograms having sharply defined peaks, the error in establishing M_{peak} far exceeds that in V_R. For asymmetrical chromatograms the choice of V_R at the peak of the chromatogram is empirical and other definitions of retention volume have been proposed. Crouzet et al. [10] proposed an average retention volume which may be defined as Σ F(V)V, with Σ F(V) normalized to unity. From theoretical considerations of the separation mechanism Van Kreveld and Van den Hoed [11] proposed that the first statistical moment could represent the mean retention volume of a chromatogram. They also indicated that the median is a useful approximation of the mean retention volume for an almost symmetrical peak and that the first moment and the median can be determined more accurately than V_R at the peak. The errors that can arise in the calibration curve from the use of V_R at the peak and of mean retention volume have been discussed [12,13] and are reviewed in Chapter 7.

2.2.1. Precision of Molecular Weight

It is generally accepted that when the reference standards have a polydispersity, defined by $\overline{M}_w/\overline{M}_n$, below 1.1, equation (2.4) will be applicable. This may be justified as follows. Polymer characterization techniques such as light scattering and membrane osmometry will give an average molecular weight with an error of about 5%, although depending on technique and molecular weight, this error could be higher [14]. It is unlikely that a reference standard will have the most precise values of both $\overline{M}_w$ and $\overline{M}_n$ simultaneously, since light scattering gives greater intensity as molecular weight increases, and the magnitude of colligative properties increases as molecular weight falls. Thus the error on the polydispersity established by absolute molecular weight techniques will be about 10%. It should be recognized that these errors are always higher than the errors involved in deriving data from a chromatogram with an instrument operating under optimum conditions. The error in determining V_R has been claimed to be as low as 0.2% [15,16], which for a typical calibration curve represented by equation (2.3) over four decades of molecular weight gives an error in molecular weight of about 1-2%. Estimates of errors in $\overline{M}_n$ and $\overline{M}_w$ determined from a chromatogram, for example with equation (2.1), have been compared among different laboratories, and in one such cooperative test the error was found to be about 6%, although the error in a single laboratory was much lower [17]. This is an extremely acceptable result (see Chapter 7).

The estimates of errors in average molecular weights computed from a W(M) distribution that has been calculated from a chromatogram may depend on whether a correction for chromatogram broadening has or has not been performed. Methods for correcting for broadening are considered in Chapter 3. We note here that it is possible to predict the correction factor $\overline{M}^*$ between an average molecular weight for an uncorrected experimental chromatogram and the true average molecular weight determined from a chromatogram corrected for dispersion effects in terms of the value of C_2 in equation (2.3) and the value of σ defined as the standard deviation of a Gaussian instrument spreading function in the correction procedure. Details of the method and the assumptions are given elsewhere [18]. The correction factors are defined by

$$\overline{M}_w^* = \frac{(\overline{M}_w)_{exp} - (\overline{M}_w)_{true}}{(\overline{M}_w)_{true}} \tag{2.5}$$

$$\overline{M}_n^* = \frac{(\overline{M}_n)_{exp} - (\overline{M}_n)_{true}}{(\overline{M}_n)_{true}} \tag{2.6}$$

Typical results for $\overline{M}^*$ for various column packings have been reported [18].

2.2.2. Reference Standards

Many calibration standards for polystyrene have been produced and may be obtained from several sources [19-22]. These reference standards have low polydispersities and cover a wide molecular weight range, from a few hundreds to several million. Reference standards having $\overline{M}_w/\overline{M}_n < 1.1$ have become available for other polymers: for example, poly(α-methyl styrene) [21], polyisoprene [19,21], polybutadiene [19, 23], polyethylene [24,25], poly(ethylene oxide) [21,22], poly(methyl methacrylate) [21], and polytetrahydrofuran [21]. However, the number of standards for each polymer is not large, so difficulties may be encountered in attempting to construct a calibration curve over a wide retention volume range. For aqueous separations monodisperse proteins are available, but the specific shapes and conformations of these macromolecules indicate that a molecular weight calibration is unlikely to be applicable to other types of polymers [26]. Reference standards for poly(ethylene oxide) [21,22] and poly(styrene sulfonate) [19,21] are also available for aqueous separations.

For other homopolymers, polymer fractionation techniques may be considered, although the experimental work is time consuming and not always efficient. Fractional precipitation invariably gives fractions having $\overline{M}_w/\overline{M}_n > 1.1$. Thus it was demonstrated that fractions of poly(dimethyl siloxane) produced by fractional precipitation had polydispersities in the range 1.1-1.2 when $\overline{M}_v$ was below 10^5 [27-29]. It is possible to lower the polydispersity further by refractionation. This may be performed for methods depending on polymer solubility with a column containing an inert nonporous support subjected to solvent and temperature gradients [30]. Fractions with quite narrow MWDs may be obtained; see, for example, the column fractionation of poly(methyl methacrylate) [31]. Preparative-scale steric exclusion chromatography may also be used for polymer fractionation [25,32-37], but fractions with sufficiently narrow distributions are obtained only by incorporating refractionation and recycle procedures into the fractionation scheme [25,33].

2.2.3. Construction of Calibration Curve

The calibration that results from using narrow distribution reference standards may appear like the curve shown in Figure 2.1. Three regions of behavior may be identified. Above molecular weight M_4 and below molecular weight M_1 no fractionation occurs because of total exclusion and total permeation, respectively, so that these values represent the limits of molecular weight that may be separated by the column packing(s). There is a region where the calibration curve may be represented by a straight line, as given by equation (2.3). The molecular weight range M_2 to M_3 and the fit to a straight line will be determined by the choice of column packings. The third region occurs for the molecular weight ranges M_1 to M_2 and M_3 to M_4, where fractionation still occurs but the calibration curve is nonlinear. The S-shaped calibration curve can be stored in a computer program and an interpolation method may be used to calculate molecular

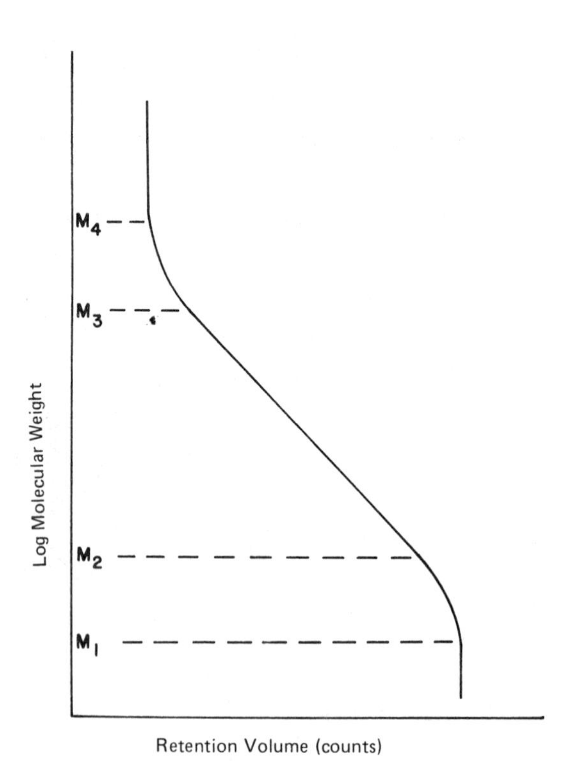

Figure 2.1. Molecular weight calibration curve. 1 count = 5 ml (see Ref. 2).

weight at particular retention volumes. The calibration curve in Figure 2.1 may be represented by a polynomial. As the S-shaped curve contains gradual slope changes, the best polynomial fit to the experimental data points gives a curve that is usually indistinguishable from the plotted experimental calibration curve. When the number of calibration standards is small, a linear interpolation between data points, to determine molecular weights at intermediate values of V, may give a physically unrealistic curve. A further problem arises because of the precision of the average molecular weights determined for the standards and the accuracy of values of M_{peak}, as discussed in Section 2.2.1. Therefore, it is then preferred to represent the set of experimental data points by a polynomial in order to generate a continuous smooth calibration curve.

Tung [38] employed orthogonal Legendre polynomials. The best fit to the experimental data by the method of moments is selected from a set of polynomials with degrees ranging from 3 to 32. This polynomial representation has the advantage of ease of differentiation, so that the slope of the calibration curve required in equation (2.1) when calculating MWD is readily accessible. Other representations of calibration curves by continuous analytical functions have been reviewed by Janča [39].

In any polynomial representation it is important to consider the precision of the molecular weight values provided for each standard. The calibration curve may be found either by standard methods of nonlinear regression with an arbitrarily selected degree of the polynomial or by an iteration method in which the degree of the polynomial is gradually increased until "an acceptable fit" between the average molecular weights of the standards calculated with the calibration function and the average molecular weights supplied from polymer characterization techniques is obtained. Clearly, little will be achieved by attempting to fit these data below the error level on the supplied average molecular weights. It may be observed that the degree of fit may be above the desired limit for some standards. As stated in Section 2.2.1, it is inevitable that some supplied average molecular weights will be more accurate than others because of the type of molecular weight technique and the inherent dependence of the accuracy of the experimental technique, either directly or inversely, on molecular weight.

For polymers with wide distributions, calibration of the tails of the chromatogram at the extremes of the retention volume range may pose problems. For the low molecular weight regime, a compound with a structure analogous to the repeating unit of the polymer or to that of an oligomer will often be available. For example, in Figure 2.2, curve BD for n-paraffins permits the extrapolation of a polyethylene calibration to short chains. At high molecular weights the tail of a chromatogram of a polydisperse sample may begin at a value of V that is below the value of V_R of the highest molecular weight reference standard. It is preferable to avoid an arbitrary extrapolation

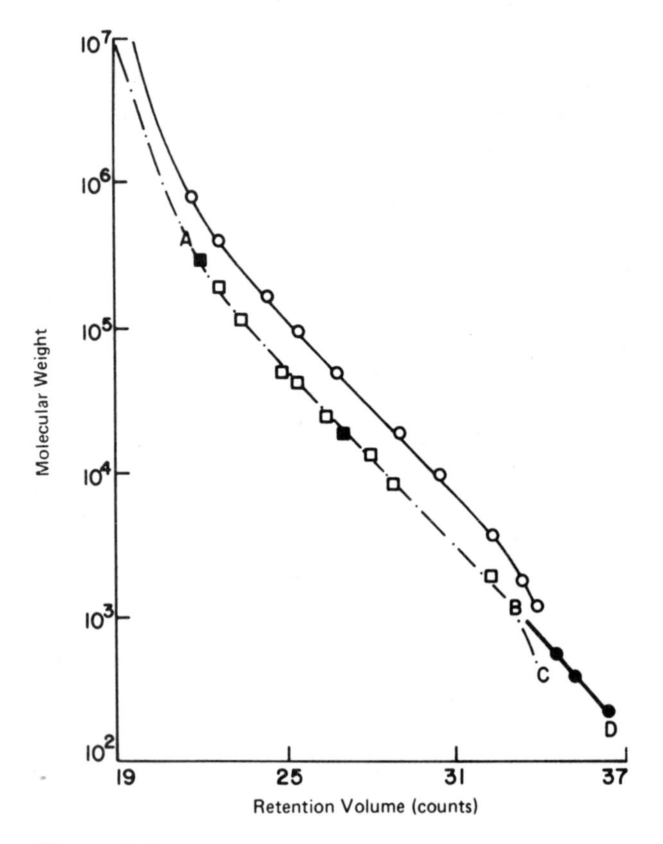

Figure 2.2. Molecular weight calibration plots for cross-linked polystyrene gels with o-dichlorobenzene at 138°C. ○, polystyrene standards; ●, n-paraffins; □■, linear polyethylene fractions; -•-, predicted polyethylene calibration curve ABC with equation (2.37) and a shift factor of log 0.466. (From Ref. 40.)

of the calibration curve to the interstitial volume. For example, let us assume that a standard having $M_{peak} = M_3$ in Figure 2.1 was the standard with the highest molecular weight. It is likely that $\overline{M}_w$ will be known for this reference standard, so that a trial-and-error method to find equation (2.2) for $V < V_R$ (corresponding to M_3) may be employed until the value of $\overline{M}_w$ calculated from the chromatogram of the standard is in reasonable agreement with the experimental value. Of course, the calculated value should be larger because of chromatogram broadening. Caution is advised when using this procedure because of the upward curvature in the calibration [41]. The extrapolation of the calibration for polyethylene to low and high V is exemplified in Figure 2.2 for $M > 2 \times 10^5$ and $M < 1800$. For routine separations with columns combined in a series arrangement, the guideline is to select cross-linked polystyrene gels that have different mean pore sizes and overlapping pore size distributions. The question which then arises is whether the calibration curve for the molecular weight range between M_2 and M_3 in Figure 2.1 is truly linear for a wide range of V.

Mori and Suzuki [42] have examined the polynomial approach to representing the calibration curve. The nth order polynomials are given by

$$V = A' + B'(\log M) + C'(\log M)^2 + D'(\log M)^3 + \cdots \quad (2.7)$$

and

$$\log M = A + BV + CV^2 + DV^3 + \cdots \quad (2.8)$$

where $A', B', C', D', \cdots$ and $A, B, C, D, \cdots$ are the coefficients of the polynomials. These coefficients were determined by a least-squares fit of the polynomial to the experimental calibration data points. For a calibration for a wide molecular weight range for cross-linked polystyrene gels, it was shown that the representation of the whole calibration curve by equation (2.3) caused serious errors in determining average molecular weights from W(M) calculated from a chromatogram. The calibration curve generally had to be represented by a third order polynomial. It may be concluded that for a column combination having overlapping pore size distributions, the calibration curve for a wide range of V is best represented by an S-shaped curve, although segments of the curve may exhibit good linearity. Similar observations have been reported for silica packings [43].

There are considerable advantages if equation (2.3) holds. First, data analysis with equation (2.1) is simplified. Second, many of the calibration methods in Section 2.3 assume a linear calibration curve. Third, corrections for chromatogram broadening and expressions for molecular weight accuracy are much easier to handle [18]. For porous inorganic packings pore size distributions may be obtained by independent methods [44-46]. Yau et al. [43] have computed theoretically the effect of pore size distribution on calibration linearity and the molecular weight separation range. The choice of function for pore size is based on experimental measurements of pore size distribution. They identify that equation (2.3) holds over four to five decades of molecular weight for a combination of two porous silicas having different mean pore sizes (known as a bimodal column system), provided that the linear portions of the calibration curves of each packing do not overlap significantly and that the pore volumes of the packings are such that these linear portions of the calibration curves are parallel. Typically, the mean pore sizes are about one decade apart and the pore volumes are the same. Experimental confirmation of the concept of the bimodal column system is

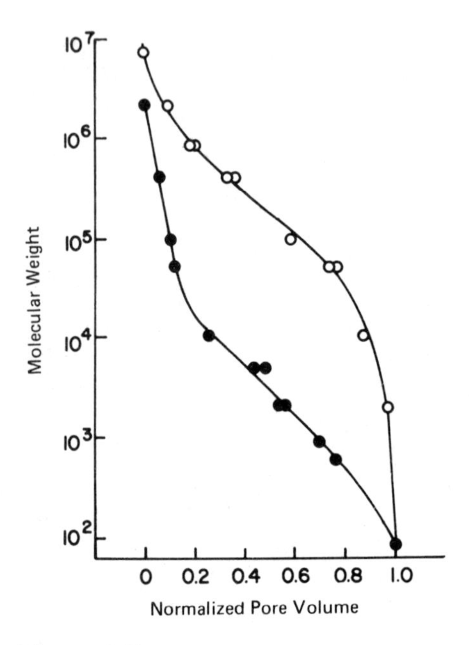

Figure 2.3. Molecular weight fractionation ranges for polystyrene for two porous silicas with pore size distributions differing by about one decade. ●, Silica PSM 60 Å; ○, silica PSM 750 Å. (From Ref. 43.)

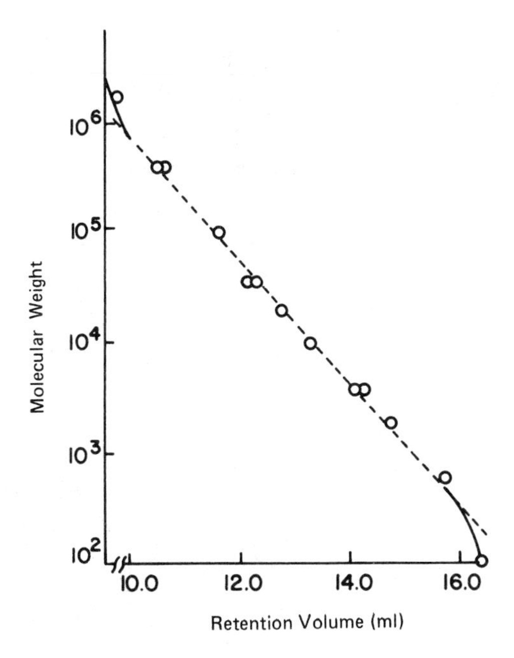

Figure 2.4. Molecular weight calibration curve for polystyrene for bimodal column system assembled with porous silicas having calibration curves in Figure 2.3. (From Ref. 43.)

demonstrated in Figures 2.3 and 2.4. The calibration curves for two appropriate porous silicas are shown in Figure 2.3., and a bimodal column system assembled with these two porous silicas gives the calibration curve in Figure 2.4.

2.3. METHODS FOR BROAD MOLECULAR WEIGHT DISTRIBUTION SAMPLES

Although more well-characterized reference materials have become widely available, many of these have broad MWDs, so that equation (2.4) is no longer applicable. It is not possible to guarantee that one of the common average molecular weights will correspond to M_{peak}, which may lead to errors in positioning the calibration curve. Reference materials with broad MWDs generally enable the calibration curve to be constructed over a wide range of V. One or more of the calibration methods in this section may then be applied. These methods may identify M_{peak} for the reference material, or average retention volumes assigned to average molecular weights of the reference material, or the calibration curve defined by equation (2.2) or (2.3). The approaches that calculate the average molecular weight and/or MWD from a chromatogram are trial-and-error methods, differing in mathematical and numerical details and in the amount of computer time required, in which the calibration curve is adjusted until an acceptable fit of the calculated molecular weight data with the experimental information on average molecular weights and/or MWD is obtained.

2.3.1. Distribution Functions

If the experimental chromatogram is fitted by a theoretical distribution function, it may be possible to determine M_{peak} easily. Berger and Shultz [47] examined the relation between M_{peak} and average molecular weights for several distribution functions assuming that equation (2.3) is valid for the range of V of the chromatogram. Their analysis shows that when the chromatogram is fitted by the Schulz-Zimm exponential distribution function, M_{peak} is given by

$$M_{peak} = \overline{M}_w \tag{2.9}$$

and when the logarithmic normal distribution function applies,

$$M_{peak} = (\overline{M}_w \overline{M}_n)^{1/2} \tag{2.10}$$

A simple test to check whether the chromatogram is fitted by a theoretical distribution function is to calculate average molecular weights from the chromatogram in order to find the values of the polydispersity indices $\overline{M}_w/\overline{M}_n$ and $\overline{M}_z/\overline{M}_w$. These ratios are the same for a logarithmic normal distribution, but the value of $\overline{M}_w/\overline{M}_n$ is larger for an exponential distribution. These trends are shown in Figure 2.5 for polysulfone and polycarbonate fractions. It appears that apart from one fraction, the distributions of the polycarbonate fractions are well represented by the exponential distribution, whereas the data for the polysulfone fractions fall between the two

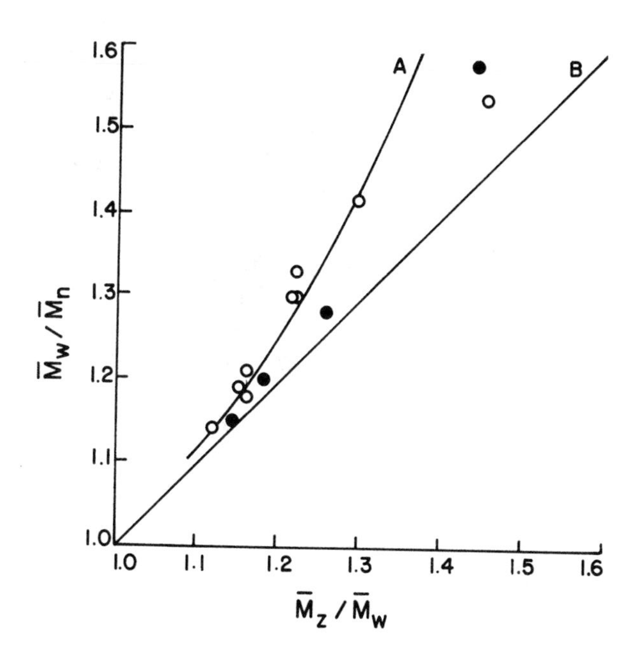

Figure 2.5. Polydispersity indices for theoretical distribution functions. A, Exponential distribution; B, logarithmic normal distribution; ●, experimental data for polysulfone fractions; ○, experimental data for polycarbonate fractions. (From Ref. 48.)

curves for the two distribution functions. This suggests that M_{peak} is given by equation (2.9) for the polycarbonate fractions and that M_{peak} lies between the values given by equations (2.9) and (2.10) for the polysulfone fractions. For fractions that are not too broad (i.e., the value of $\bar{M}_w/\bar{M}_n$ is between 1.1 and 1.2), M_{peak} will be very close to $\bar{M}_v$ when M_{peak} is between $(\bar{M}_n\bar{M}_w)^{1/2}$ and $\bar{M}_w$ [48]. For poly(dimethyl siloxane) fractions having $\bar{M}_w/\bar{M}_n$ in the range 1.1 to 1.2, it was shown that the chromatograms were approximated by logarithmic normal distributions and that $\bar{M}_v$ may be assumed to be M_{peak} [28]. The use of $\bar{M}_v$ of fractions for establishing the calibration curve has also been advocated for poly(vinyl chloride) [49] and for poly(α-methyl styrene), polypropylene, and polyethylene [50].

Many experimental chromatograms appear symmetrical, or pseudo-symmetrical with skewed tails, and equation (2.10), which was first proposed for polystyrene standards [51-53], has been applied to broad distribution samples, including polybutadiene [51] and branched polystyrene [54]. A range of well-characterized reference materials has been produced by preparative-scale steric exclusion chromatography [36] at the National Physical Laboratory, United Kingdom, and calibration data have been reported for polypropylene ($\bar{M}_w = 10^4$-62.5×10^4) [55], poly(vinyl chloride) ($\bar{M}_n = 3 \times 10^4$-$17.5 \times 10^4$) [56,57], linear poly(vinyl acetate) ($\bar{M}_n = 1.6 \times 10^4$-$1.1 \times 10^6$) [58], and poly(vinyl alcohol) ($\bar{M}_n = 2 \times 10^4$-$25 \times 10^4$) [59]. It was demonstrated that reference materials of polypropylene [55], poly(vinyl chloride) [56,57], and poly(vinyl acetate) [58] had chromatograms that could be approximated by the logarithmic normal function, so that equation (2.10) was valid for finding M_{peak}.

2.3.2. Molecular Weight Distributions

For some samples it may be difficult to assess whether the MWD is accurately represented by a theoretical distribution function. The possible correction of the chromatogram for broadening may complicate this assessment. In principle, greater precision should be obtained when the whole MWD is known from an independent experimental investigation. It is clear from equation (2.1) that if the distributions I(M) and W(M) have been determined independently for a sample, equation (2.2) or (2.3) may be calculated for the chromatogram of the same sample. This was first attempted by Cantow et al. [60] with a polyisobutene sample whose MWD had been well characterized by chromatographic fractionation depending on polymer solubility. Their method involved integrating the experimental chromatogram and comparing the normalized weight fractions with the known cumulative weight distribution I(M). When the two distributions are superimposed, the dependence of log M on V could be calculated and was shown to be nonlinear. Van Dijk et al. [61] have established by this superposition procedure a calibration curve with dextran reference materials having known I(M).

Polyolefins invariably have wide MWDs. A standard reference material, No. 1475, supplied by the National Bureau of Standards, has been recommended for establishing the calibration curve for linear polyethylene. The distribution I(M) is supplied with the sample. Values of I(V) are determined from an experimental chromatogram of the reference material in order to find the cumulative weight distribution as a function of V. Wild and coworkers [37,62] then correlated the two distributions with the calibration curve represented by the polynomial given by equation (2.8), in which values of the coefficients A, B, C, and D were determined. These workers reported that this method was very effective since a continuous calibration is obtained and because the chromatogram of a reference material with a broad distribution minimizes corrections for chromatogram broadening. The use of standard reference material No. 1475 is also recommended by the results of Swartz et al. [63], who used the superposition procedure for cumulative weight distributions. They showed that this method gave a calibration curve for linear polyethylene in reasonable agreement with a curve produced by the method of Balke and coworkers [64], which is discussed in Section 2.3.3 (see Figure 2.6).

Theoretical MWDs may be predicted from a detailed consideration of the mechanism and kinetics of the polymerization process. If a polymer is prepared under carefully controlled experimental conditions such that the assumptions involved in the calculation of the theoretical distribution are valid, the dependence of W(M) and I(M) on M can then be determined. Consequently, it is possible to apply the MWD method even though the experimental distribution has not been estimated. In practice, the calculation of W(M) is facilitated if average molecular weights have been determined for the broad distribution sample by osmometry, viscometry, and light scattering. The procedure was first described by Weiss and Cohn-Ginsberg [65] for broad distribution samples of poly(methyl methacrylate). The weight fractions determined independently for a sample were calculated by assuming that W(M) corresponds to a generalized exponential distribution and knowing the experimental values of $\overline{M}_n$ and $\overline{M}_w$ (or $\overline{M}_n$ and $\overline{M}_v$). A comparison of these weight fractions with experimental weight fractions from a chromatogram of the sample enabled the cali-

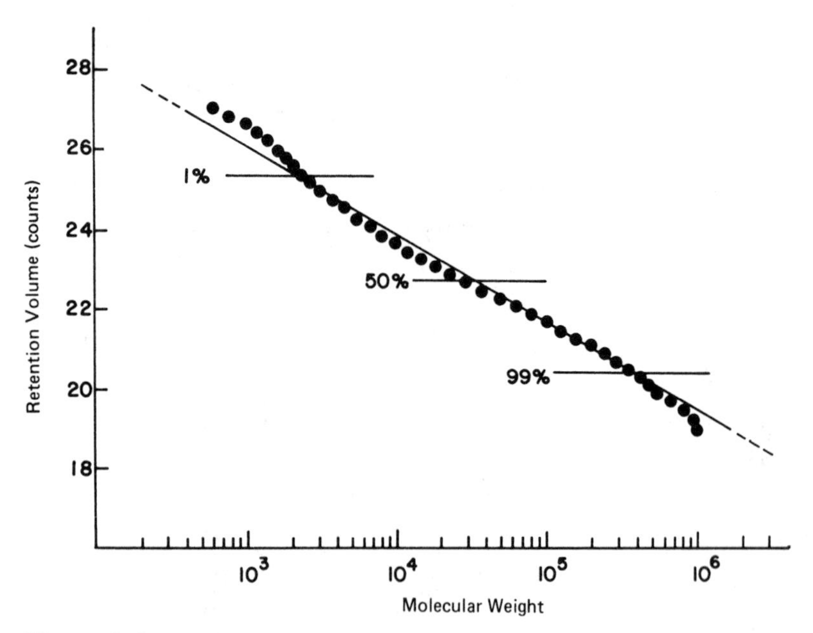

Figure 2.6. Calibrations for the reference material NBS 1475 of linear polyethylene. ●●●, By the molecular weight distribution method showing the retention volumes corresponding to the cumulative elution peak areas of 1, 50, and 99%; ——, by the BHLP method. (From Ref. 63, © John Wiley & Sons, Inc.)

bration curve to be found by the superposition procedure. Swartz et al. [63] have shown that the representation of a nylon 66 sample by a most probable distribution, calculated from experimental values of $\overline{M}_n$ and $\overline{M}_w$, enables the calculation of the calibration curve by matching W(M) with the chromatogram.

2.3.3. Average Molecular Weights

In view of the assumptions involving the application of theoretical distributions in Section 2.3.1 and the problems of obtaining accurate information on W(M) and I(M) for the methods in Section 2.3.2, methods involving the moments of the MWD have been more widely used. In the simplest example, when equation (2.3) is assumed to be valid, a broad distribution sample with two known average molecular weights, or alternatively two samples each having one known average molecular weight, may be employed to find the parameters C_1 and C_2. If this sample has a chromatogram with a well-defined peak, a value for M_{peak} may be assigned with equation (2.3). For a broad reference material the known average molecular weights are not likely to be identical to M_{peak}. Therefore, when the parameters C_1 and C_2 have been determined, an average retention volume corresponding to a given average molecular weight may be calculated for the broad distribution sample. Consequently, the information provided may be utilized in several ways: as a plot of the calibration curve according to equation (2.3) for the retention volume range of the chromatogram of the broad distribution sample, as a plot of M_{peak} versus V_R, or as a plot of average molecular weight versus the appropriate value of retention volume. In practice, the operation of methods involving moments may be quite complex, and it is intended here to summarize the principles of the more important procedures.

Average Retention Volumes

This method was first examined for fractions of poly(butene-1) by Ring and Holtrup [49]. They began with an experimental plot of log $\bar{M}_v$ versus V_R, and calculated values of the viscosity average molecular weight were then determined from the chromatograms of the fractions. The calibration plot according to equation (2.2) was then altered by a trial-and-error method until the experimental and calculated average molecular weights were in good agreement. The average retention volumes corresponding to $\bar{M}_n$, $\bar{M}_v$, and $\bar{M}_w$ were identified in order to plot the correct calibration curve for a series of fractions. Purdon and Mate [66] proposed a similar method. They commenced by constructing a calibration curve with narrow distribution polystyrene standards. This curve was used with the chromatogram of a broad distribution sample to calculate values of $\bar{M}_n$ and $\bar{M}_w$ from which number- and weight-average retention volumes were assigned. The calibration data were then replotted employing the true experimental values of $\bar{M}_n$ and $\bar{M}_w$ for the sample and the corresponding assigned retention volumes, and the whole calibration curve was generated for a series of well-characterized broad distribution samples. The method was exemplified with samples of polyisoprene, polybutadiene, poly(vinyl chloride), and several copolymers containing butadiene [66]. It is possible that further adjustments to the calibration curve may be necessary to obtain close agreement between the experimental and calculated average molecular weights. Bombaugh et al. [67] constructed a preliminary calibration curve for reference materials of dextran with known $\bar{M}_w$, and this curve together with a chromatogram permitted the calculation of $\bar{M}_n$ and $\bar{M}_w$. Number- and weight-average retention volumes were then assigned to these calculated average molecular weights from the preliminary calibration curve. The true calibration curve was constructed from a semilogarithmic plot of the experimentally determined average molecular weights and these average retention volumes. These results illustrated that the dextran samples are quite polydisperse. The broad distributions of reference materials of dextran were investigated by Basedow and coworkers [68], who prepared narrow distribution fractions by preparative steric exclusion chromatography. Ogawa and Inaba [69] also recommended that the calibration curve should be constructed with data for $\bar{M}_n$ and $\bar{M}_w$ of the reference materials and the corresponding number-average and weight-average retention volumes. They assumed that the chromatograms of their reference materials were fitted by the logarithmic normal distribution function.

A more detailed treatment of the average retention volume approach has been proposed by Szewczyk [70]. He represented the molecular weight calibration curve by polynomials according to equations (2.7) and (2.8). A series of equations was then defined in terms of the polynomial coefficients, an average molecular weight, and the corresponding average retention volume $\bar{V}$. Thus, in the case of the number average, the variables are $\bar{M}_n$ and $\bar{V}_n$. The series of equations permit the search of the polynomial coefficients by a successive approximation method which is also used to find the coefficients in the equations

$$\bar{V}_n = q(\log \bar{M}_n) = q\{-\log [\Sigma W(V)10^{-p(V)}]\} \quad (2.11)$$

$$\bar{V}_v = q(\log \bar{M}_v) = q\{a^{-1} \log [\Sigma W(V)10^{ap(V)}]\} \quad (2.12)$$

$$\bar{V}_w = q(\log \bar{M}_w) = q\{\log [\Sigma W(V)10^{p(V)}]\} \quad (2.13)$$

where q and p denote polynomials according to equations (2.7) and (2.8), respectively; W(V) is the weight fraction of polymer having retention volume V, which is estimated from a chromatogram; and a is the Mark-Houwink exponent [see equation (2.29)]. The approximation procedure is continued until acceptable agreement between the experimental and calculated values of $\bar{M}_n$, $\bar{M}_v$, and $\bar{M}_w$ is obtained, when the corresponding average retention volumes are defined. This procedure is useful for reference materials having a unimodal distribution for a wide polydispersity range [71,72]. Corrections for chromatogram broadening may be incorporated [71]. This method [72] was shown to give good results for a sample of poly(vinyl chloride) and fractions of poly(propylene oxide) when compared with other methods based on average molecular weights and on universal calibration (Section 2.4.3). Only one or two polydisperse samples may be required to establish the calibration curve over a wide range of V. A further merit of all the methods in this section is that the calibration relation according to equation (2.2) may be nonlinear.

Search Methods

The more common approach of using broad distribution samples for calibration was first described by Almin [73], Frank et al. [74], and Balke and coworkers [64]. These procedures assume that the sample has known average molecular weights and that the calibration curve is represented by equation (2.3). Then, with the chromatogram for the reference sample, the constants C_1 and C_2 in this equation are found by a trial-and-error method. Almin [73] presented a theoretical analysis in which a specific MWD was assumed and a correction for chromatogram broadening was necessary. Unfortunately, his method was not tested with experimental data. Here, the other two methods, denoted FWW [74] and BHLP [64], will be surveyed, and the modification proposed by Yau et al. [75], referred to as the YSB method, will be discussed.

FWW method: This is a general approach involving a graphical procedure that finds a pair of values of C_1 and C_2 compatible with the following two equations:

$$(C_1)_n = \log \bar{M}_n + \log \int_0^\infty \exp (C_2V)F(V)\, dV - \log \int_0^\infty F(V)\, dV \tag{2.14}$$

$$(C_1)_w = \log \bar{M}_w + \log \int_0^\infty F(V)\, dV - \log \int_0^\infty \exp (-C_2V)F(V)\, dV \tag{2.15}$$

The calibration curve over a wide range of retention volumes is then obtained by determining the chromatograms for a series of well-characterized fractions. A linear calibration is obtained for each fraction over its retention volume range. The complete calibration curve is then obtained by combining calibration curves for all the fractions. The procedure has been demonstrated with polyethylene fractions [74], samples of poly(2-vinyl pyridine) [76], and reference materials of polypropylene [55] and poly(vinyl chloride) [56, 57]. Frank et al. [74] considered that in their graphical procedure it was not essential to correct for chromatogram broadening, as their results suggested that its effect was small even when experimental broadening contributes significantly to the width of the chromatogram.

Having defined the constants C_1 and C_2 in equation (2.3) by using equations (2.14) and (2.15), it is evident that M_{peak} at V_R and the average retention volumes corresponding to $\bar{M}_n$ and $\bar{M}_w$ may be defined. These results may be used in previous methods for plotting the calibration curve. It is not necessary to use a graphical method to find C_1 and C_2, which can be calculated by an iteration procedure [76,77]. The FWW method [74] may be extended to fractions for which values of $\bar{M}_n$ and $\bar{M}_v$ are available [77]. Equation (2.14) is then used with

$$(C_1)_v = \log \bar{M}_v - a^{-1} \log \int_0^\infty F(V) \exp (-aC_2V)\, dV + a^{-1} \log \int_0^\infty F(V)\, dV \tag{2.16}$$

to find unique values of C_1 and C_2 for the retention volume range of a broad distribution fraction. Examples of defining the constants C_1 and C_2 with equations (2.14) and (2.16) have been reported for polyethylene fractions [77] and samples of poly(γ-benzyl-L-glutamate) [76].

BHLP method: Balke and coworkers [64] used a search routine to determine the linear calibration curve. To aid the comparison between the BHLP and YSB methods in the following section, the nomenclature adopted by Yau et al. [75] will be employed. This necessitates an alternative definition of equation (2.3) representing the linear calibration curve, as follows:

$$\log M = \log D_1 - D_2 V \quad (2.17)$$

In the BHLP method an iterative method is required to find the best fit between the known experimental values of $\overline{M}_n$ and $\overline{M}_w$ and the calculated values from the chromatogram of the broad distribution sample so that the values of D_1 and D_2 may be deduced for the retention volume range of the chromatogram. The errors involved in this method have been considered by Pollock et al. [78]. The general equations containing D_1 and D_2 are

$$\overline{M}_n = \frac{D_1}{\Sigma\ F(V)\ \exp\ (D_2 V)} \quad (2.18)$$

$$\overline{M}_w = \Sigma\ F(V) D_1\ \exp\ (-D_2 V) \quad (2.19)$$

The iteration method to find the two coefficients in equation (2.17) may be time consuming. Loy [79] showed that the method could be simplified so that D_1 and D_2 could be calculated more quickly. He derived an expression for the polydispersity of the broad distribution reference material, and from equations (2.18) and (2.19) this is given by

$$\frac{\overline{M}_w}{\overline{M}_n} = [\Sigma\ F(V)\ \exp\ (-D_2 V)][\Sigma\ F(V)\ \exp\ (D_2 V)] \quad (2.20)$$

Equation (2.20) contains only one coefficient instead of two, so that from experimental values of $\overline{M}_w$, $\overline{M}_n$, and $F(V)$, D_2 may be calculated by a single-variable search routine. It is then possible to find the value of D_1 with equation (2.18) or (2.19). Results from the procedure employing equation (2.20) have been reported by Chaplin et al. [80] for polydisperse samples of poly(vinyl chloride). They concluded that this calibration method was reliable as long as the assumption of a linear calibration curve is reasonable. Several well-characterized polydisperse samples may be necessary to calibrate accurately a wide range of V. Other iteration methods have been described by Crouzet and coworkers [81], giving results for samples of polyethylene and polyisobutene, and by Malawer and Montana [82].

It was suggested that the BHLP method could be readily modified for nonlinear calibration curves [64]. McCrackin [83] represented the calibration curve by the first three terms in equation (2.8), neglecting higher terms. His method requires polydisperse samples having known values of $\overline{M}_n$, $\overline{M}_v$, or $\overline{M}_w$. McCrackin [83] then derives the coefficients A, B, and C in equation (2.8), assuming that D is zero, by a procedure involving considerable calculation. The same representation of a nonlinear calibration curve was used by Chaplin and coworkers [80,84]. The method of Loy [79] involving equation (2.20) may be extended to incorporate the calibration coefficients A, B, and C [84]. A search routine is then applied to find these coefficients. The results of Chaplin and coworkers [80,84] for samples of poly(vinyl chloride) and polystyrene suggest that significant errors may arise with search methods involving a nonlinear calibration curve. They conclude that this method is less reliable than calculation of the calibration curve by a universal calibration method (see Section 2.4.3). Andersson [85] has examined the least-squares approximation of the calibration curve by the method of cubic splines. He proposes that this method should be preferred when a nonlinear calibration curve must be calculated from the experimental data of the average molecular weights and the chromatogram for the broad distribution reference material.

YSB method: The BHLP method does not include a correction for chromatogram broadening. The values of D_1 and D_2 derived with equations (2.18), (2.19), and (2.20) are the true values only when chromatogram broadening is not significant. These equations will provide apparent values of D_1 and D_2 only if chromatogram broadening cannot be neglected [78]. Several of the initial applications of the BHLP method involved corrections for chromatogram broadening, although precise details of the correction procedure were not stated. Thus Whitehouse [86] described a search method for the linear calibration curve for samples of linear polyethylene having broad distributions, reporting that the effects of chromatogram broadening were taken into account. Swartz et al. [63] included a correction for chromatogram broadening in the BHLP method giving a linear calibration (see Figure 2.6) in good agreement with the calibration method described in Section 2.3.2 for polydisperse samples having well-defined distributions.

Yau et al. [75] proposed a formulation of the BHLP method which included a correction for chromatogram broadening. The relevant equations are

$$\overline{M}_n = \frac{D_1 \exp\ [(D_2\sigma)^2/2]}{\Sigma\ F(V)\ \exp\ (D_2V)} \tag{2.21}$$

$$\overline{M}_w = \left[\exp\ \frac{-(D_2\sigma)^2}{2}\right] \Sigma\ F(V) D_1\ \exp\ (-D_2V) \tag{2.22}$$

where σ is the peak standard deviation caused by column dispersion. It is assumed that chromatogram broadening of a single species is symmetrical and obeys a Gaussian function and that σ does not depend on V (i.e., is independent of M) (for more details, see Section 3.4.1). A search procedure is then employed with equations (2.21) and (2.22) in order to find the constants D_1 and D_2 in equation (2.17). The procedure may be simplified by deriving an equation for polydispersity analogous to equation (2.20). From equations (2.21) and (2.22) Pollock et al. [78] derive

$$\frac{\overline{M}_w}{\overline{M}_n} = \exp\ -(D_2\sigma)^2 [\Sigma\ F(V)\ \exp\ (-D_2V)][\Sigma\ F(V)\ \exp\ (D_2V)] \tag{2.23}$$

The value of σ may be evaluated in a separate experiment from the chromatogram for a narrow distribution polystyrene standard. However, Hamielec and Omorodion [87] noted that when equations (2.21) and

(2.22) are multiplied, σ vanishes. In the resulting expression values of D_1 and D_2 may be obtained with chromatograms for two broad distribution reference materials having known experimental data for $\overline{M}_n$ and $\overline{M}_w$. After searching for D_1 and D_2, it is possible to derive σ for each reference material directly from equation (2.21) or (2.22). Hamielec and Omorodion [87] illustrated the procedure with dextran samples. A more detailed correction procedure for chromatogram broadening has been investigated by Yau and coworkers [88] and may be found in Chapter 3. The shape of the chromatogram of a single species is defined in terms of σ and a peak skew parameter τ which must be determined experimentally. Equations (2.21) and (2.22) are easily extended to incorporate τ for a chromatogram represented by an exponentially modified Gaussian function.

A comparison of the YSB and BHLP methods was performed by Yau et al. [75]. They employed a single polydisperse polystyrene having a bimodal broad distribution. The calibrations established by the two methods are shown in Figure 2.7 together with the direct calibration curve plotted for narrow distribution polystyrene standards. Although only one polydisperse reference material is used, the YSB method gives a satisfactory determination of the calibration curve despite the assumptions regarding chromatogram broadening inherent in equations (2.21) and (2.22). The deviation of the curve derived by the BHLP method from the other curves in Figure 2.7 depends on the extent of chromatogram broadening. Yau et al. [75] proposed therefore that the BHLP method is best applied when the reference material and the polymer to be characterized have a similar MWD. Pollock et al. [78] have considered the errors that may arise in the BHLP and YSB methods. In addition to errors arising from neglect of the skewing contribution to chromatogram broadening and from the mismatch of the distributions of the reference material and the unknown samples, they identified errors in the experimental values of $\overline{M}_n$ and $\overline{M}_w$ from techniques such as osmometry and light scattering, in the replication of chromatograms both for the reference material and the unknown sample, and in the evaluation of σ. From the total error analysis, Pollock

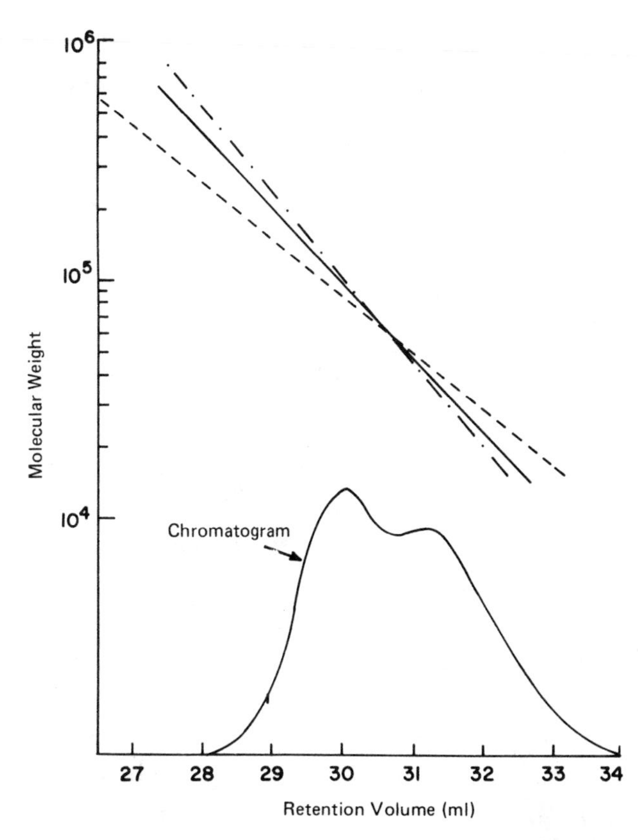

Figure 2.7. Calibrations for broad distribution sample of polystyrene having bimodal chromatogram. ——, YSB method; ----, BHLP method; -•-•-, curve for narrow distribution polystyrene standards. (From Ref. 75, © John Wiley & Sons, Inc.)

et al. [78] concluded that the replication error was the most important and that errors arising from imprecise values of $\overline{M}_n$ and $\overline{M}_w$ could be very significant. Therefore, any attempt to improve the calibration method by incorporating a correction procedure for chromatogram broadening may not be worthwhile when experimental errors are high. An extensive discussion of the errors involved in search methods for establishing calibration curves with dextran reference materials has been presented by Nilsson and Nilsson [8].

Search methods involving average molecular weights for a broad distribution reference material have proved to be a convenient and popular calibration procedure. The three methods described illustrate the basic principles for determining the calibration curve by equation (2.3) or (2.17). The methods are capable of further development in terms of mathematical and numerical detail, and therefore in the amount of computation required. Vrijbergen et al. [89] proposed an iteration method requiring a broad distribution sample for which two average molecular weights must be known and assuming that equation (2.3) is valid for the sample. This method was found to be most precise when chromatograms are corrected for broadening. A calibration curve is found for each broad distribution sample, and the whole calibration curve for a wide range of V is determined by a further iteration procedure with all the samples. Highest accuracy is obtained when these samples have chromatograms with overlapping ranges of V. The method has been illustrated with dextran reference materials having known values of $\overline{M}_n$ and $\overline{M}_w$ [89,90]. These results show that care is required in calibrating with dextran samples, which are widely used for aqueous studies, because the samples are quite polydisperse with $\overline{M}_w/\overline{M}_n > 1.4$ and have branched chains. McCrackin [83] examined reference materials with narrow MWD and with known values of both $\overline{M}_n$ and $\overline{M}_w$. A correction for chromatogram broadening was included in this method, assuming that broadening is represented by a Gaussian function. For polystyrene standards chromatogram broadening was reported to be small and independent of molecular weight. Tsvetkovskii et al. [91] also described a detailed iteration method for the evaluation of D_1 and D_2 in equation (2.17), incorporating a correction for chromatogram broadening.

2.4. UNIVERSAL CALIBRATION

In the absence of well-characterized reference materials of the polymer requiring analysis for part or all of a wide retention volume range, a calibration curve can be obtained provided that solute size is controlling separation. The mechanism must be steric exclusion when theory predicts that the elution behavior of all polymers can be represented by a universal size parameter. Thus a single calibration curve plotted in terms of size then permits the establishment of a molecular weight calibration for a given polymer provided that the relation between molecular size and molecular weight is known for that polymer. This relation will vary from polymer to polymer because of the dependence of molecular size on the structure of the monomer unit in the chain and on chain flexibility. The universal calibration curve may be readily established with narrow distribution polystyrene standards for organic separations and with narrow distribution poly(ethylene oxide) standards and/or well-characterized dextran reference materials for aqueous separations.

Several theoretical treatments (see Chapter 1) relate the distribution coefficient, and therefore V, to solute size. The initial theories of steric exclusion considered simple geometrical models (for references, see Ref. 92). The first thermodynamic theory for a flexible coil polymer in a separation operating at equilibrium was proposed by Casassa [93] and developed further in subsequent papers [94-96]. His theoretical treatment considered the radius of gyration of a random coil polymer as the size parameter. Giddings and coworkers [97] in the theoretical treatment of the separation of rigid biopolymers proposed that elution is determined by the mean molecular projection independent of molecular geometry. Casassa [98] has shown that the concept of the mean external length is also applicable in his treatment of random coil polymers.

2.4.1. Hydrodynamic Volume

Studies of the dilute solution viscosity of synthetic polymers indicate that the polymer molecule in solution can be represented as an equivalent hydrodynamic sphere [99]. Then the intrinsic viscosity $[\eta]$ is defined according to the Einstein viscosity relation

$$[\eta]M = 0.025 N_A V_h \tag{2.24}$$

where V_h is the hydrodynamic volume of the equivalent sphere and N_A is Avogadro's number. If hydrodynamic volume, or a size parameter related to this volume such as the radius or diameter of the hydrodynamic sphere, controls the separation, a plot of $[\eta]M$ versus V

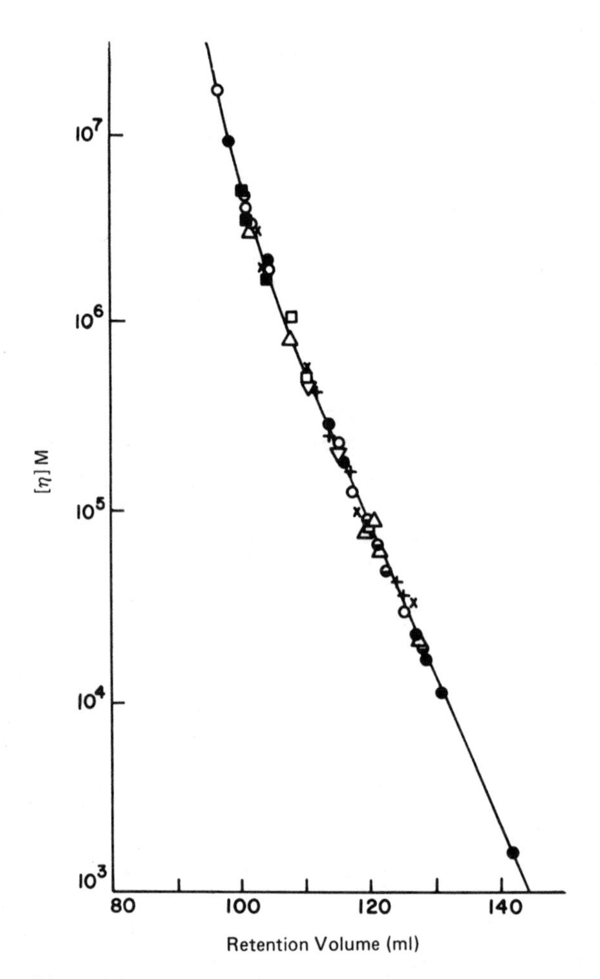

Figure 2.8. Hydrodynamic volume universal calibration curve for cross-linked polystyrene gels with tetrahydrofuran. ●, Linear polystyrene; ○, branched polystyrene (comb type); +, branched polystyrene (star type); △, branched block copolymer of styrene/methyl methacrylate; ×, poly(methyl methacrylate); ◒, poly(vinyl chloride); ▽, graft copolymer of styrene/methyl methacrylate; ■, poly(phenyl siloxane); □, polybutadiene. (From Ref. 100, © John Wiley & Sons, Inc.)

will be the same for all polymers. Experimental evidence for the validity of hydrodynamic volume, as defined by [η]M, for universal calibration was presented by Grubisic et al. [100], who studied homopolymers and copolymers having various chemical and geometrical structures. Their data are shown on a semilogarithmic plot in Figure 2.8.

The dimensions of flexible coil polymers are defined by the root-mean-square end-to-end distance $(\langle r^2 \rangle)^{1/2}$ and the root-mean-square radius of gyration $(\langle s^2 \rangle)^{1/2}$. Flory and Fox [99,101] extended equation (2.24) to

$$[\eta]M = \Phi_0(\langle r^2 \rangle)^{3/2} = \Phi_0'(\langle s^2 \rangle)^{3/2} \tag{2.25}$$

where Φ_0 and Φ_0' are universal constants. Therefore, the hydrodynamic radius R_e given by

$$V_h = \frac{4\pi R_e^3}{3} \tag{2.26}$$

is directly proportional to an average linear dimension of the polymer molecule, e.g. $(\langle r^2 \rangle)^{1/2}$ or $(\langle s^2 \rangle)^{1/2}$. Consequently, these molecular parameters should be just as satisfactory as $[\eta]M$ for universal calibration.

The results in Figure 2.8 were obtained with cross-linked polystyrene gels and tetrahydrofuran as eluant. The first report in 1966 of the calibration method based on hydrodynamic volume [102] initiated studies of universal calibration by many workers. Much of the early work with cross-linked polystyrene gels was confined to eluants that are good solvents for linear polystyrene [28]. Calibration data were presented in terms of hydrodynamic volume, hydrodynamic radius, radius of gyration, and end-to-end distance, as reviewed by Coll [103]. Some workers preferred to calculate molecular parameters with the modification of equation (2.25) proposed by Ptitsyn and Eizner [104] for the mean-square end-to-end distance

$$(\langle r^2 \rangle)^{3/2} = \frac{[\eta]M}{\Phi(\varepsilon)} \tag{2.27}$$

where

$$\Phi(\varepsilon) = \Phi_0(1 - 2.63\varepsilon + 2.86\varepsilon^2) \tag{2.28}$$

and ε is a parameter depending on polymer-solvent interaction. A measure of polymer-solvent interaction is provided by the exponent a in the dilute solution viscosity Mark-Houwink equation

$$[\eta] = KM^a \tag{2.29}$$

where K and a are constants for a given polymer-solvent-temperature system. The parameters ε and a are related by

$$2a = 1 + 3\varepsilon \tag{2.30}$$

Eluants such as tetrahydrofuran (THF), chloroform (CHL), 1,2,4-trichlorobenzene (TCB), and o-dichlorobenzene (ODCB), having values of a in the range 0.67 to 0.80 as shown in Table 2.1, will be classified as *good* solvents for polystyrene. Theta solvent is that for which a = 0.5.

In any assessment of the universal calibration method, separations must be performed such that contributions to elution from polymer-gel interactions are effectively absent. It is assumed that polystyrene standards in a good solvent do not display preferential affinity for the mobile phase or the stationary phase because the eluant is very compatible with the cross-linked gel. Consequently, it is presumed that the separation of linear polystyrene standards by the primary steric exclusion mechanism is not modified by secondary partition and adsorption mechanisms. The other polymers selected for testing universal calibration parameters should also not participate in secondary separation mechanisms. Consequently, nonpolar homopolymers should be preferred for separations with cross-linked polystyrene gels, and the eluants THF, CHL, TCB and ODCB should be good solvents for these homopolymers. The linear polymers given in Table 2.2 all have a polarity, as judged by the solubility parameter δ (see Ref.

Table 2.1. Solution Viscosity Constants for Polystyrene Standards

Solvent	Temperature (°C)	K	a	References
Tetrahydrofuran	25	1.16×10^{-4}	0.73	105
Chloroform	30	4.9×10^{-5}	0.794	27
1,2,4-Trichlorobenzene	135	1.21×10^{-4}	0.707	50, 106
o-Dichlorobenzene	138	1.38×10^{-4}	0.7	28, 40

Table 2.2. Universal Calibration with Hydrodynamic Volume for Linear Polymers with Cross-Linked Polystyrene Gels

Polymer	Solvent	Temperature (°C)	K	a	References
Polyethylene	TCB	140	3.95×10^{-4}	0.726	37, 50, 107, 108
Polyethylene	ODCB	138	5.06×10^{-4}	0.7	40
Polypropylene	TCB	135	1.37×10^{-4}	0.75	50
Polypropylene	ODCB	135	2.42×10^{-4}	0.707	55
Polyisobutylene	TCB	35, 150	—	0.69[a]	109
Polybutadiene	THF	25	$4.03\text{-}4.57 \times 10^{-4}$	0.693	110
Polyisoprene	THF	25	1.77×10^{-4}	0.735	110
Poly(dimethyl siloxane)	CHL	30	5.4×10^{-5}	0.77	27
Poly(α-methylstyrene)	TCB	135	1.61×10^{-4}	0.672	50

[a]Value calculated from solution viscosity data for polyisobutylene in TCB at 150°C from Ref. 111.

112), below that of polystyrene, and clearly the values of a indicate that the eluants may be regarded as good solvents. Universal calibration data presented in the references quoted in Table 2.2 confirm the validity of the hydrodynamic volume method. For a given eluant, polymer-solvent interactions as judged by the values of a in Tables 2.1 and 2.2 are similar for polystyrene and the other linear polymers. Consequently, the universal calibration plots with parameters defined in equations (2.24) to (2.27), are equivalent.

Dawkins et al. [28] recognized that it is not possible to distinguish between the various size parameters with eluants that are good solvents for all the polymers. They reported calibration data for two polymers having different polymer-solvent interactions. However, even for poly(dimethyl siloxane) in ODCB at 138°C, when a = 0.57, and polystyrene (a = 0.7), it was indicated that an unambiguous assessment of the true universal calibration parameter could not be made. In order to obtain a large difference in polymer-solvent interaction, measurements must be performed for a polymer in a theta solvent and compared with data for polystyrene in a good solvent. Dawkins and Hemming [29] reported calibration results for polystyrene ($a = 0.73$) and poly(dimethyl siloxane) ($a = 0.5$) in ODCB at 87°C. They concluded that universal calibration based on hydrodynamic volume or on dimensions given by equations (2.24) to (2.26), was valid, whereas molecular dimensions derived from equation (2.27) did not give a universal calibration when the difference in the values of a for the two polymers is large. Because of the different solution viscosity expressions for deducing values of $(\langle r^2 \rangle)^{1/2}$ and $(\langle s^2 \rangle)^{1/2}$, it appears preferable to regard $[\eta]M$, or a linear hydrodynamic dimension, as the universal calibration parameter, although no ambiguity will arise when the eluant has a value of a which is about the same for all polymers.

The experimental results discussed in the preceding two paragraphs provide considerable support for the universal calibration method proposed by Grubisic et al. [100], so that at a given retention volume the relation

$$\log [\eta]_p M_p = \log [\eta]_{ps} M_{ps} \tag{2.31}$$

will apply. Here p refers to a polymer requiring analysis and subscript ps to polymer standards, which will be polystyrene standards for organic eluants such as those in Table 2.1 and will be poly(ethylene oxide) standards and/or dextran reference materials for aqueous eluants. The precision of molecular weight and the polydispersity of a reference material, which were discussed in Sections 2.2 and 2.3, must also be considered in experiments aimed at testing equation (2.31) and in calibration methods for finding M_p from equation (2.31) (see Section 2.4.3). Boni et al. [106] recognized that the number-average molecular weight should be used for M in equation (2.24) for polydisperse samples. The calculation of universal calibration parameters involving the number-average molecular weight has also been discussed by other workers [113-115]. Consequently, V_h and R_e are number-average size parameters for polydisperse samples. However, the number-average size, or $\bar{M}_n$, of a sample is generally not the average at the peak of a chromatogram at V_R, which often lies between the weight average and the geometric mean of the number average and the weight average [see equations (2.9) and (2.10) and the experimental results cited in Section 2.3]. Much of the data in the literature claiming the validity of the universal calibration plot has involved a plot of ln $[\eta]\bar{M}_w$ versus V_R. Dawkins and Hemming [29] examined calibration plots of ln $[\eta]\bar{M}_w$ versus V_R and ln $[\eta]\bar{M}_n$ versus $\bar{V}_n$ for poly(dimethyl siloxane) fractions with $\bar{M}_v \simeq \bar{M}_w$. It was observed that these two plots produced the same calibration curve. Similar conclusions are observed in the results reported for samples of cellulose nitrate and polyoxypropylene [116]. Therefore, the preferred method involves $\bar{M}_n$, but determination of $\bar{V}_n$ may be difficult, whereas the use of $\bar{M}_w$ is less satisfactory from a theoretical viewpoint but may be regarded as a correction for the polydispersity of a sample when used in a plot with V_R at the peak of a chromatogram whose determination is often straightforward. It is necessary to emphasize that fractions which have been carefully characterized must be used for testing universal calibration plots. The omission by Gilding et al. [117] of a discussion on how the polydispersity and the accuracy of the molecular weights of their fractions of polyisobutylene might influence calibration plots makes it difficult to judge their claim that molecular dimensions derived with equation (2.27) should be preferred for universal calibration. They reported calibration data for polystyrene standards (a = 0.744) and for narrow distribution fractions of polyisobutylene (a = 0.5) in benzene at 25°C. Even for this difference in polymer-solvent interaction, fractions having high molecular weight should be preferred since a falls to 0.5 for short chains in good solvents [48]. Consequently, a unique definition of the universal size parameter can be established only with high polymers whose fractions are generally quite polydisperse so that accurate assignment of M_{peak} will be necessary. The complementary discussion of this problem is given in Chapter 7.

2.4.2. Limitations

Polymer Geometry

In view of the theoretical support for a steric exclusion mechanism in which elution is determined by the mean external length for any molecular model [98], it is of interest to determine whether the hydrodynamic volume calibration method is valid for other molecular geometries apart from linear chains. The data in Figure 2.8 include branched polymers, and two reports during 1967 suggested that the hydrodynamic volume plot was also applicable for universal calibration for branched (or low density) and linear polyethylene [118,119]. The separation of polyethylene containing long-chain branching has been studied thoroughly [120], and almost without exception, hydrodynamic volume has been found to be a valid universal calibration parameter. Many of the methods developed for the determination of molecular weight and the degree of long-chain branching in branched polymers depend on equation (2.31) being correct for both linear and branched polymers [121]. On the other hand, experimental results for highly branched polystyrene indicate that the hydrodynamic volume method is incorrect for universal calibration [54,122,123]. A theoretical analysis with elution determined by the mean molecular projection indicates that $[\eta]M$ should be a useful universal calibration parameter for polymers that are linear and have a limited amount of long-chain branching [98]. Ambler [124] has discussed these deviations and recommended a correction to the hydrodynamic volume procedure with the empirical equation

$$\log\ [\eta]_b M_b = \log\ ([\eta]_\ell M_\ell)(2 - g^{0.5}) \tag{2.32}$$

where the subscript b refers to a polymer with long-chain branching and subscript ℓ to the linear polymer. Here g is the ratio of the mean-square radii of gyration of the branched and linear polymers having the same molecular weight and may be calculated from theoretical expressions for the influence of the type and degree of branching on molecular dimensions.

Equation (2.24) may be extended to molecules having a nonspherical geometry. Simha [125] proposed the equation

$$[\eta]M = \gamma N_A V_h \tag{2.33}$$

where the parameter γ is constant (0.025) for spheres but varies with the molecular weight of nonspherical polymers. For a prolate ellipsoid having semimajor and semiminor axes x and y, γ is given by

$$\gamma = 0.14\left(\frac{x}{y}\right)^{1.8} \qquad (2.34)$$

when the axial ratio is large [126]. Equating the volumes and lengths of a prolate ellipsoid and a rigid rod permits the determination of γ in equation (2.33) for rigid rod polymers. Casassa [98] has noted that when the mean external length of a molecule controls elution the values of $[\eta]M$ in equations (2.25) and (2.33) coincide only for $x/y = 33$. This suggests a limited molecular size range in which the hydrodynamic volume plot should be universal for both random coil and rodlike polymers. Theory therefore predicts that hydrodynamic volume will not be a universal calibration parameter for all types of polymers.

Since it follows from equation (2.33) that the hydrodynamic behavior is markedly affected by molecular shape, elution experiments with polymers having different geometries represent a critical test of any universal calibration method. The rodlike conformation of poly(γ-benzyl-L-glutamate) (PBLG) in N,N-dimethylformamide (DMF) is well documented [126]. Grubisic et al. [127] examined the separation behavior of polystyrene standards and samples of PBLG in dimethylformamide with cross-linked polystyrene gels. They proposed that equation (2.31) was valid for both flexible coil and rigid rod polymers. However, their samples of PBLG had relatively broad MWD, and the identification of M_{peak} was not discussed. Furthermore, it is now established that DMF is not entirely satisfactory for the establishment of universal calibrations with polystyrene standards separating on cross-linked polystyrene gels because V for polystyrene is higher than expected from a steric exclusion mechanism. Dawkins and Hemming [76] examined separations with dimethylacetamide as solvent at 80°C. It was established that dimethylacetamide is a good solvent for polystyrene at 80°C and that PBLG has a rigid rod conformation and does not associate in dimethylacetamide at 80°C. The samples of PBLG were well characterized so that M_{peak} values were readily established. It was confirmed that equation (2.31) applies to both flexible coil and rigid rod polymers separating with cross-linked polystyrene gels, as shown in Figure 2.9. Furthermore, the elution data were interpreted

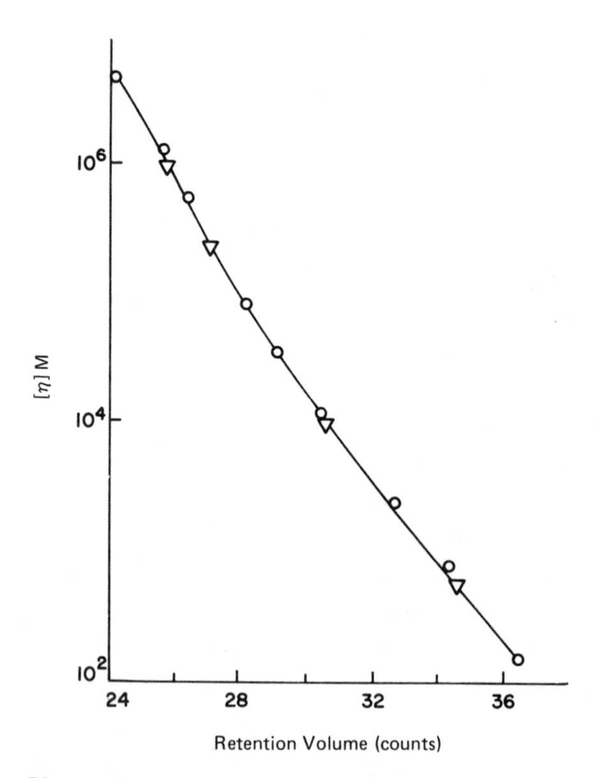

Figure 2.9. Universal calibration plot for cross-linked polystyrene gels with N,N-dimethylacetamide at 80°C. ○, Polystyrene standards; ▽, poly(γ-benzyl-L-glutamate). (From Ref. 76, © IPC Business Press Ltd.)

in terms of the theoretical treatment of Giddings and co-workers [97] for separations with packings assumed to have pores with a parallel plane geometry. The polystyrene molecules were represented as solid equivalent spheres with diameter $2R_e$. Good agreement between experimental results and theoretical prediction was obtained suggesting that the parallel plane model is a useful representation of pore geometry. This is consistent with the conclusions of Casassa and Tagami [94] for other experimental data.

The plots of log $[\eta]M$ versus V_R in Figures 2.8 and 2.9 permit the determination of a molecular weight calibration curve (see Section 2.4.3) which is then used with a chromatogram in the calculation of a MWD. The theoretical treatment of solute separation by a steric exclusion mechanism by Giddings and co-workers [97] predicts that log (distribution coefficient) is linearly related to the mean molecular projection of a solute. Consequently, calibration data should be correlated by a plot of log V_R versus mean molecular projection. This semilogarithmic plot has been examined by Hester and Mitchell [128], who employed $[\eta]M$ to represent solute size. Although this plot is justified theoretically, the inverse plot in Figure 2.8 is more useful from a practical standpoint, since the presentation of distributions in terms of log M facilitates the comparison of samples having wide distributions.

Polymer Concentration

The intrinsic viscosity defined in equation (2.24) is obtained by extrapolating experimental data for solution viscosity to infinite dilution. Calibration methods must be performed at finite polymer concentrations, and in the earliest experiments with cross-linked polystyrene gels [129,130], a shift of V_R to higher V was demonstrated as the concentration of injected polymer increased. This concentration effect is discussed in Chapter 4 and a listing of references is given elsewhere [39]. The shift to high V becomes more pronounced as the molecular weight of the polymer increases, as the polydispersity of the polymer is reduced, and as polymer-solvent interaction increases; that is, the concentration effect is less important for a theta solvent than for a good solvent [16,130-134].

Interpretations of experimental data for the concentration effect in terms of theoretical treatments of dilute polymer solutions have been proposed [135-138]. These theories consider that the dimensions of flexible coil polymers decrease with increasing polymer concentration in solution. Consequently, an increase in polymer concentration is accompanied by a decrease in V_h and an increase in V_R. The degree of expansion and contraction of a flexible coil polymer is determined by the excluded volume effect, which is represented by a linear expansion factor α [99]. Polymer solution theory predicts that α decreases as polymer-solvent interaction and polymer molecular weight decrease and as polymer concentration increases. For solutions at infinite dilution, equation (2.25) may be extended to

$$[\eta]M = \Phi_0 \alpha^3 (\langle r_0^2 \rangle)^{3/2} \tag{2.35}$$

where $\langle r_0^2 \rangle$ is the unperturbed mean-square end-to-end distance of the polymer in a theta solvent when α is unity [99]. In a good solvent the molecular dimensions increase with $\alpha > 1.0$. Because of the dependence of α on polymer concentration, Rudin and Hoegy [139] proposed that the hydrodynamic volume should be estimated at finite polymer concentration for universal calibration plots. However, this is not necessarily a straightforward procedure. First, there is a difference between the concentration of polymer in the injected solution and in the eluant at the column outlet [138,140,141]. Second, the variation of polymer dimensions with polymer concentration in bulk solution may not apply to a polymer molecule in a pore [142]. Third, a shift in V_R with polymer concentration may result from other phenomena, such as the viscosity of the polymer solution in the mobile phase and secondary exclusion owing to occupancy of a pore by another polymer molecule [143-145]. In view of all these observations, which are discussed in Chapter 4, the rigorous application of hydrodynamic volume for universal calibration should involve experimental conditions selected so that the concentration effect is insignificant. However, for polymers of fairly high molecular weight in a good solvent, a shift in V_R may be unavoidable. It is recommended that experimental data for the concentration effect should be extrapolated to find the value of V_R corresponding to infinite dilution [146], and this value together with V_h may be employed in a universal calibration plot.

Solvent

The results cited in Tables 2.1 and 2.2 are for polymers in good solvents. The eluants are all good solvents for polystyrene with a_{ps} in the range 0.67-0.80 and have a solubility parameter similar to the value $\delta = 8.6$ $[(J \cdot cm^3)^{1/2}]$ for polystyrene [112]. Highly polar polymers will dissolve only in solvents that have a solubility parameter much larger than δ = 18.6. Thus DMF, which is widely employed for separations of polar polymers with cross-linked polystyrene gels has δ = 24.7. Very polar eluants are generally poor solvents for polystyrene, and we shall define a *poor solvent* as an eluant with a in the range 0.50-0.65.

The validity of the hydrodynamic volume calibration concept for polystyrene in various solvents has been investigated for cross-linked polystyrene gels [147-154]. A good example of the displacement that can occur for polystyrene in a poor solvent is shown in Figure 2.10. As discussed in Chapter 4, the displacement of plots of log [η]M versus V_R for polystyrene to high V_R arises from polymer retardation

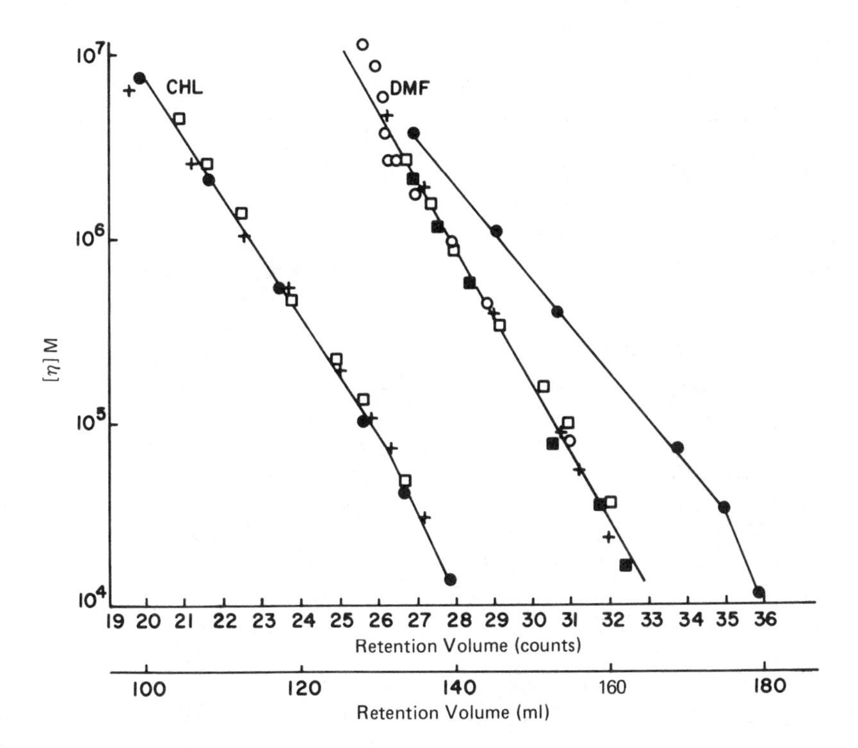

Figure 2.10. Universal calibration plots for cross-linked polystyrene gels with chloroform and N,N-dimethylformamide at 20°C. ●, Polystyrene; □ + ■, copolymers of styrene and acrylonitrile; ○, polyacrylonitrile; (From Ref. 148.)

due to secondary adsorption and partition mechanisms resulting from polymer-gel interactions.

Several polar polymers having $\delta > 18.6$ are included in Figure 2.8. For the eluants in Table 2.1 and for other good solvents for linear polystyrene, it is not easy to predict whether secondary mechanisms will occur for a polar polymer separating on cross-linked polystyrene gels. A comprehensive listing of universal calibration studies demonstrates that equation (2.31) is applicable to a wide range of polymers, including polymers containing polar groups [39]. However, the possibility of specific polymer-gel interactions should always be considered.

For example, Dubin et al. [155] demonstrated that retardation with cross-linked polystyrene gels was dependent on polymer type for polar polymers in DMF, as shown in Figure 2.11. Displacement of the hydrodynamic volume plot for cellulose nitrate ($\delta = 21.7$) in acetone ($\delta = 20.3$) [156] and for polyamides in m-cresol ($\delta = 20.9$) and o-chlorophenol [157,158], among others, has been reported (see Chapter 4). Thus it is recommended that δ(eluant) > δ(polymer) in order to minimize polymer-gel interactions [152].

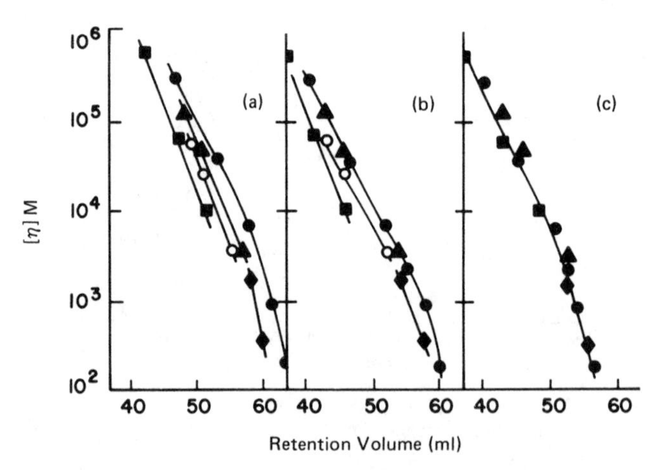

Figure 2.11. Universal calibration plots for cross-linked polystyrene gels (a), silanized porous glass (b), and porous glass (c), with N,N-dimethylformamide. ●, Polystyrene; ▲, poly(methyl acrylate); o, poly(vinyl pyrrolidone); ■, poly(p-nitrostyrene); ♦, poly(ethylene oxide). (From Ref. 155, © John Wiley & Sons, Inc.)

Inorganic Packings

Porous silica and porous glass packings have always appeared attractive for polymer fractionation because of their excellent thermal and mechanical stability and well-defined pore size distribution. However, because of the active surface sites, polymer retardation, and even irreversible polymer adsorption, is always possible [159]. The first attempt to study universal calibration for inorganic packings was performed by Moore and Arrington [160], who separated polystyrene with porous glass using a binary liquid mixture as eluant consisting of methyl ethyl ketone (δ = 19.0) and isopropanol (δ = 23.5). It may be assumed that isopropanol, because of its considerable affinity for the surface of the porous glass, will tend to reduce the adsorption of polystyrene. Moore and Arrington [160] also performed separations with benzene (δ = 18.8) as eluant, showing that polystyrene had a higher value of V_R than for the binary liquid mixture. Although benzene is a good solvent for polystyrene, retardation occurs because polystyrene and benzene have a similar affinity for the pore surface.

The plots of log (hydrodynamic volume) for polystyrene in an eluant formed from a similar binary liquid mixture are always displaced to low V_R with respect to plots for polystyrene in single liquids which are good solvents for polystyrene [161-166]. Single solvents such as 1,2-dichloroethane (δ = 20.1), benzene (δ = 18.8), CHL (δ = 19.0), methyl ethyl ketone (MEK) (δ = 19.0), and carbon tetrachloride (δ = 17.6) all have a polarity similar to that of polystyrene, so that polystyrene can compete for surface adsorption sites. When the nonsolvent component in the binary liquid mixture is less polar than the solvent component (e.g., elutions of polystyrene with a mixture of n-heptane and MEK) [163-165], the plot of log (hydrodynamic volume) is displaced to high V_R with respect to the plot for polystyrene in MEK. An increase in the fraction of n-heptane reduces the polarity of the binary liquid mixture, which also becomes a poorer solvent and eventually a theta solvent. Both reduced eluant polarity and polymer-solvent interaction will facilitate polymer-gel interactions [159,162,165,167]. Even though hydrodynamic volume may appear to be a universal size parameter [e.g., the results of Kotera et al.

[168] for porous glass with polystyrene in diethylmalonate (theta solvent) and THF (gcod solvent)], the separations may be influenced by specific eluant-packing interactions. A discussion of these secondary adsorption and partition mechanisms is presented in Chapter 4.

In view of the experimental observations for polystyrene, the separation behavior of polar polymers with inorganic packings is likely to be influenced by secondary mechanisms. Thus Nakamura and Endo [169] reported that polystyrene and poly(ethylene oxide) in THF at 60°C followed the same plot of log $[\eta]M$ versus V_R for cross-linked polystyrene gels, whereas the plot for poly(ethylene oxide) was displaced to high V_R for porous glass, the same shift reported by Figueruelo et al. [170,171] for poly(methyl methacrylate) with silica as column packing.

Anomalous retention volumes have been observed for some polar polymers in pure DMF both for inorganic packings and cross-linked polystyrene gels. Both charged and nonionic polymers, in particular polar polymers containing nitrogen, may behave as polyelectrolytes in DMF [172-178]. The polar chains associate in DMF and the aggregate formation is prevented in the presence of salt, which interacts with both polymer and solvent [172,179,180]. However, the addition of salt will tend to make the eluant an even poorer solvent for polystyrene, thus facilitating interaction effects with the pore surface when universal calibration based on hydrodynamic volume will not be satisfactory. For more details, see Chapter 4.

Silica packings having well-defined surface-bonded phases have been reported [181-184]. For some polymer-solvent systems secondary mechanisms will occur, and partition effects can produce separations in which retention volume rises as molecular weight increases (see Chapter 4).

Aqueous Separations

In the last several years there has been considerable interest in the separation of water-soluble polymers with microparticulate inorganic packings [185,186]. The validity of the calibration method based on hydrodynamic volume may depend on polymer type and eluant composition. For polyelectrolytes it is recognized that the ionic strength of the eluant must be sufficiently high to minimize interactions between charges on the surface of the column packing and ionic groups on the polyelectrolyte [187]. Secondary mechanisms arising from ion exchange, ion exclusion, ion inclusion, or adsorption are then reduced. The addition of salt to the eluant is advantageous not only for suppressing these secondary mechanisms but also for lowering electrostatic repulsions along the polyelectrolyte chain. Consequently, the hydrodynamic volume of the polyelectrolyte is lowered, which decreases the incidence of polymer concentration effects, which can be more severe than for hydrophobic polymers in organic media [186]. The plot in Figure 2.12 of log $[\eta]M$ versus V_R for polystyrene sodium sulfonate separating on porous glass is shown to be universal at two different concentrations of sodium sulfate in the eluant [188]. This universal plot has been confirmed for porous silica by Rinaudo and coworkers [189,190], who recommended that the ionic strength of the eluant should exceed 0.05 M.

Dextran reference materials are widely used for calibrating columns for the separation of water-soluble polymers. Calibration data for dextrans are in good agreement with the universal calibration curve for polystyrene sodium sulfonate in Figure 2.12. It is expected that nonionic hydrophilic polymers should be less influenced by eluant ionic strength than polyelectrolytes. Some experimental data do suggest that chromatographic results for dextrans are not influenced significantly by the presence of electrolytes, so that water may be used as eluant [90,187,188,191]. However, other workers have indicated that low concentrations of salt should be present in the aqueous eluant because secondary electrostatic effects arising from negative charges on the dextran chain may occur [192,193].

For silica packings electrostatic effects between polymer and pore surface may arise from negative charges generated by the dissociation of silanol groups in aqueous media. For anionic polyelectrolytes, electrostatic repulsion occurs, generating a secondary ion exclusion mechanism, so that the retention volume of the polyelectrolyte is reduced in the absence of salt in the eluant. Cationic

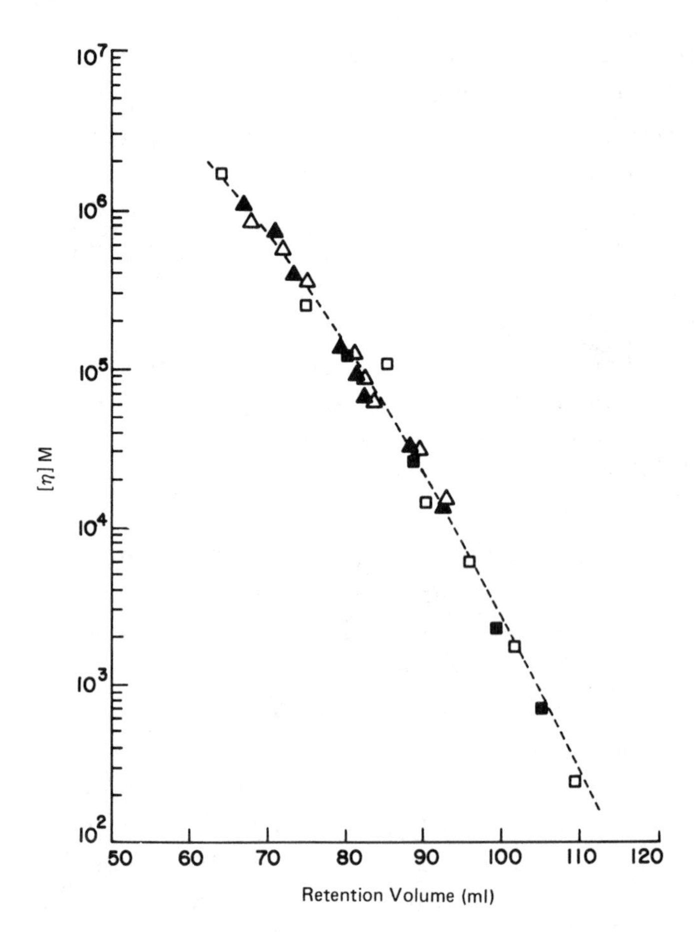

Figure 2.12. Universal calibration for porous glass with aqueous sodium sulfate at salt concentrations 0.2 M (open symbols) and 0.8 M (filled symbols). □■, Polystyrene sodium sulfonate; △▲, dextran. (From Ref. 188, © John Wiley & Sons, Inc.)

polyelectrolytes therefore tend to generate attractive interactions. It has been reported that cationic polyelectrolytes may be separated with porous glass to which a bonded phase containing a quaternary ammonium group was attached [194]. However, plots of log (hydrodynamic volume) for dextrans and samples of poly(2-vinyl pyridine) in eluants consisting of acidic salt solutions did not superimpose. Not all silanol groups may dissociate in aqueous media, and undissociated silanol groups may hydrogen bond with the polymer. If hydrogen bonding leading to higher V_R is suspected, the addition of urea or guanidine hydrochloride may reduce the interaction [186]. Because of the secondary mechanisms resulting from silanol groups, silicas with bonded phases (e.g., resulting from the reaction with γ-glycidoxypropyltrimethoxysilane) have been advocated [186]. However, two problems remain. First, not all silanol groups are reacted, and second, adsorption or partition of the polymer with the bonded phase may occur at high ionic strength of the eluant via a hydrophobic interaction [195,196]. It is emphasized therefore that the validity of a universal calibration method for a water-soluble polymer will depend on a careful choice of eluant composition after considering the nature of possible polymer-packing interactions.

2.4.3. Molecular Weight Equations

As long as a plot of log $[\eta]M$ versus V_R is valid, a molecular weight calibration curve established with polystyrene standards permits the calculation of the molecular weight calibration M_p for a second polymer with equation (2.31). At a given V_R the calibration curves are related by

$$\log M_p - \log M_{ps} = \log \frac{[\eta]_{ps}}{[\eta]_p} \qquad (2.36)$$

It follows that polymers will have different molecular weight calibrations unless the molecular weight dependence of $[\eta]$ is the same for both polymers. Equation (2.36) permits the determination of M_p, when the dependence of $[\eta]_{ps}$ and $[\eta]_p$ on retention volume has been established. Experimentally, the right-hand side of equation (2.36) is obtained by measuring $[\eta]$ with an on-line viscometer detector. These measurements are discussed in Chapter 5.

The simple relationship between $[\eta]$ and M given by equation (2.29) may be substituted into equation (2.36) for *linear homopolymers* to give after rearrangement

$$\log M_p - \frac{1 + a_{ps}}{1 + a_p} \log M_{ps} = \frac{1}{1 + a_p} \log \frac{K_{ps}}{K_p} \qquad (2.37)$$

where K_{ps}, a_{ps} and K_p, a_p are the Mark-Houwink constants for polystyrene and the second polymer. Mark-Houwink constants may be determined by classical methods involving preparative fractionation of the polymer and determination of the intrinsic viscosities and molecular weights of the isolated fractions. Much of the data given in Tables 2.1 and 2.2 involved such procedures. Problems may arise when the fractions are polydisperse since the hydrodynamic volume is a number-average parameter. It is then convenient to consider that values of K and a should correspond to those for a monodisperse polymer [86]. For polydisperse fractions that are characterized by light scattering in order to find K and a, Whitehouse [86] proposed that equation (2.29) should be modified to

$$[\eta] = K\bar{M}_w(1 - \Delta) \tag{2.38}$$

where Δ is a function of polydispersity. Mark-Houwink constants are available in the literature [112,197,198]. However, for a wide molecular weight range, a unique pair of values of K and a may not apply [199] because of the molecular weight dependence of α in equation (2.35) [99]. It is advisable to use Mark-Houwink constants only over the molecular weight range of the fractions employed in establishing equation (2.29).

An alternative method for determining Mark-Houwink constants was reported by Weiss and Cohn-Ginsberg [200]. This method requires a well-characterized polydisperse polymer having two known experimental values of $[\eta]$, $\bar{M}_w$, and $\bar{M}_n$ which for a normalized distribution of the polymer are defined by

$$[\eta] = \sum_i W_i[\eta]_i \tag{2.39}$$

$$\bar{M}_w = \sum_i W_iM_i \tag{2.40}$$

$$\bar{M}_n = \frac{1}{\sum_i (W_i/M_i)} \tag{2.41}$$

where W_i is the weight fraction of component i having molecular weight M_i and intrinsic viscosity $[\eta]_i$. A universal calibration curve determined experimentally with polystyrene standards is assumed to be valid, and Weiss and Cohn-Ginsberg [200] define a parameter J by

$$J_i = [\eta]_i M_i \tag{2.42}$$

for each component i in the polymer. With equations (2.29) and (2.42), equations (2.39) to (2.41) may be expressed as

$$[\eta] = K_p^{1/(a_p+1)} \sum_i W_i J_i^{a_p/(a_p+1)} \tag{2.43}$$

$$\bar{M}_w = K_p^{-1/(a_p+1)} \sum_i W_i J_i^{1/(a_p+1)} \tag{2.44}$$

$$\bar{M}_n = \frac{K_p^{-1/(a_p+1)}}{\sum_i (W_i/J_i^{1/(a_p+1)})} \tag{2.45}$$

Each of these equations contains the unknown constants K_p and a_p. A typical experimental procedure involves the experimental determination of $[\eta]$, $\bar{M}_n$ (e.g., by osmometry) and a chromatogram, in order to define W_i, for the polydisperse polymer. Combination of equations (2.43) and (2.45) gives

$$[\eta]\bar{M}_n = \frac{\sum_i W_i J_i^{a_p/(a_p+1)}}{\sum_i (W_i/J_i^{1/(a_p+1)})} \tag{2.46}$$

which can be solved by trial-and-error methods in order to find a_p [200]. The value of K_p is then obtained from equation (2.43) or (2.45). Alternatively, equations (2.44) and (2.45) could be employed if $\bar{M}_w$ is known from light-scattering measurements, but this combination is likely to be less satisfactory than equation (2.46) because $[\eta]$ can be determined much more accurately than $\bar{M}_n$ and $\bar{M}_w$. The combination of equations (2.43) and (2.44) is not recommended because errors in molecular weight determinations will be important in the trial-and-error calculations when values of $\bar{M}_v$ and $\bar{M}_w$ are close together. Since viscometry is the most accurate characterization technique, an alternative method is to use only equation (2.43) with two polydisperse samples having different intrinsic viscosities $[\eta]_1$ and $[\eta]_2$ and known chromatograms to define W_{i1} and W_{i2}. Combining equation (2.43) for the two samples, it can be shown that

$$\frac{[\eta]_1}{[\eta]_2} = \frac{\sum_i W_{i1} J_i^{a_p/(a_p+1)}}{\sum_i W_{i2} J_i^{a_p/(a_p+1)}} \tag{2.47}$$

which may be solved for a_p. The value of K_p is then found with equation (2.43) for one of the polydisperse samples.

Methods based on equations (2.46) and (2.47) have been examined [72,115,201-209]. Morris [201] reported values of K_p and a_p for samples of styrene-butadiene copolymers, but the agreement between values derived from equations (2.46) and (2.47) was poor. The use of equation (2.47) for samples of butyl rubber produced values of K_p and a_p that depended on sample selection [201]. Samay et al. [206] compared Mark-Houwink constants determined by the classical approach involving polymer fractionation and by the method proposed by Weiss and Cohn-Ginsberg [200]. Six different methacrylate homopolymers and copolymers in THF were examined. Samay et al. [206] concluded that values of K_p and a_p deduced by equations (2.46) and (2.47) are not very accurate. Dobbin et al. [208] also indicated that equation (2.47) produced less reliable values of K_p and a_p for poly(methyl methacrylate) in THF than those produced by the classical approach. These workers also reported that equation (2.46) is not satisfactory because the low molecular weight tail of a chromatogram which has a pronounced effect on the calculation of $\overline{M}_n$ is influenced by chromatogram broadening. Hamielec and Omorodion [87] have shown how corrections for chromatogram broadening may be incorporated into the method, giving results for samples of poly(vinyl chloride) and dextran. In summary, although methods based on equations (2.46) and (2.47) require relatively little experimental effort, the results for K_p and a_p tend to be less reliable than values determined by the classical approach involving fractionation.

Methods for determining K_p and a_p, which are variations on the procedure proposed by Weiss and Cohn-Ginsberg [200], have been examined. Zhongde et al. [210] employed equation (2.47) with the simplification that the chromatogram was represented by a Gaussian distribution, thus facilitating the calculation of W_i. Reasonable values for K_p and a_p for polybutadiene in THF were obtained. Mahabadi and O'Driscoll [211] utilized a molecular weight calibration curve M_{ps} established with polystyrene standards rather than the universal calibration curve defined by equation (2.42). These two calibrations are related by equations (2.31) and (2.37). It is convenient to define equation (2.37) by

$$M_p = \kappa(M_{ps})^{\beta} \tag{2.48}$$

where

$$\kappa = \left(\frac{K_{ps}}{K_p}\right)^{1/(1+a_p)} \tag{2.49}$$

and

$$\beta = \frac{1 + a_{ps}}{1 + a_p} \tag{2.50}$$

Mahabadi and O'Driscoll [211] show that equation (2.46) then becomes

$$\frac{[\eta]\bar{M}_n}{K_{ps}} = \frac{\sum_i W_i M_{psi}^{\beta a_p}}{\sum_i W_i / M_{psi}^{\beta}} \tag{2.51}$$

where W_i is obtained from the chromatogram for the polymer p having known values of $[\eta]$ and $\bar{M}_n$ and M_{psi} is "the molecular weight of component i" from the calibration curve established with polystyrene standards. Trial-and-error methods then permit the determination of a_p from equation (2.51), and the value of K_p is obtained from relations in terms of M_{psi} corresponding to equations (2.43) and (2.45). Chaplin and coworkers [80,84] have reported reasonable values of K_p and a_p for poly(methyl methacrylate), poly(vinyl chloride), and polybutadiene in THF. A further modification has been described by Mori [212], who proposes the use of a polydisperse polymer with known experimental values of $\overline{M}_w$ and $\overline{M}_n$. It is assumed that the calibration curve M_{ps} has been determined so that with equation (2.48) it follows that equations (2.40) and (2.41) may be transformed to

$$\overline{M}_w = \sum_i W_i \kappa M_{psi}^{\beta} \tag{2.52}$$

$$\overline{M}_n = \frac{1}{\sum_i W_i / \kappa M_{psi}^{\beta}} \tag{2.53}$$

By trial-and-error or search methods, it is then possible to match the experimental value of $\overline{M}_n$ with the computed value from the chromatogram, and likewise for $\overline{M}_w$, in order to find the correct values of κ and β in equations (2.52) and (2.53). Since a_{ps} and K_{ps} are known for most SEC solvents, equations (2.49) and (2.50) can be solved to determine a_p and K_p. Mori [212] demonstrated that the determination of κ and β was influenced by chromatogram broadening. This method did not yield very accurate values for K_p and a_p for poly(methyl methacrylate), poly(vinyl chloride), and poly(vinyl acetate) in THF [212], and methods based on equation (2.47) appear to be preferable. Dobbin et al. [208,209] have shown how equation (2.47) may be extended to provide values of K_p and a_p for a polymer in a solvent that is not the SEC eluant. This appears to be a useful modification for separations involving eluants which are nonsolvents for polystyrene [117,213].

The use of equation (2.37) is illustrated in Figure 2.13 for polystyrene standards and poly(dimethyl siloxane) fractions in ODCB at 87°C. Since polystyrene is in a good solvent and poly(dimethyl siloxane) is in a theta solvent, it follows from equation (2.37) that the calibration curves log M_{ps} and log M_p will be nonparallel. The calculated curve for log M_p from equation (2.37) in Figure 2.13 is in good agreement with the experimental data for poly(dimethyl siloxane) fractions. It is evident from equation (2.25) that equation

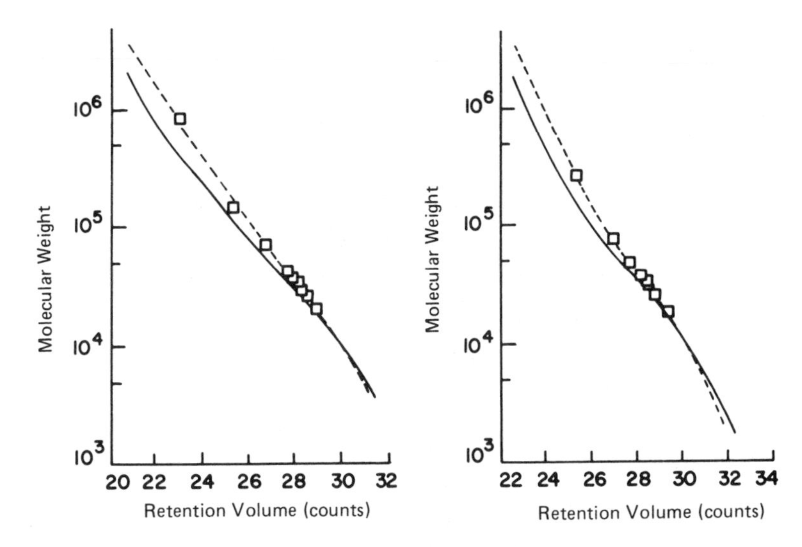

Figure 2.13. Molecular weight calibration plots for cross-linked polystyrene gels with o-dichlorobenzene at 87°C. □, poly(dimethyl siloxane) fractions; ——, M_{ps} calibration for polystyrene standards; ---, M_p calibration for poly(dimethyl siloxane) calculated with equation (2.37). (From Ref. 29 by permission of the publishers, Huthig and Wepf Verlag, Basel.)

(2.37) will apply for universal calibration for size parameters defined by $[\eta]M$, $(\langle s^2 \rangle)^{1/2}$, or $(\langle r^2 \rangle)^{1/2}$. However, if $(\langle r^2 \rangle)^{1/2}$ [or $(\langle s^2 \rangle)^{1/2}$], as defined by equation (2.27), is assumed to be the universal calibration parameter, the relation for calculating the calibration curve M_p from M_{ps} is given by [214].

$$\log M_p - \frac{1 + a_{ps}}{1 + a_p} \log M_{ps} = \frac{1}{1 + a_p} \log \frac{K_{ps}}{K_p} \frac{\Phi(\varepsilon)_p}{\Phi(\varepsilon)_{ps}} \qquad (2.54)$$

where values of $\Phi(\varepsilon)$ are calculated with equations (2.28) and (2.30). For the polymers having $a_{ps} = 0.73$ and $a_p = 0.5$ in Figure 2.13, M_p calibration curves calculated with equations (2.37) and (2.54) will be different. Dawkins and Hemming [29] indicated that their data were more consistent with equation (2.37) than with equation (2.54). Gilding et al. [117], who examined polystyrene standards ($a_{ps} = 0.74$) and polyisobutylene fractions ($a_p = 0.5$) in benzene at 25°C, preferred equation (2.54). Equations (2.37) and (2.54) become very similar when good solvents for both polymers are used, so that it is not possible to distinguish between these two equations when $a_{ps} \sim a_p$ [50,103].

The experimental results reviewed in Section 2.4.2 and in Chapter 4 clearly indicate that eluants which are good solvents should be preferred for polymer separations with cross-linked polystyrene gels. Polymers will then have similar polymer-solvent interactions. It follows from equation (2.37) that the M_{ps} and M_p calibrations will be parallel when $a_{ps} = a_p$. The right-hand side of equation (2.37) can then be considered as a shift factor. Although the Mark-Houwink constants may not be known, a relation analogous to equation (2.37) may still be employed. For two polymers having $a_{ps} = a_p$, it can be shown that K is related to K_θ (when a = 0.5) [77,215], as defined by

$$K_\theta = \Phi_0 \left(\frac{\langle r_0^2 \rangle}{M} \right)^{3/2} \tag{2.55}$$

from equation (2.35), where $\langle r_0^2 \rangle / M$ is a constant for a polymer and is independent of molecular weight and retention volume ($\langle r_0^2 \rangle$ being unperturbed end-to-end distance). It can be shown that equation (2.37) may then be simplified to

$$\log M_p - \log M_{ps} = \frac{1}{1 + a_{ps}} \log \frac{K_{\theta_{ps}}}{K_{\theta_p}} \tag{2.56}$$

The results in Figure 2.2 for polystyrene ($a_{ps} = 0.7$) and polyethylene ($a_p = 0.7$) in ODCB at 138°C illustrate parallel calibrations, and the shift factor calculated with equation (2.37) is log 0.466, whereas a value of log 0.477 is obtained with equation (2.56) [215]. Equation (2.56) has the merit that values of a_{ps} have now been determined for many SEC eluants (e.g., Table 2.1), and data for K_θ have been tabulated for many polymers [112]. An alternative expression to equation (2.56) may be obtained by considering equation (2.35). When two polymers have similar polymer-solvent interactions, their values for

α^3 are very similar, so $\langle r_0^2 \rangle$ should be as satisfactory as $[\eta]M$ for universal calibration [27]. The M_{ps} and M_p calibrations are then related by [216]

$$\log M_p - \log M_{ps} = \log \left(\frac{\langle r_0^2 \rangle}{M} \right)_{ps} \left(\frac{M}{\langle r_0^2 \rangle} \right)_p \tag{2.57}$$

Equation (2.57) when written in terms of K_θ

$$\log M_p - \log M_{ps} = \frac{2}{3} \log \frac{K_{\theta_{ps}}}{K_{\theta_p}} \tag{2.58}$$

is a special case of equation (2.56) with $a_{ps} = 0.5$. The usefulness of equation (2.57) has been demonstrated [40,48,77,215,216]. Funt and Hornof [217] have provided support for universal calibration involving a shift factor when $a_{ps} \sim a_p$.

The application of equation (2.37) to a wide molecular weight range that includes short chains may lead to errors. Theory predicts that α falls to unity for short chains [99], so that from equations (2.29), (2.35), and (2.55), a must tend to 0.5 for short chains even in good solvents. This behavior has been confirmed experimentally for a number of polymers for the molecular weight range 1000-10,000 [218]. Consequently, equations (2.36), (2.37), and (2.56) to (2.58) are then identical. Experiments by Donkai et al. [219] have shown that equation (2.57) is valid for six types of polymers in THF for the molecular weight range 282-72,500. Therefore, when changes in K and a as a function of molecular weight may be difficult to assess for low polymers, equations (2.57) and (2.58) may be preferred to equation (2.37).

In the initial work with cross-linked polystyrene gels, calibration plots of log (projected extended chain length)(log L_{max}) versus V_R were constructed with polystyrene standards [2,220]. From this plot of log L_{max} versus V_R a molecular weight calibration for a particular polymer can be obtained with the Q factor, defined as the molecular weight per unit Angstrom assuming a fully extended molecule.

Polymer molecules in solution correspond to flexible coils rather than extended chains. Consequently, an L_{max} calibration will not be correct for a wide range of polymers. Calibration curves plotted in terms of L_{max} may be used to define exclusion limits for cross-linked polystyrene gels [2,220] and the plot is still used by manufacturers of these gels, thus partly explaining the application of the Q factor for calibration by some workers.

If L_{max} is assumed to be a universal calibration parameter, a calibration curve M_p is related to the experimental calibration established with polystyrene standards by

$$\log M_p - \log M_{ps} = \log \frac{Q_p}{Q_{ps}} \tag{2.59}$$

Although equation (2.59) would be acceptable for poly(methyl methacrylate) and poly(dimethyl siloxane) [77], it is incorrect for polyethylene and polyisoprene [77,216]. Heller and Moacanin [221] observed that three vinyl aromatic polymers, having similar values for K and a, followed the same curve on an L_{max} plot. However, it is evident that polymers such as poly(methyl methacrylate) and poly(dimethyl siloxane) in CHL at 30°C having $K_p \sim K_{ps}$ and $a_p \sim a_{ps}$ follow the molecular weight calibration established with polystyrene standards [27,77], so that universal calibration is unnecessary. The same conclusion is reached by inspecting equation (2.37). A more general condition for the validity of universal calibration based on L_{max} was proposed by Dawkins [77], who suggested that polystyrene and the polymer p should have the same value of $Q \langle r_0^2 \rangle / M$, that is,

$$Q_p \left(\frac{\langle r_0^2 \rangle}{M} \right)_p = Q_{ps} \left(\frac{\langle r_0^2 \rangle}{M} \right)_{ps} \tag{2.60}$$

This criterion was verified for polystyrene and polycarbonate having different values of K_{ps} and K_p [48]. Even though equation (2.60) may be valid, equation (2.59) will only be accurate when polymers have similar polymer-solvent interactions (i.e., $a_p \sim a_{ps}$). Since $\langle r_0^2 \rangle$ follows from a more realistic model of a polymer chain in solution, equations (2.56) to (2.58) will apply to a wider range of polymers than will equation (2.59).

ACKNOWLEDGMENTS

This chapter was written during the tenure of a Visiting Professorship in Polymer Science at the Institute of Materials Science in The University of Connecticut. The author wishes to thank Professors R. M. Fitch, J. F. Johnson, and L. V. Azaroff for the invitation to hold this professorship and for providing financial support.

REFERENCES

1. Moore, J. C., J. Polym. Sci., A, *2*, 835, 1964.
2. Maley, L. E., J. Polym. Sci., C, *8*, 253, 1965.
3. Peebles, L. H., *Molecular Weight Distributions in Polymers*, Wiley-Interscience, New York, 1971.
4. Ouano, A. C., Rubber Chem. Technol., *54*, 535, 1981.
5. Rodriguez, F., and Clark, O. K., Ind. Eng. Chem., Prod. Res. Dev., *5*, 118, 1966.
6. Pickett, H. E., Cantow, M. J. R., and Johnson, J. F., J. Appl. Polym. Sci., *10*, 917, 1966.
7. Yau, W. W., and Fleming, S. W., J. Appl. Polym. Sci., *12*, 2111, 1968.
8. Nilsson, G., and Nilsson, K., J. Chromatogr., *101*, 137, 1974.
9. Dawkins, J. V., Br. Polym. J., *4*, 87, 1972.
10. Crouzet, P., Martens, A., and Mangin, P., J. Chromatogr. Sci., *9*, 525, 1971.
11. Van Kreveld, M. E., and Van den Hoed, N., J. Chromatogr., *83*, 111, 1973.
12. Ivory, C. F., and Bratzler, R. L., J. Chromatogr., *198*, 354, 1980.
13. Vander Linden, C., Polymer, *21*, 171, 1980.
14. Billingham, N. C., *Molar Mass Measurements in Polymer Science*, Kogan Page, London, 1976.
15. Boni, K. A., Sliemers, F. A., and Stickney, P. B., Polym. Prepr., *8*, 446, 1967.
16. Boni, K. A., Sliemers, F. A., and Stickney, P. B., J. Polym. Sci., A2, *6*, 1567, 1968.
17. Adams, H. E., Ahad, E., Chang, M. S., Davis, D. B., French, D. M., Hyer, H. J., Law, R. D., Simkins, R. J. J., Stuchbury, J. E., and Tremblay, M., J. Appl. Polym. Sci., *17*, 269, 1973.
18. Yau, W. W., Kirkland, J. J., Bly, D. D., and Stoklosa, H. J., J. Chromatogr., *125*, 219, 1976.
19. Pressure Chemical Co., 3419 Smallman Street, Pittsburgh, PA 15201.

20. Waters Associates, Inc., Maple Street, Milford, MA 01757.
21. Polymer Laboratories Ltd., Essex Road, Church Stretton, Shropshire, England.
22. Toyo Soda Mfg. Co. Ltd., Toso Building, Tokyo, Japan.
23. Phillips Petroleum Company, Bartlesville, OK 74003.
24. Polysciences, Inc., Paul Valley Industrial Park, Warrington, PA 18976.
25. Peyrouset, A., Prechner, R., Panaris, R., and Benoit, H., J. Appl. Polym. Sci., *19*, 1363, 1975.
26. Ackers, G. K., Adv. Protein Chem., *24*, 343, 1970.
27. Dawkins, J. V., J. Macromol. Sci., B, *2*, 623, 1968.
28. Dawkins, J. V., Maddock, J. W., and Coupe, D., J. Polym. Sci., A2, *8*, 1803, 1970.
29. Dawkins, J. V., and Hemming, M., Makromol. Chem., *155*, 75, 1972.
30. Tung, L. H., *Fractionation of Synthetic Polymers: Principles and Practices,* Marcel Dekker, New York, 1977.
31. Dawkins, J. V., and Peaker, F. W., Eur. Polym. J., *6*, 209, 1970.
32. Barlow, A., Wild, L., and Roberts, T., J. Chromatogr., *55*, 155, 1971.
33. Peyrouset, A., and Panaris, R., J. Appl. Polym. Sci., *16*, 315, 1972.
34. Montague, P. G., and Peaker, F. W., J. Polym. Sci., Polym. Symp., *43*, 277, 1973.
35. Cooper, A. R., Hughes, A. J., and Johnson, J. F., J. Appl. Polym. Sci., *19*, 435, 1975.
36. Vaughan, M. F., and Francis, M. A., J. Appl. Polym. Sci., *21*, 2409, 1977.
37. Barlow, A., Wild, L., and Ranganath, R., J. Appl. Polym. Sci., *21*, 3319, 1977.
38. Tung, L. H., Sep. Sci., *5*, 339, 1970.
39. Janča, J., Adv. Chromatogr., *19*, 37, 1981.
40. Dawkins, J. V., and Maddock, J. W., Eur. Polym. J., *7*, 1537, 1971.
41. Ambler, M. R., Fetters, L. J., and Kesten, Y., J. Appl. Polym. Sci., *21*, 2439, 1981.
42. Mori, S., and Suzuki, T., J. Liq. Chromatogr., *3*, 343, 1980.
43. Yau, W. W., Ginnard, C. R., and Kirkland, J. J., J. Chromatogr., *149*, 465, 1978.
44. Cooper, A. R., Bruzzone, A. R., Cain, J. H., and Barrall, E. M., J. Appl. Polym. Sci., *15*, 571, 1971.
45. Cooper, A. R., and Johnson, J. F., J. Appl. Polym. Sci., *15*, 2293, 1971.
46. Longman, G. W., Wignall, G. D., Hemming, M., and Dawkins, J. V., Colloid Polym. Sci., *252*, 298, 1974.
47. Berger, H. L., and Shultz, A. R., J. Polym. Sci., A, *3*, 3643, 1965.
48. Dawkins, J. V., Maddock, J. W., and Nevin, A., Eur. Polym. J., *9*, 327, 1973.
49. Ring, W., and Holtrup, W., Makromol. Chem., *103*, 83, 1967.
50. Coll, H., and Gilding, D. K., J. Polym. Sci., A2, *8*, 89, 1970.
51. Hazell, J. E., Prince, L. A., and Stapelfeldt, H. E., J. Polym. Sci., C, *21*, 43, 1967.

52. Duerksen, J. H., and Hamielec, A. E., J. Appl. Polym. Sci., *12*, 2225, 1968.
53. Cooper, A. R., Johnson, J. F., and Bruzzone, A. R., Eur. Polym. J., *9*, 1393, 1973.
54. Pannell, J., Polymer, *13*, 277, 1972.
55. Atkinson, C. M. L., and Dietz, R., Makromol. Chem., *177*, 213, 1976.
56. Atkinson, C. M. L., Dietz, R., and Green, J. H. S., J. Macromol. Sci., B, *14*, 101, 1977.
57. Atkinson, C. M. L., and Dietz, R., Polymer, *18*, 408, 1977.
58. Atkinson, C. M. L., and Dietz, R., Eur. Polym. J., *15*, 21, 1979.
59. Atkinson, C. M. L., Dietz, R., and Francis, M. A., Polymer, *21*, 891, 1980.
60. Cantow, M. J. R., Porter, R. S., and Johnson, J. F., J. Polym. Sci., A1, *5*, 1391, 1967.
61. Van Dijk, J. A. P. P., Henkens, W. C. M., and Smit, J. A. M., J. Polym. Sci., Polym. Phys. Ed., *14*, 1485, 1976.
62. Wild, L., Ranganath, R., and Ryle, T., J. Polym. Sci., A2, *9*, 2137, 1971.
63. Swartz, T. D., Bly, D. D., and Edwards, A. S., J. Appl. Polym. Sci., *16*, 3353, 1972.
64. Balke, S. T., Hamielec, A. E., LeClair, B. P., and Pearce, S. L., Ind. Eng. Chem., Prod. Res. Dev., *8*, 54, 1969.
65. Weiss, A. R., and Cohn-Ginsberg, E., J. Polym. Sci., A2, *8*, 148, 1970.
66. Purdon, J. R., and Mate, R. D., J. Polym. Sci., A1, *6*, 243, 1968.
67. Bombaugh, K. J., Dark, W. A., and Little, J. N., Anal. Chem., *41*, 1337, 1969.
68. Basedow, A. M., Ebert, K. H., Ederer, H., and Hunger, H., Makromol. Chem., *177*, 1501, 1976.
69. Ogawa, T., and Inaba, T., J. Appl. Polym. Sci., *20*, 2101, 1976.
70. Szewczyk, P., Polymer, *17*, 90, 1976.
71. Szewczyk, P., J. Polym. Sci., Polym. Symp., *68*, 191, 1980.
72. Szewczyk, P., J. Appl. Polym. Sci., *26*, 2727, 1981.
73. Almin, K. E., Polym. Prepr., *9*, 727, 1968.
74. Frank, F. C., Ward, I. M., and Williams, T., J. Polym. Sci., A2, *6*, 1357, 1968.
75. Yau, W. W., Stoklosa, H. J., and Bly, D. D., J. Appl. Polym. Sci., *21*, 1911, 1977.
76. Dawkins, J. V., and Hemming, M., Polymer, *16*, 554, 1975.
77. Dawkins, J. V., Eur. Polym. J., *6*, 831, 1970.
78. Pollock, M. J., MacGregor, J. F., and Hamielec, A. E., J. Liq. Chromatogr., *2*, 895, 1979.
79. Loy, B. R., J. Polym. Sci., Polym. Chem. Ed., *14*, 2321, 1976.
80. Chaplin, R. P., Haken, J. K., and Paddon, J. J., J. Chromatogr., *171*, 55, 1979.
81. Crouzet, P., Hauteville, A., Lucet, M., and Martens, A., Analusis, *4*, 450, 1976.
82. Malawer, E. G., and Montana, A. J., J. Polym. Sci., Polym. Phys. Ed., *18*, 2155, 1980.
83. McCrackin, F. L., J. Appl. Polym. Sci., *21*, 191, 1977.

84. Chaplin, R. P., and Ching, W., J. Macromol. Sci., A, *14*, 257, 1980.
85. Andersson, L., J. Chromatogr., *216*, 23, 1981.
86. Whitehouse, B. A., Macromolecules, *4*, 463, 1971.
87. Hamielec, A. E., and Omorodion, S. N. E., Am. Chem. Soc. Symp. Ser., *138*, 183, 1980.
88. Yau, W. W., Kirkland, J. J., and Bly, D. D., *Modern Size-Exclusion Liquid Chromatography,* Wiley-Interscience, New York, 1979, p. 302.
89. Vrijbergen, R. R., Soeteman, A. A., and Smit, J. A. M., J. Appl. Polym. Sci., *22*, 1267, 1978.
90. Soeteman, A. A., Roels, J. P. M., Van Dijk, J. A. P. P., and Smit, J. A. M., J. Polym. Sci., Polym. Phys. Ed., *16*, 2147, 1978.
91. Tsvetkovskii, I. B., Valuev, V. I., and Shlyakhter, R. A., Vysokomol. Soedin., A, *19*, 2637, 1977.
92. Determann, H., *Gel Chromatography,* Springer-Verlag, Berlin, 1969, Chap. 3.
93. Casassa, E. F., J. Polym. Sci., B, *5*, 773, 1967.
94. Casassa, E. F., and Tagami, Y., Macromolecules, *2*, 14, 1969.
95. Casassa, E. F., Sep. Sci., *6*, 305, 1971.
96. Casassa, E. F., J. Phys. Chem., *75*, 3929, 1971.
97. Giddings, J. C., Kucera, E., Russell, C. P., and Myers, M. N., J. Phys. Chem., *72*, 4397, 1968.
98. Casassa, E. F., Macromolecules, *9*, 182, 1976.
99. Flory, P. J., *Principles of Polymer Chemistry,* Cornell University Press, Ithaca, N.Y., 1953.
100. Grubisic, Z., Rempp, P., and Benoit, H., J. Polym. Sci., B, *5*, 753, 1967.
101. Flory, P. J., and Fox, T. G., J. Am. Chem. Soc., *73*, 1904, 1951.
102. Benoit, H., Grubisic, Z., Rempp, P., Decker, D., and Zilliox, J. G., J. Chim. Phys., *63*, 1507, 1966.
103. Coll, H., Sep. Sci., *5*, 273, 1970.
104. Ptitsyn, O. B., and Eizner, Yu. E., Sov. J. Tech. Phys. (English transl.), *4*, 1020, 1960.
105. Spychaj, T., Lath, D., and Berek, D., Polymer, *20*, 437, 1979.
106. Boni, K. A., Sliemers, F. A., and Stickney, P. B., J. Polym. Sci., A2, *6*, 1579, 1968.
107. Wagner, H. L., and Hoeve, C. A. J., J. Polym. Sci., Polym. Phys. Ed., *11*, 1189, 1973.
108. Cervenka, A., Makromol. Chem., *170*, 239, 1973.
109. Cantow, M. J. R., Porter, R. S., and Johnson, J. F., J. Polym. Sci., A1, *5*, 987, 1967.
110. Kraus, G., and Stacy, C. J., J. Polym. Sci., A2, *10*, 657, 1972.
111. Cooper, A. R., and Bruzzone, A. R., J. Polym. Sci., Polym. Phys. Ed., *11*, 1423, 1973.
112. Brandrup, J., and Immergut, E. H., *Polymer Handbook,* Wiley-Interscience, New York, 1975.
113. Nichols, E., Polym. Prepr., *12*, 828, 1971.
114. Hamielec, A. E., and Ouano, A. C., J. Liq. Chromatogr., *1*, 111, 1978.

115. Kolinsky, M., and Janča, J., J. Polym. Sci., Polym. Chem. Ed., *12*, 1181, 1974.
116. French, D. M., and Nauflett, G. W., J. Liq. Chromatogr., *4*, 197, 1981.
117. Gilding, D. K., Reed, A. M., and Askill, I. N., Polymer, *22*, 505, 1981.
118. Wild, L., and Guliana, R. J., J. Polym. Sci., A2, *5*, 1087, 1967.
119. Drott, E., 4th Int. Semin. GPC, Miami Beach, Fla., 1967.
120. Drott, E. E., in *Liquid Chromatography of Polymers and Related Materials* (Chromatographic Science Series, Vol. 8, J. Cazes, ed.), Marcel Dekker, New York, 1977.
121. Scholte, Th. G., and Meijerink, N. L. J., Br. Polym. J., *9*, 133, 1977.
122. Kato, T., Itsubo, A., Yamamoto, Y., Fujimoto, T., and Nagasawa, M., Polym. J., *7*, 123, 1975.
123. Ambler, M. R., and McIntyre, D., J. Polym. Sci., Polym. Lett. Ed., *13*, 589, 1975.
124. Ambler, M. R., J. Appl. Polym. Sci., *21*, 1655, 1977.
125. Simha, R., J. Phys. Chem., *44*, 25, 1940.
126. Tanford, C., *Physical Chemistry of Macromolecules*, Wiley, New York, 1961, p. 335.
127. Grubisic, Z., Reibel, L., and Spach, G., C.R. Acad. Sci., Ser. C, *264*, 1690, 1967.
128. Hester, R. D., and Mitchell, R. H., J. Polym. Sci., Polym. Chem. Ed., *18*, 1727, 1980.
129. Waters, J. L., Polym. Prepr., *6*, 1061, 1965.
130. Cantow, M. J. R., Porter, R. S., and Johnson, J. F., J. Polym. Sci., B, *4*, 707, 1966.
131. Lambert, A., Polymer, *10*, 213, 1969.
132. Berek, D., Bakos, D., Soltes, L., and Bleha, T., J. Polym. Sci., Polym. Lett. Ed., *12*, 277, 1974.
133. Bleha, T., Bakos, D., and Berek, D., Polymer, *18*, 897, 1977.
134. Kato, Y., and Hashimoto, T., J. Appl. Polym. Sci., *18*, 1239, 1974.
135. Rudin, A., J. Polym. Sci., A1, *9*, 2587, 1971.
136. Rudin, A., and Wagner, R. A., J. Appl. Polym. Sci., *20*, 1483, 1976.
137. Mahabadi, H. K., and Rudin, A., Polym. J., *11*, 123, 1979.
138. Bleha, T., Mlynek, J., and Berek, D., Polymer, *21*, 798, 1980.
139. Rudin, A., and Hoegy, H. W., J. Polym. Sci., A1, *10*, 217, 1972.
140. Janča, J., Polym. J., *12*, 405, 1980.
141. Vander Linden, C., and Van Leemput, R., Macromolecules, *11*, 1237, 1978.
142. Anderson, J. L., and Brannon, J. H., J. Polym. Sci., Polym. Phys. Ed., *19*, 405, 1981.
143. Janča, J., Anal. Chem., *51*, 637, 1979.
144. Janča, J., J. Chromatogr., *170*, 309, 1979.
145. Janča, J., Pokorný, S., Bleha, M., and Chiantore, O., J. Liq. Chromatogr., *3*, 953, 1980.
146. Elsdon, W. L., Goldwasser, J. M., and Rudin, A., J. Polym. Sci., Polym. Lett. Ed., *19*, 483, 1981.

147. Otocka, E. P., and Hellman, M. Y., J. Polym. Sci., Polym. Lett. Ed., *12*, 331, 1974.
148. Kranz, D., Pohl, U., and Baumann, H., Angew. Makromol. Chem., *26*, 67, 1972.
149. Dubin, P. L., and Wright, K. L., Polym. Prepr., *15*, 673, 1974.
150. Mencer, H. J., and Grubisic-Gallot, Z., J. Liq. Chromatogr., *2*, 649, 1979.
151. Dawkins, J. V., and Hemming, M., Makromol. Chem., *176*, 1795, 1975.
152. Dawkins, J. V., and Hemming, M., Makromol. Chem., *176*, 1777, 1975.
153. Dawkins, J. V., and Hemming, M., Makromol. Chem., *176*, 1815, 1975.
154. Iwama, M., and Homma, T., Kogyo Kagaku Zasshi (J. Chem. Soc. Jpn., Ind. Chem. Sect.), *74*, 277, 1971.
155. Dubin, P. L., Koontz, S., and Wright, K. L., J. Polym. Sci., Polym. Chem. Ed., *15*, 2047, 1977.
156. Meyerhoff, G., Makromol. Chem., *134*, 129, 1970.
157. Ede, P. S., J. Chromatogr. Sci., *9*, 275, 1971.
158. Walsh, E. K., J. Chromatogr., *55*, 193, 1971.
159. Dawkins, J. V., in *Chromatography of Synthetic and Biological Polymers*, Vol. 1 (R. Epton, ed.), Ellis Horwood, Chichester, England, 1978, p. 30.
160. Moore, J. C., and Arrington, M. C., 3rd Int. Semin. GPC, Geneva, 1966.
161. Swenson, H. A., Kaustinen, H. M., and Almin, K. E., J. Polym. Sci., B, *9*, 261, 1971.
162. Berek, D., Bakos, D., Bleha, T., and Soltes, L., Makromol. Chem., *176*, 391, 1975.
163. Bakoš, D., Bleha, T., Ozima, A., and Berek, D., J. Appl. Polym. Sci., *23*, 2233, 1979.
164. Campos, A., and Figueruelo, J. E., Makromol. Chem., *178*, 3249, 1977.
165. Campos, A., Soria, V., and Figueruelo, J. E., Makromol. Chem., *180*, 1961, 1979.
166. Spychaj, T., Lath, D., and Berek, D., Polymer, *20*, 1108, 1979.
167. Cooper, A. R., and Johnson, J. F., J. Appl. Polym. Sci., *13*, 1487, 1969.
168. Kotera, A., Furusawa, K., and Okamoto, K., Rep. Prog. Polym. Phys. Jpn., *16*, 69, 1973.
169. Nakamura, K., and Endo, R., J. Appl. Polym. Sci., *26*, 2657, 1981.
170. Figueruelo, J. E., Soria, V., and Campos, A., in *Liquid Chromatography of Polymers and Related Materials II* (Chromatographic Science Series, Vol. 13, J. Cazes and X. Delamare, eds.), Marcel Dekker, New York, 1980, p. 49.
171. Figueruelo, J. E., Soria, V., and Campos, A., Makromol. Chem., *182*, 1525, 1981.
172. Dubin, P. L., J. Liq. Chromatogr., *3*, 623, 1980.
173. Rochas, C., Domard, A., and Rinaudo, M., Polymer, *20*, 76, 1979.
174. Harwood, D. D., and Fellers, J. F., Macromolecules, *12*, 693, 1979.

175. Siebourg, W., Lunberg, R. D., and Lenz, R. W., Macromolecules, *13*, 1013, 1980.
176. Cha, C. Y., J. Polym. Sci., B, *7*, 343, 1969.
177. Hann, N. D., J. Polym. Sci., Polym. Chem. Ed., *15*, 1331, 1977.
178. Booth, C., Forget, J. L., Georgii, I., Li, W. S., and Price, C., Eur. Polym. J., *16*, 255, 1980.
179. Coppola, G., Fabbri, P., Pallesi, B., and Bianchi, U., J. Appl. Polym. Sci., *16*, 2829, 1972.
180. Kenyon, A. S., and Motthus, E. H., Appl. Polym. Symp., *25*, 57, 1974.
181. Lecourtier, J., Audebert, R., and Quivoron, C., J. Liq. Chromatogr., *1*, 367, 1978.
182. Lecourtier, J., Audebert, R., and Quivoron, C., J. Liq. Chromatogr., *1*, 479, 1978.
183. Lecourtier, J., Audebert, R., and Quivoron, C., Macromolecules, *12*, 141, 1979.
184. Monrabel, B., in *Liquid Chromatography of Polymers and Related Materials III* (Chromatographic Science Series, Vol. 19, J. Cazes, ed.), Marcel Dekker, New York, 1981, p. 79.
185. Cooper, A. R., and Van Derveer, D. S., J. Liq. Chromatogr., *1*, 693, 1978.
186. Barth, H. G., J. Chromatogr. Sci., *18*, 409, 1980.
187. Cooper, A. R., and Matzinger, D. P., J. Appl. Polym. Sci., *23*, 419, 1979.
188. Spatorico, A. L., and Beyer, G. L., J. Appl. Polym. Sci., *19*, 2933, 1975.
189. Rochas, C., Domard, A., and Rinaudo, M., Eur. Polym. J., *16*, 135, 1980.
190. Rinaudo, M., Desbrieres, J., and Rochas, C., J. Liq. Chromatogr., *4*, 1297, 1981.
191. Van Dijk, J. A. P. P., Roels, J. P. M., and Smit, J. A. M., in *Liquid Chromatography of Polymers and Related Materials II* (Chromatographic Science Series, Vol. 13, J. Cazes and X. Delamare, eds.), Marcel Dekker, New York, 1980, p. 95.
192. Buytenhuys, F. A., and Van der Maeden, F. P. E., J. Chromatogr., *149*, 489, 1978.
193. Omorodion, S. N. E., Hamielec, A. E., and Brash, J. L., J. Liq. Chromatogr., *4*, 41, 1981.
194. Talley, C. P., and Bowman, L. M., Anal. Chem., *51*, 2239, 1979.
195. Schmidt, D. E., Giese, R. W., Conron, D., and Karger, B. L., Anal. Chem., *52*, 177, 1980.
196. Dawkins, J. V., and Yeadon, G., Faraday Symp., *15*, 127, 1980.
197. Evans, J. M., Polym. Eng. Sci., *13*, 401, 1973.
198. Evans, J. M., and Maisey, L. J., in *Industrial Polymers: Characterization by Molecular Weight* (J. H. S. Green and R. Dietz, eds.), Transcripta Books, London, 1973, p. 89.
199. Ambler, M. R., J. Appl. Polym. Sci., *25*, 901, 1980.
200. Weiss, A. R., and Cohn-Ginsberg, E., J. Polym. Sci., B, *7*, 379, 1969.
201. Morris, M. C., J. Chromatogr., *55*, 203, 1971.
202. Braun, G., J. Appl. Polym. Sci., *15*, 2321, 1971.

203. Spatorico, A. L., and Coulter, B., J. Polym. Sci., Polym. Phys. Ed., *11*, 1139, 1973.
204. Ambler, M. R., J. Polym. Sci., Polym. Chem. Ed., *11*, 191, 1973.
205. Kato, Y., Takamatsu, T., Fukuda, M., and Hashimoto, T., J. Appl. Polym. Sci., *21*, 577, 1977.
206. Samay, G., Kubin, M., and Podesva, J., Angew. Makromol. Chem., *72*, 185, 1978.
207. Balke, S. T., and Patel, R. D., Am. Chem. Soc. Symp. Ser., *138*, 149, 1980.
208. Dobbin, C. J. B., Rudin, A., and Tchir, M. F., J. Appl. Polym. Sci., *25*, 2985, 1980.
209. Dobbin, C. J. B., Rudin, A., and Tchir, M. F., J. Appl. Polym. Sci., *27*, 1081, 1982.
210. Zhongde, X., Mingshi, S., Hadjchristides, N., and Fetters, L. J., Macromolecules, *14*, 1591, 1981.
211. Mahabadi, H. F., and O'Driscoll, K. F., J. Appl. Polym. Sci., *21*, 1283, 1977; *22*, 297, 1978.
212. Mori, S., Anal. Chem., *53*, 1813, 1981.
213. Provder, T., Woodbrey, J. C., and Clark, J. H., Sep. Sci., *6*, 101, 1971.
214. Coll, H., and Prusinowski, L. R., J. Polym. Sci., B, *5*, 1153, 1967.
215. Dawkins, J. V., Eur. Polym. J., *13*, 837, 1977.
216. Dawkins, J. V., Denyer, R., and Maddock, J. W., Polymer, *10*, 154, 1969.
217. Funt, B. L., and Hornof, V., J. Appl. Polym. Sci., *15*, 2439, 1971.
218. Bianchi, U., and Peterlin, A., J. Polym. Sci., A2, *6*, 1759, 1968.
219. Donkai, N., Nakazawa, A., and Inagaki, H., Bull. Inst. Chem. Res. Kyoto Univ., *48*, 79, 1970.
220. Moore, J. C., and Hendrickson, J. G., J. Polym. Sci., C, *8*, 233, 1965.
221. Heller, J., and Moacanin, J., J. Polym. Sci., B, *6*, 595, 1968.

—3—

CORRECTION FOR AXIAL DISPERSION

ARCHIE E. HAMIELEC / *McMaster University, Hamilton, Ontario, Canada*

3.1. INTRODUCTION

This chapter is concerned with the phenomenon of axial dispersion in SEC and has the objective of presenting a practical methodology for its detection and correction.

The detector response at retention volume V for a polydispersed polymer sample is due also to neighboring species, both smaller and larger in size. In other words, the detector cell contains polymer solute with species that differ in size in the mobile phase. As discussed in Chapters 1 and 2, the basic size parameter appropriate for SEC of hard-sphere, rigid rod, and flexible, random-coil solutes is the radius of gyration [1-5]. The deviation from a monodispersed size distribution for solute in the detector cell is referred to as axial dispersion, band broadening, peak broadening, and many other terms. The various sources of axial dispersion have been adequately summarized elsewhere [5,6], and discussed in Chapter 1. Ouano and Barker [7] have modeled the band broadening process in the columns and have identified two characteristic times that affect the extent of dispersion and the shape of the instrumental spreading function. The instrumental spreading function is the normalized detector response for a monodispersed solute. These times are the time required for the solute to diffuse into and back out of a packing pore (t_D) and the time for the pulse of solute in the interstitial volume to move around the packing particle downstream (t_F). Under ideal SEC operation, $t_D << t_F$, local equilibrium between solute in the pores and interstitial volume will be approached closely and the instrumental spreading function will be Gaussian and relatively narrow (small variance about the mean). As t_D lengthens relative to t_F, the spreading function broadens and eventually becomes skewed toward longer retention volumes. One therefore might expect that for large polymer solutes with small diffusion coefficients, high mobile-phase flow rates, and deep pores, skewed instrumental spreading functions would result. This has been shown experimentally [8,9]. Skewed spreading functions can also be caused by column overloading with polymer solute [10]. Polymer coil size is reduced, with increasing solute concentrations with the reduction greatest for the larger molecules [11] (see Chapter 4 for detailed discussion). This affects the molecular weight calibration curve at the high molecular weight end [12] and gives skewed spreading functions [13].

The detector response cannot reflect the true amount of a particular sized species in a polymer sample because peak broadening cannot be eliminated and sufficient resolution is not obtained to give separate peaks for individual species. It is therefore important to interpret detector responses or chromatograms to provide information which has been corrected for peak broadening and can be used to calculate molecular weight distribution (MWD) and averages for the polymer sample. To this end, some general relationships for a general detector and simple and complex polymers will be provided. A simple polymer is one for which a unique relationship exists between size (radius of gyration) in the mobile phase and molecular weight. Examples of simple polymers include linear homopolymers and linear homogeneous copolymers. Most other polymers are complex. Examples include polymers with long-chain branching, heterogeneous copolymers, and blends of linear homopolymers and homogeneous copolymers.

3.2. GENERAL RELATIONSHIPS

3.2.1. General Detector

The response for a general detector (called the chromatogram), F(V), at retention volume V is given by

$$F(V) = \int_0^\infty W(V,Y)\, dY \tag{3.1}$$

where

$$W(V,Y) = \beta(Y)C(V,Y)\overline{M}_k^p(Y) \tag{3.1a}$$

for complex polymers and for simple polymers

$$W(V,Y) = \beta(Y)C(V,Y)M^p(Y) \tag{3.1b}$$

and where W(V,Y) dV dY is the area under the detector response in the retention volume range V - V + dV due to species with mean retention volume in the range Y - Y + dY. C(V,Y) is the mass concentration in V - V + dV due to species with Y - Y + dY. $\overline{M}_k(Y)$ is the kth average molecular weight of species with Y - Y + dY (when k = 1, one is referring to $\overline{M}_n$, k = 2 to $\overline{M}_w$, and so on). $\beta(Y)$ is the detector response factor for species with Y - Y + dY. p defines the detector type. When p = 0, one is considering a mass concentration detector and when p = 1, the light-scattering photometer, and so on. For the special case of simple polymers, $\overline{M}_k^p(Y)$ can be replaced by $M^p(Y)$, where M(Y) is the molecular weight calibration curve. For complex polymers, the detector response factor may not be the same for solute having the same radius of gyration and mean retention volume. For example, a branched and linear chain of the same size may have significantly different response factors, and this could also be true for heterogeneous copolymers (see Chapter 7).

3.2.2. Simple Polymers

Properties of Detector Cell Contents

Molecular weight averages of the detector cell contents are given by

$$\bar{M}_k(V,Uc) = \frac{\int_0^\infty W(V,Y)\ \beta(Y)^{-1}\ M(Y)^{k-p-1}\ dY}{\int_0^\infty W(V,Y)\ \beta(Y)^{-1}\ M(Y)^{k-p-2}\ dY} \tag{3.2}$$

where $\bar{M}_k(V,Uc)$ is the kth average molecular weight of the detector cell contents and Uc stands for uncorrected for dispersion. In the absence of peak broadening,

$$\bar{M}_k(V,Uc) = M(V) \tag{3.2a}$$

the molecular weight calibration curve. The intrinsic viscosity of the detector cell contents is given by

$$[\eta](V,Uc) = \frac{K \int_0^\infty W(V,Y)\ \beta(Y)^{-1}\ M(Y)^{a-p}\ dY}{\int_0^\infty W(V,Y)\ \beta(Y)^{-1}\ M(Y)^{-p}\ dY} \tag{3.3}$$

where K and a are Mark-Houwink constants. The dependence of the response factor on retention volume would generally not be strong except perhaps for very low molecular weight polymers, and thus two approximations could be used. First, $\beta(Y)$ can be replaced by $\beta(V)$ with small error and the next step would be to neglect altogether the dependence of β on retention volume. With either approximation, equations (3.2) and (3.3) take the form

$$\bar{M}_k(V,Uc) = \frac{\int_0^\infty W(V,Y)\ M(Y)^{k-p-1}\ dY}{\int_0^\infty W(V,Y)\ M(Y)^{k-p-2}\ dY} \tag{3.2b}$$

$$[\eta](V,Uc) = \frac{K \int_0^\infty W(V,Y)\ M(Y)^{a-p}\ dY}{\int_0^\infty W(V,Y)\ M(Y)^{-p}\ dY} \tag{3.3a}$$

Whole Polymer Properties

When β is constant

$$\bar{M}_k(c) = \frac{\int_0^\infty W(V)\ M(V)^{k-p-1}\ dV}{\int_0^\infty W(V)\ M(V)^{k-p-2}\ dV} \tag{3.4}$$

$$\bar{M}_k(Uc) = \frac{\int_0^\infty F(V)\ M(V)^{k-p-1}\ dV}{\int_0^\infty F(V)\ M(V)^{k-p-2}\ dV} \tag{3.5}$$

$$[\eta](c) = K \int_0^\infty W(V)\ M(V)^{a-p}\ dV \tag{3.6}$$

For the case of a mass concentration detector ($p = 0$), $\bar{M}_n(c)$ and $\bar{M}_w(c)$, and the distribution of the whole polymer corrected for peak broadening, are given by

$$\bar{M}_n(c) = \frac{\int_0^\infty F(V)\ dV}{\int_0^\infty F(V)\ M_n(V,Uc)} \tag{3.7}$$

$$\bar{M}_w(c) = \frac{\int_0^\infty F(V)\bar{M}_w(V,Uc)\ dV}{\int_0^\infty F(V)\ dV} \tag{3.8}$$

$$W(M) = \frac{-W(V)}{dM/dV} \tag{3.9a}$$

$$W(\ln M) = \frac{-W(V)}{d\ \ln M/dV} \tag{3.9b}$$

where $W(M)\ dM$ is the weight fraction of polymer with molecular weight in the range $M - M + dM$ and $W(\ln M)\ d\ \ln M$ is the weight fraction of polymer in the range $\ln M - \ln M + d\ \ln M$. $W(V)$ is the normalized detector response corrected for peak broadening. For the case of a variable response factor $\beta(V)$ and a mass concentration detector ($p = 0$), the normalized detector response is found using

$$W(V) = \frac{\beta(V)^{-1}\ H(V)}{\int_0^\infty \beta(V)^{-1}\ H(V)\ dV} \tag{3.10}$$

where $H(V)$ is the unnormalized detector response corrected for peak broadening. When the response factor is independent of retention volume, the detector response is normalized by simply dividing each chromatogram height by the total area under the chromatogram. The

MWD on a log scale [equation (3.9b)] is a more convenient representation as the entire molecular weight range can be conveniently displayed graphically.

3.2.3. Complex Polymers

Properties of Detector Cell Contents

A complex polymer has an infinite number of molecular weight calibration curves, one for each molecular weight average. Under conditions of perfect resolution, the detector cell contains polymer species having the same radius of gyration but possibly very different copolymer composition and molecular weights. The molecular weight calibration curves are $\bar{M}_n(V)$, $\bar{M}_w(V)$, and so on, with the detector cell contents having molecular weight averages $\bar{M}_n(V,Uc)$, $\bar{M}_w(V,Uc)$, and so on, and intrinsic viscosity $[\eta](V,Uc)$.

Whole Polymer Properties

Molecular weight averages for the whole polymer are given by ($p = 0$ and β constant)

$$\bar{M}_k(c) = \frac{\int_0^\infty W(V)\ \bar{M}_k^{k-1}(V)\ dV}{\int_0^\infty W(V)\ \bar{M}_k^{k-2}(V)\ dV} \tag{3.11}$$

$$[\eta](c) = \frac{\int_0^\infty W(V)\ [\eta](V)\ dV}{\int_0^\infty W(V)\ dV} \tag{3.12}$$

Similar relationships for $\bar{M}_k(Uc)$ and $[\eta](Uc)$ are found by simply replacing $W(V)$ with $F(V)$.

3.2.4. Generalization of Universal Molecular Weight Calibration

Benoit's universal molecular weight calibration [14] involving the product $[\eta](V)\ M(V)$ is valid for simple polymers but must be generalized for complex polymers. This generalization was made by Hamielec and Ouano [15] and the development follows. Under conditions of perfect resolution, all of the species in the detector cell have the same hydrodynamic volume and therefore

$$J_{ps}(V) = J_1(V) = J_2(V) = \cdots = J_i(V) \tag{3.13}$$

where $J_i(V) = [\eta]_i(V)M_i(V)$ and subscript ps stands for polystyrene the reference polymer. The intrinsic viscosity of the detector cell contents is given by

$$[\eta](V) = J_1 \frac{W_1}{M_1} + J_2 \frac{W_2}{M_2} + \cdots + J_i \frac{W_i}{M_i} \tag{3.14}$$

Substituting equation (3.13) in (3.14) gives

$$[\eta](V) = J_{ps}(V) \left(\frac{W_1}{M_1} + \frac{W_2}{M_2} + \cdots + \frac{W_i}{M_i} \right) \tag{3.15}$$

Now

$$\bar{M}_n(V) = \left(\frac{W_1}{M_1} + \frac{W_2}{M_2} + \cdots + \frac{W_i}{M_i} \right)^{-1} \tag{3.16}$$

Hence

$$[\eta](V)\bar{M}_n(V) = ([\eta]M(V))_{ps} \tag{3.17}$$

An SEC viscosity detector system gives a measure of $[\eta](V,Uc)$. Correction for peak broadening gives $[\eta](V)$. Application of the generalized universal molecular weight calibration curve gives $\bar{M}_n(V)$. An inverse correction for peak broadening provides $\bar{M}_n(V,Uc)$ and integration using equation (3.7) gives $\bar{M}_n(c)$. In summary, a viscosity detector together with the universal molecular weight calibration curve can provide a measure of $\bar{M}_n$ for the whole polymer.

3.2.5. Principle of Universal Peak Broadening Calibration

Polymer solutes having different composition but the same radius of gyration would have diffusion coefficients that do not greatly differ and would also experience about the same packing pore volume. One might therefore expect that peak broadening parameters would be the same for these polymer solutes. In other words, polystyrene standards could be used to estimate peak broadening parameters for a particular column combination. These peak broadening parameters could then be used for other polymer types. A limited number of experiments have tended to confirm this principle of universal peak broadening calibration [16].

3.3. INTEGRAL EQUATION FOR AXIAL DISPERSION

All rigorous methods of correcting for peak broadening use the following integral equation as the basis:

$$F(V) = \int_0^\infty W(Y)G(V,Y)\, dY \tag{3.18}$$

where $G(V,Y)$ is the normalized detector response for a solute of uniform size with mean retention volume Y. $G(V,Y)$ will be called the instrumental spreading function or just the spreading function. $W(Y)\, dY$ is the area under the detector response due to solute with mean retention volume in the range $Y - Y + dY$. $W(Y)$ or $W(V)$ are again chromatograms corrected for peak broadening. The use of this integral equation or continuous form rather than the more basic discrete form is justified on the basis that a polymer sample has many species differing in molecular weight by a small increment (the molecular weight of the repeat unit). When dealing with oligomers it may be necessary to use the discrete form

$$F(V) = \sum_{i=1}^{n} A_i G_i(V) \tag{3.19}$$

where n is the number of species, A_i the area under the detector response due to species i, and $G_i(V)$ the instrumental spreading function for species i. Equation (3.18) is a Fredholme integral equation of the first kind and has been used extensively in various science and engineering applications. Tung [17] was the first to apply it to SEC, and for this reason it is often referred to as Tung's integral equation.

It is clear from the integral equation that

$$W(V,Y) = W(Y)G(V,Y) \tag{3.20}$$

and that

$$W(Y) = \int_0^\infty W(V,Y)\ dV \tag{3.21}$$

Equations (3.20) and (3.21) will be later used to derive correction equations for peak broadening in the detector cell.

There is some question about the applicability of the integral equation for complex polymers. The main point is whether the instrumental spreading function, G(V,Y), is the same for species having the same mean retention volume. For example, with a heterogeneous copolymer, would all polymer molecules having the same radius of gyration have a spreading function with the same shape parameters? To date, no experimental data that could be used to address this concern have been published. Until evidence to the contrary becomes available, it will be assumed that the shape parameters of G(V,Y) are unique functions of the retention volume. In the practical application of the integral equation to correct for peak broadening, a number of assumptions should have reasonable validity. A basic assumption is that the normalized instrumental spreading function is independent of concentration of all solute species. This is equivalent to the principle of linear superposition. It has already been mentioned that column overloading with polymer solute can lead to skewed chromatograms as a consequence of skewed spreading function. This can be a serious limitation with high molecular weight polymers, where even the molecular weight calibration curve shifts with polymer solute concentration [12]. Another basic assumption is that the detector response is linear in concentration.

3.3.1. Uniform Gaussian Spreading Function

The instrumental spreading function is uniform when its shape parameters are independent of retention volume; in other words, the shape parameters are the same for solute species of different size. The integral equation now takes the form of a convolution integral:

$$F(V) = \int_0^\infty W(Y)G(V - Y)\, dY \tag{3.22}$$

This limiting form should be valid for polymer samples with relatively narrow MWD and for samples with broad MWD when peak broadening corrections are small.

Bilateral Laplace transformation of the convolution integral gives [9,18]

$$\bar{F}(s) = \bar{W}(s)\bar{G}(s) \tag{3.23}$$

This algebraic equation will be later used to develop corrections for peak broadening for the whole polymer molecular weight averages and intrinsic viscosity.

For the case of a uniform Gaussian spreading function, equation (3.22) takes the form

$$F(V) = \frac{1}{\sqrt{2\pi\sigma^2}} \int_0^\infty W(Y) \exp\left[-\frac{(V - Y)^2}{2\sigma^2}\right] dY \tag{3.22a}$$

where σ^2, the variance of the spreading function, is independent of retention volume.

3.3.2. Nonuniform Gaussian Spreading Function

The integral equation takes the form

$$F(V) = \int_0^\infty W(Y) \frac{1}{\sqrt{2\pi\sigma(Y)^2}} \exp\left[-\frac{(V - Y)^2}{2\sigma(Y)^2}\right] dY \tag{3.24}$$

When peak broadening is not excessive, an excellent approximation of equation (3.24) is given by

$$F(V) = \frac{1}{\sqrt{2\pi\sigma(V)^2}} \int_0^\infty W(Y) \exp\left[-\frac{(V - Y)^2}{2\sigma(V)^2}\right] dY \tag{3.24a}$$

Equation (3.24a) is the basis for the derivation of correction equations for peak broadening in the detector cell to be made later in the chapter.

3.3.3. General Spreading Function

Provder and Rosen [9] have proposed the use of a general statistical shape function to account for deviations of the instrumental spreading function from the Gaussian shape. It has the form

$$G(x) = \phi(x) + \sum_{n=3}^{\infty} \frac{(-1)^n A_n \phi^n(x)}{n!} \tag{3.25}$$

where

$$\phi(x) = \frac{1}{\sqrt{\pi}} \exp\left[\frac{-x^2}{2}\right]$$

$$x = \frac{V}{\sigma}$$

and $\sigma^n(x)$ denotes the nth order derivative. The coefficients A_n are functions of μ_n, the nth order moments about the mean retention volume μ_1 of the normalized detector response for a single species. The first two coefficients are of direct statistical significance and also represent the most useful terms in the infinite series for SEC application.

$$A_3 = \frac{\mu_3}{\mu_2^{3/2}} \tag{3.25a}$$

$$A_4 = \frac{\mu_4}{\mu_2^2 - 3} \tag{3.25b}$$

where μ_2 is the variance and is an equivalent symbol to σ^2. The coefficient A_3 provides an absolute statistical measure of skewness (when $\mu_3 = 0$ the spreading function is symmetrical about the mean retention volume μ_1 or Y). When $\mu_3 > 0$, skewing is toward longer retention volumes. The coefficient A_4 provides a statistical measure of flattening or kurtosis. When $A_4 > 0$ the shape function is taller and slimmer than a Gaussian, and so forth. For SEC application a truncated form of the series with $A_5 = 0$ and $A_6 = 10A_3^2$ and $A_7 = A_8 = \cdots = 0$ was employed [9]. Tung and Runyon [16] used the following simple form to fit skewed chromatograms:

$$G(V,Y) = [1 + f(V - Y)] \frac{1}{\sqrt{2\pi\sigma^2}} \exp\left[-\frac{(V - Y)^2}{2\sigma^2}\right] \tag{3.26}$$

Yau et al. [5] have used an exponentially modified Gaussian [19]. Insufficient information is available to compare these proposed shape functions and recommend one as most suitable for SEC application (see Section 1.3.4).

3.4. ANALYTICAL SOLUTION OF THE INTEGRAL EQUATION: MOLECULAR WEIGHT AVERAGES AND INTRINSIC VISCOSITY OF THE WHOLE POLYMER

The first solutions of this kind were based on the use of bilateral Laplace transformations [9,18] and Tung's integral equation with a uniform instrumental spreading function and a linear molecular weight calibration curve. Yau et al. [20] obtained similar solutions for the case of a uniform spreading function and a linear molecular weight calibration curve but used a more direct solution technique by focusing on peak broadening in the detector cell (for details, see Section 2.3.3). The latter approach was used by Hamielec et al. [21,22] to generalize these solutions for a nonuniform Gaussian spreading function and a nonlinear molecular weight calibration curve.

3.4.1. Uniform Spreading Function and Linear Molecular Weight Calibration Curve

Gaussian Spreading Function

For a mass concentration detector (p = 0) and a linear molecular weight calibration curve it can readily be shown that

$$\frac{\bar{M}_k(c)}{\bar{M}_k(Uc)} = \frac{\bar{W}((k - 1)D_2)/\bar{W}((k - 2)D_2)}{\bar{F}((k - 1)D_2)/\bar{F}((k - 2)D_2)} = \frac{\bar{G}((k - 2)D_2)}{\bar{G}((k - 1)D_2)} \tag{3.27}$$

where the molecular weight calibration curve is given by

$$M(V) = D_1 \exp(-D_2 V) \tag{3.28}$$

where D_1, $D_2 > 0$ and constant and

$$\frac{[\eta](c)}{[\eta](Uc)} = \frac{1}{\bar{G}(aD_2)} \tag{3.29}$$

where a is the Mark-Houwink exponent. For a uniform Gaussian spreading function the correction equations follow:

$$\frac{\bar{M}_k(c)}{\bar{M}_k(Uc)} = \exp\left[\frac{(3 - 2k)(D_2\sigma)^2}{2}\right] \tag{3.27a}$$

$$\frac{[\eta](c)}{[\eta](Uc)} = \exp\left[-\frac{(aD_2\sigma)^2}{2}\right] \tag{3.29a}$$

These equations, which correct molecular weight averages and the intrinsic viscosity of the whole polymer, are consistent with our knowledge of resolution [23]. This idea was later further developed and promoted by Yau et al. [5]. D_2, the slope of the molecular weight calibration curve, gives a measure of peak separation and σ^2 the variance of the Gaussian instrumental spreading function, a measure of peak broadening. Clearly, as the resolution increases the arguments of the exponentials decrease and the corrections for dispersion approach zero. The smallest correction is for the whole polymer intrinsic viscosity and is a result of a Mark-Houwink exponent which is less than unity. It is interesting to note that when k = 3/2 (3/2 molecular weight average), the correction for dispersion vanishes. Apparently, the use of this noninteger molecular weight average is not widespread.

These correction equations have been generalized to include complex polymers [21] with D_2 replaced by D_{2k} and aD_2 replaced by $\bar{D}_2$, where

$$\bar{M}_k(V) = D_{1k} \exp(-D_{2k} V) \tag{3.30a}$$

and

$$[\eta](V) = \bar{D}_1 \exp(-\bar{D}_2 V) \tag{3.30b}$$

It is now assumed that for a complex polymer the calibration curves for the molecular weight averages are linear and also that the intrinsic viscosity calibration curve is linear.

These correction equations will now be generalized to include a general detector as defined by equations (3.1a) and (3.1b). Equation (3.27) now becomes

$$\frac{\bar{M}_k(c)}{\bar{M}_k(Uc)} = \frac{\bar{G}((k - p - 2)D_2)}{\bar{G}((k - p - 1)D_2)} \tag{3.27b}$$

with the correction equations taking the form

$$\frac{\bar{M}_k(c)}{\bar{M}_k(Uc)} = \exp\left[\frac{[3 - 2(k - p)](D_2\sigma)^2}{2}\right] \tag{3.27c}$$

$$\frac{[\eta](c)}{[\eta](Uc)} = \exp\left[-\frac{[(a - p)D_2\sigma]^2}{2}\right] \tag{3.29b}$$

Correction equations for a general detector should find use when multidetector systems are employed. For example, when a DRI/LALLSP detector system is used, equations (3.27c) and (3.29b) with $p = 1$ could be used with the LALLSP detector response to correct for dispersion. In this case, $\bar{M}_k(Uc)$ and $[\eta](Uc)$ should be calculated using the LALLSP response. Equations (3.27c) and (3.29b) can be used with complex polymers by making the appropriate changes to D_2 and $(a - p)D_2$.

General Uniform Spreading Function

Provder and Rosen [9] used the general spreading function, equation (3.25), and bilateral Laplace transformations to obtain the following correction equations for a mass concentration detector ($p = 0$):

$$\frac{\bar{M}_k(c)}{\bar{M}_k(Uc)} = \exp\left[\frac{(3-2k)(D_2\sigma)^2}{2}\right] \frac{1 + \sum_{n=3}^{\infty} (A_n/n!)[(2-k)D_2\sigma]^n}{1 + \sum_{n=3}^{\infty} (A_n/n!)[(1-k)D_2\sigma]^n} \tag{3.31}$$

$$\frac{[\eta](c)}{[\eta](Uc)} = \exp\left[-\frac{(aD_2\sigma)^2}{2}\right] \frac{1}{1 + \sum_{n=3}^{\infty} (A_n/n!)(-aD_2\sigma)^n} \tag{3.32}$$

These correction equations reduce to those derived earlier for a uniform Gaussian spreading function. These solutions apply for a linear molecular weight calibration curve. Yau et al. [5] employed a uniform exponentially modified Gaussian spreading function and obtained the following correction equations:

$$\frac{\bar{M}_n(c)}{\bar{M}_n(Uc)} = \exp\left[\frac{(D_2\sigma)^2}{2}\right] \left[\frac{1}{1 - D_2\tau} \exp(-D_2\tau)\right] \tag{3.33a}$$

For $D_2\tau < 1$

$$\frac{\bar{M}_w(c)}{\bar{M}_w(Uc)} = \exp\left[-\frac{(D_2\sigma)^2}{2}\right] [(1 + D_2\tau)\exp(-D_2\tau)] \tag{3.33b}$$

For $D_2\tau > -1$

$$\frac{\bar{M}_z(c)}{\bar{M}_z(Uc)} = \exp\left[-\frac{3(D_2\sigma)^2}{2}\right] \left[\frac{1 + 2D_2\tau}{1 + D_2\tau} \exp(-D_2\tau)\right] \tag{3.33c}$$

The parameter τ accounts for a skewed spreading function.

Balke and Hamielec [8] suggested the use of the following semi-empirical correction equations, where SK is a factor that accounts for skewing:

$$\frac{\bar{M}_n(c)}{\bar{M}_n(Uc)} = \left[1 + \frac{SK}{2}\right] \exp\left[\frac{(D_2\sigma)^2}{2}\right] \tag{3.34a}$$

$$\frac{\bar{M}_w(c)}{\bar{M}_w(Uc)} = \left[1 + \frac{SK}{2}\right] \exp\left[-\frac{(D_2\sigma)^2}{2}\right] \tag{3.34b}$$

When skewing is not excessive, all of these correction equations are equivalent. The most serious deficiency of these methods which attempt to correct for skewing is the use of uniform skewing parameters in the spreading function. It is well known that low- and intermediate-sized polymer solutes have Gaussian spreading functions, whereas it is only the largest polymer solute molecules that have significantly skewed spreading functions. In other words, when solving Tung's integral equation, one should employ a skewing parameter in the spreading function which depends on mean retention volume Y.

3.4.2. Nonuniform Spreading Function and Nonlinear Molecular Weight Calibration Curve

Gaussian Spreading Function

Correction equations for this case are based on solutions for peak broadening in the detector cell [21,22].

$$\frac{\bar{M}_k(V,Uc)}{M(V)} = \frac{F(V - (k-1)D_2(V)\sigma(V)^2)}{F(V - (k-2)D_2(V)\sigma(V)^2)} \exp\left[\frac{2k-3}{2}\,[D_2(V)\sigma(V)]^2\right] \tag{3.35a}$$

$$\bar{M}_k(c) = \frac{\int_0^\infty F(V)\, M_k^{k-1}(V,Uc)\, dV}{\int_0^\infty F(V)\, M_k^{k-2}(V,Uc)\, dV} \tag{3.35b}$$

and

$$\frac{[\eta](V,Uc)}{[\eta](V)} = \frac{F(V - aD_2(V)\sigma(V)^2)}{F(V)} \exp\left[\frac{[aD_2(V)\sigma(V)]^2}{2}\right] \tag{3.36a}$$

$$[\eta](c) = \frac{\int_0^\infty F(V)[\eta](V,Uc)\, dV}{\int_0^\infty F(V)\, dV} \tag{3.36b}$$

Equations (3.35a) and (3.36a) apply for simple polymers and a mass concentration detector. The full derivation for a general detector is given later in the chapter. To obtain corrected whole polymer molecular weight averages and intrinsic viscosity, equations (3.35a) and (3.36a) are substituted into equations (3.35b) and (3.36b) and the integrations performed. $D_2(V)$ and $\sigma(V)$ should be available through prior calibration.

Another valid approach is to correct the detector response for peak broadening using the analytical solution [22].

$$W(V) = F(V) \frac{\sigma(V)}{\bar{\sigma}(V)} \exp\left[-\frac{[V - \bar{Y}(V)]^2}{2\bar{\sigma}(V)^2}\right] \tag{3.37}$$

where

$$\bar{Y}(V) = V + \frac{1}{D_2(V)} \ln \frac{F(V + D_2(V)\sigma(V)^2)}{\sqrt{F(V - D_2(V)\sigma(V)^2)F(V + D_2(V)\sigma(V)^2)}} \tag{3.37a}$$

$$\bar{\sigma}(V)^2 = \sigma(V)^2 + \frac{1}{D_2(V)^2} \ln \frac{F(V - D_2(V)\sigma(V)^2)F(V + D_2(V)\sigma(V)^2)}{F(V)^2} \tag{3.37b}$$

One then uses the usual integration procedures with W(V) to calculate the whole polymer properties corrected for peak broadening. The full derivation of this analytical solution for W(V) is given later in this chapter. Again $D_2(V)$ and $\sigma(V)$ values should be available from prior calibration. The quantities $\bar{Y}(V)$ and $\bar{\sigma}(V)^2$ are the mean retention volume and variance of the distribution of different-sized species in the detector cell, W(V,Y). A more complete definition is given later. To date, analytically derived correction equations for dispersion for the case of a nonuniform nonsymmetric instrumental spreading function have not been published.

3.4.3. Application of Correction Equations for Peak Broadening

To illustrate the usual magnitude of peak broadening corrections, calculations have been made for a mass concentration detector ($p = 0$)

Table 3.1. Corrections for Peak Broadening for a Mass-Concentration Detector (p = 0) and for a Light-Scattering Photometer Detector (LALLSP) (p = 1)--Simple Polymers, Uniform Gaussian Spreading Function, and Linear Molecular Weight Calibration Curve

$(D_2\sigma)$	$[\eta]$ (%)	$\overline{M}_n$ (%)	$\overline{M}_w$ (%)	$\overline{M}_z$ (%)	$\overline{M}_{z+1}$ (%)
		Mass-concentration detector (p = 0)			
0.25	1.5	-3.1	3.2	9.8	16.9
0.50	6.3	-11.8	13.3	45.5	86.8
1.0	27.8	-39.3	64.8	--	--
		Light-scattering photometer detector (p = 1)			
0.25	0.3	-9.1	-2.9	3.2	9.9
0.50	1.1	-31.5	-11.4	13.4	45.6
1.0	16.1	--	-39.4	65.0	--

and for a LALLSP detector (p = 1). The results of these calculations for a simple polymer and for the case where the spreading function is uniform and Gaussian and the molecular weight calibration curve is linear are tabulated in Table 3.1.

An examination of Table 3.1 clearly indicates that the magnitude of the corrections differ markedly with detector type and molecular weight average. To obtain reliable higher averages such as $\overline{M}_z$, it is better to integrate the LALLSP detector response to find $\overline{M}_z(\mathrm{Uc})$. It is interesting to note that $\overline{M}_w(\mathrm{Uc})$ values obtained with the mass concentration detector and with the LALLSP detector bracket $\overline{M}_w(\mathrm{c})$ and thus provide upper and lower bounds. When $\sigma(V)$ values are not available, this information could be useful.

Figure 3.1 illustrates experimental corrections for peak broadening for a uniform Gaussian spreading function showing the symmetrical corrections to $\overline{M}_n$ and $\overline{M}_w$ and the dependence on number of columns in series, mobile-phase flow rate, and residence time in the SEC [24].

Figure 3.2 shows the dependence of the standard deviation of a uniform Gaussian spreading function on the size of packing particles

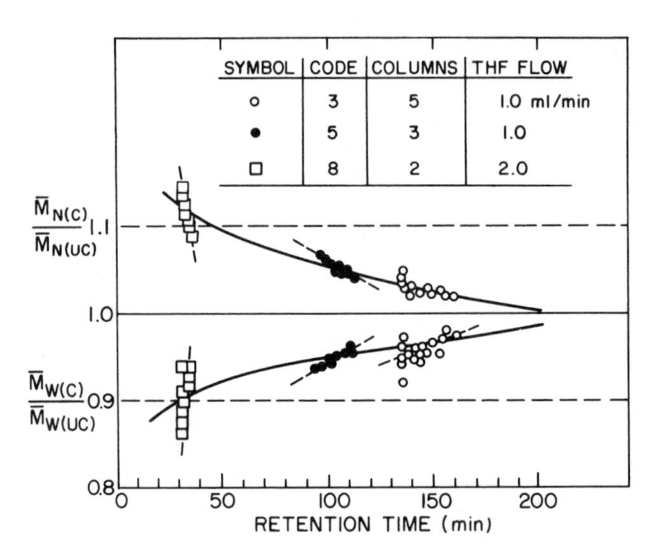

Figure 3.1. Experimental corrections [$\overline{M}_n(c)/\overline{M}_n(Uc)$ and $\overline{M}_w(c)/\overline{M}_w(Uc)$] for the whole polymer made using a uniform Gaussian spreading function showing symmetry in the corrections for $\overline{M}_n$ and $\overline{M}_w$ and the dependence on retention time or magnitude of σ for different column combinations. (From Ref. 24.)

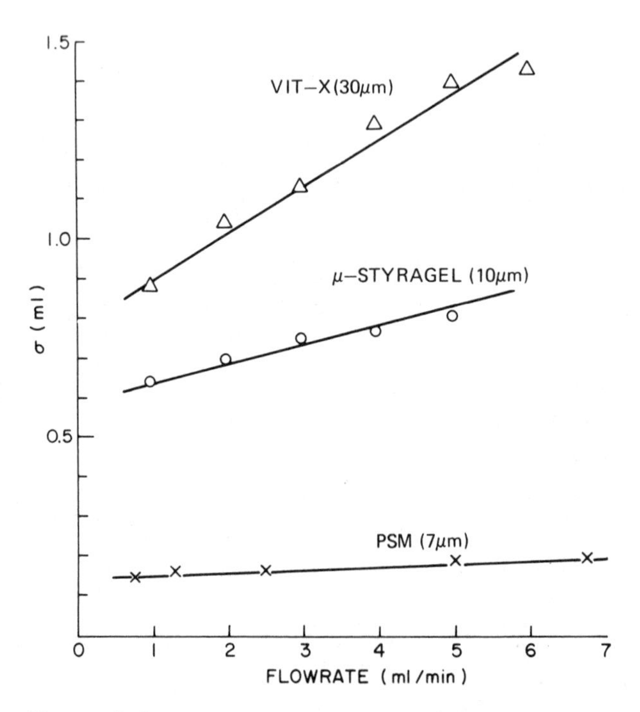

Figure 3.2. Experimental σ values for different-sized packing and mobile-phase flow rates measured for a narrow MWD polystyrene standard (MV - 97,200). (From Ref. 13.)

and mobile-phase flow rate [13]. The small variances observed for small packing particles suggest that with these micropackings, skewing of single species chromatograms due to slow pore diffusion even at high molecular weight levels in the columns would be small, if not negligible.

Balke and Hamielec [8] were the first to publish a systematic investigation of skewing in SEC. It was observed that skewing could result when SEC operation gave large peak broadening and little peak separation and also when the columns are overloaded with polymer solute [8]. Figures 3.3 to 3.5 show different measurements which clearly illustrate the effects of skewed instrumental spreading functions. Figure 3.3 shows that skewing occurs mainly for the higher molecular weight species. The excessive skewing is a consequence of excessive peak broadening due to the use of too few columns with a high mobile-phase flow rate. Figure 3.4 shows MWDs measured

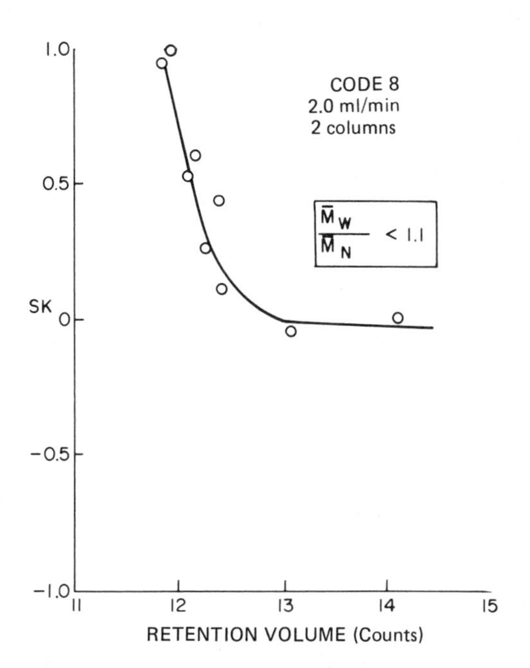

Figure 3.3. Experimental skewing factors (SK) for different narrow MWD polystyrene standards (different retention volumes) for a high mobile-phase flow rate and a column combination with modest peak separation. (From Ref. 8.)

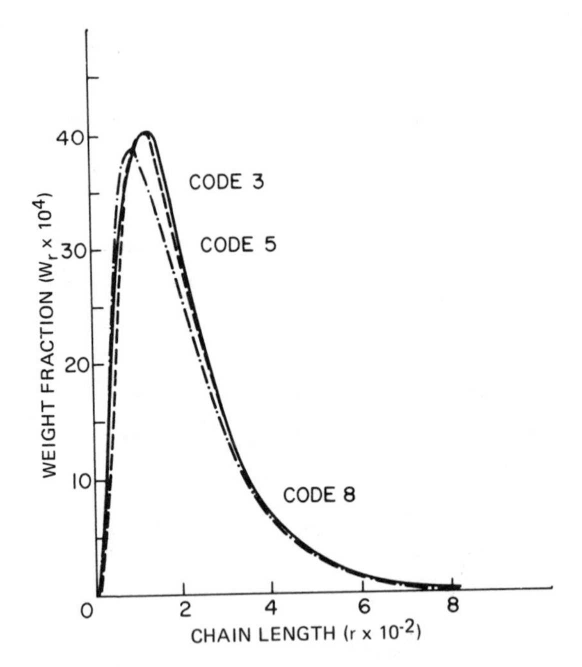

Figure 3.4. Experimental molecular weight distributions for the same polymer sample measured using SEC with different column combinations. [Code 8 had significant skewing factors (SK).] (From Ref. 27.)

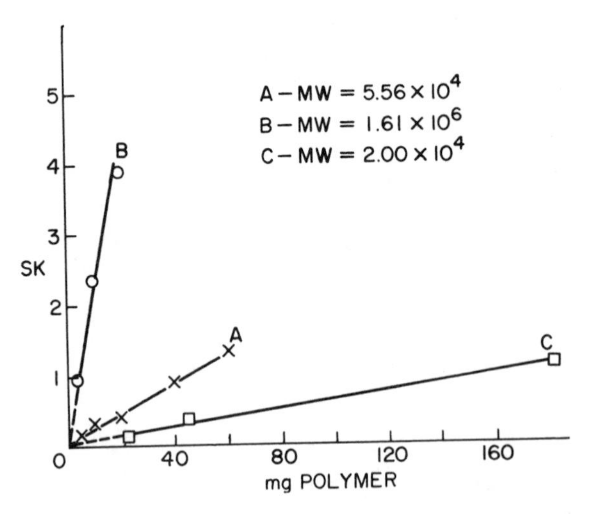

Figure 3.5. Experimental skewing factors (SK) showing the effects of weight of polymer solute injected for different molecular weight polyisobutenes. (From Ref. 25.)

for the same polymer sample. Minimal skewing was observed for codes 3 and 5. The shift toward lower molecular weights for code 8 is a direct result of a skewed instrumental spreading function and is consistent with the SK values in Figure 3.3. Figure 3.5 shows the effect of column overloading. High solute concentrations result in a shift in the molecular weight calibration curve and skewed instrumental spreading functions. The SK values in Figure 3.5 are a result of the combined effects of skewing and shift in the molecular weight calibration curve. The SK values were calculated using molecular weight data after Cantow et al. [25]. The data are tabulated in Table 3.2. These molecular weight averages were calculated with the same molecular weight calibration curve at all polymer solute loadings.

Table 3.2. Skewing of the Instrumental Spreading Function and Shift of the Molecular Weight Calibration Curve--Effect of Polymer Solute Loading on Molecular Weight Averages

Sample weight (mg)	$\overline{M}_n$(Uc)	$\overline{M}_w$(Uc)
	Polyisobutene A	
5	21,300	53,600
10	18,800	50,800
20	17,900	51,000
40	14,400	43,500
60	14,800	39,800
	Polyisobutene B	
4	468,000	1,080,000
10	332,000	698,000
20	208,000	470,000
	Polyisobutene C	
23	11,500	20,600
45	9,900	19,200
181	7,300	15,100

3.5. NUMERICAL SOLUTION OF THE INTEGRAL EQUATION: CORRECTED DETECTOR RESPONSE

The earlier numerical methods of solving the integral equation are not completely satisfactory for all SEC operational conditions [26, 27]. For instance, when corrections for broadening are significant, severe oscillations appear in the corrected detector response. Such oscillations are caused not only by detector noise but are also a result of error propagation in the mathematical methods employed. This is easily verified by applying these mathematical methods to synthesized chromatograms in which detector noise is excluded. Recent works on numerical solution techniques have concentrated on reducing error propagation, computer storage, and computation time [28-30]. The ultimate aim, of course, is to find an analytical solution of the integral equation. This has recently been done for the case of a nonuniform Gaussian spreading function [22]. The details of this analytical solution are given later in the chapter. It remains to generalize this analytical solution for a general instrumental spreading function.

3.5.1. Criteria to Be Satisfied

The evaluation of a numerical technique must be based on certain practical criteria. Among the more important of these are minimum oscillations in the absence of detector noise at high levels of peak broadening for narrow chromatograms (an effective smoothing procedure for the raw detector response is essential); accurate recovery of corrected $\bar{M}_n$, $\bar{M}_w$, and $\bar{M}_z$ emphasizes the accuracy of the low and high molecular weight tails of the corrected detector response; additionally, the methods should be designed for small computer storage and computation times.

3.5.2. Effective Numerical Methods

The methods of Chang and Huang [28], Ishige et al. [29], and Vozka and Kubin [30] are among the best methods published to date. In particular, the method of Ishige et al. [29] has been comprehensively evaluated by Kato and Hashimoto [31] and has been widely used in SEC and hydrodynamic chromatography (HDC) [32].

3.6. DISPERSION IN THE DETECTOR CELL

Early attempts to solve Tung's integral equation focused on the detector response (raw and corrected), instrumental spreading function, whole polymer molecular weight averages, and intrinsic viscosity of the whole polymer. With the advent of absolute detector systems such as DRI/LALLSP, a need arose to examine dispersion in the detector cell and develop quantitative correction equations for peak broadening locally. The first attempt in this direction was by Yau et al. [20], who considered a uniform Gaussian spreading function and a linear molecular weight calibration curve. This solution was generalized to include a nonuniform Gaussian and a nonlinear molecular weight calibration curve [21], and more recently an analytical solution of Tung's integral equation was obtained based on this generalized solution [22].

3.6.1. General Relationships

The bivariate distribution function $W(V,Y)$ gives a measure of dispersion in the detector cell. When the variance with respect to Y about Y mean approaches zero, perfect resolution is approached, with all solute species in the detector cell having the same size. It is clear that a knowledge of $W(V,Y)$ permits a direct integration to obtain $W(Y)$ or more directly via equation (3.20) when the instrumental spreading function is known.

When peak broadening is not excessive, $W(V,Y)$ is narrow and can be adequately approximated by a Gaussian distribution of the form

$$W(V,Y) = \frac{F(V)}{\sqrt{2\pi\bar{\sigma}(V)^2}} \exp\left[-\frac{[Y - \bar{Y}(V)]^2}{2\bar{\sigma}(V)^2}\right] \tag{3.38}$$

where $F(V)$ is a normalization factor.

3.6.2. Corrections for Dispersion for General Detector: Nonuniform Gaussian Spreading Function and Nonlinear Molecular Weight Calibration Curve

Simple Polymers

The nonlinear molecular weight calibration curve can be represented as

$$M(Y) = D_1(V) \exp [-D_2(V)Y] \tag{3.39}$$

It can readily be shown for the case of a constant detector response factor that

$$\frac{\bar{M}_k(V,Uc)}{M(V)} = \frac{F(V - (k - p - 1)D_2(V)\sigma(V)^2)}{F(V - (k - p - 2)D_2(V)\sigma(V)^2)} \exp \left[\frac{[2(k - p) - 3][D_2(V)\sigma(V)]^2}{2} \right] \tag{3.40}$$

For a mass-concentration detector such as DRI, set p = 0 and for a LALLSP detector set p = 1 to obtain

$$\frac{\bar{M}_k(V,Uc)}{M(V)} = \frac{F(V - (k - 2)D_2(V)\sigma(V)^2)}{F(V - (k - 3)D_2(V)\sigma(V)^2)} \exp \left[\frac{(2k - 5)[D_2(V)\sigma(V)]^2}{2} \right] \tag{3.40a}$$

It should be emphasized that when using equation (3.40a) the LALLSP detector response should be used to evaluate F(V) quantities. It is of interest to compare corrections for $\bar{M}_w$ using both detector responses. For p = 0 (mass-concentration detector response)

$$\frac{\bar{M}_w(V,Uc)}{M(V)} = \frac{F_M(V - D_2(V)\sigma(V)^2)}{F_M(V)} \exp \left[\frac{[D_2(V)\sigma(V)]^2}{2} \right] \tag{3.40b}$$

Local peak broadening corrections for $\bar{M}_w(V,Uc)$ using equation (3.40b) for micro and macropackings [22] are given in Table 3.3. The smaller σ for micropackings is compensated for by steeper detector response giving corrections that are about the same. For p = 1 (LALLSP detector response)

Table 3.3. Corrections for Peak Broadening in the Detector Cell for $\bar{M}_w(V,Uc)$

Micropacking[a]			Macropacking[b]		
V (ml)	$F(V) \times 10^3$	$\frac{\bar{M}_w(V,Uc)}{M(V)}$	V (ml)	$F(V) \times 10^3$	$\frac{\bar{M}_w(V,Uc)}{M(V)}$
27	16.0	0.745	56	3.6	0.740
28	64.0	0.853	60	14.4	0.846
29	136.3	0.937	64	44.1	0.995
30	186.6	1.002	68	44.1	0.995
31	187.7	1.055	72	46.1	1.041
32	152.1	1.084	76	38.9	1.077
33	105.9	1.109	80	28.2	1.105
34	66.4	1.127	84	18.4	1.122
35	38.6	1.139	88	11.1	1.134
36	21.3	1.153	92	6.3	1.147

[a]Parameters employed: $D_2 = 0.357\ \text{ml}^{-1}$, $\sigma = 0.70$ ml.

[b]Parameters employed: $D_2 = 8.62 \times 10^{-2}\ \text{ml}^{-1}$, $\sigma = 2.90$ ml.

$$\frac{\bar{M}_w(V,Uc)}{M(V)} = \frac{F_L(V)}{F_L(V + D_2(V)\sigma(V)^2)} \exp\left[-\frac{[D_2(V)\sigma(V)]^2}{2}\right] \tag{3.41}$$

where the subscripts M and L on F(V) are to indicate detector response for a mass-concentration detector and for a LALLSP detector, respectively. The use of these correction equations for both detectors is illustrated later in the chapter.

Complex Polymers

The molecular weight calibration curves for a complex polymer can be represented as

$$\bar{M}_k(Y) = D_{1k}(V) \exp\left[-D_{2k}(V)Y\right] \tag{3.39a}$$

and the correction equations now take the form

$$\frac{\overline{M}_k(V,Uc)}{\overline{M}_k(V)} = \frac{F(V-(k-p-1)D_{2k}(V)\sigma(V)^2)}{F(V-(k-p-2)D_{2k}(V)\sigma(V)^2)} \exp\left[\frac{[2(k-p)-3][D_{2k}(V)\sigma(V)]^2}{2}\right] \tag{3.40c}$$

3.7. ANALYTICAL SOLUTION OF THE INTEGRAL EQUATION: CORRECTED DETECTOR RESPONSE

3.7.1. Nonuniform Gaussian Spreading Function

The basis for an analytical solution of the integral equation for the corrected detector response is W(V,Y) given by equation (3.38). When the instrumental spreading function is Gaussian, equation (3.37) is valid. To complete the analytical solution a relationship for $\overline{Y}(V)$ and $\overline{\sigma}(V)$ in terms of V and $\sigma(V)$ is required. This is obtained by calculating $M_n(V,Uc)$ and $M_w(V,Uc)$ using W(V,Y) given by equation (3.38) and equating these values to the correction equations obtained for a nonuniform Gaussian spreading function, equation (3.40) with k = 1 and 2. The resulting relationships follow:

$$\overline{\sigma}(V)^2 = \sigma(V)^2 + \frac{1}{D_2^2(V)} \ln \frac{F(V+(1+p)D_2(V)\sigma(V)^2)F(V-(1-p)D_2(V)\sigma(V)^2)}{F(V+pD_2(V)\sigma(V)^2)^2} \tag{3.42a}$$

$$\overline{Y}(V) = V + \frac{1}{2D_2(V)} \ln \frac{F(V+(1+p)D_2(V)\sigma(V)^2)}{F(V-(1-p)D_2(V)\sigma(V)^2)} \times \frac{F(V+pD_2(V)\sigma(V)^2)}{F(V+(1+p)D_2(V)\sigma(V)^2)F(V-(1-p)D_2(V)\sigma(V)^2)} \tag{3.42b}$$

It can be shown with a Taylor series expansion that the dependence of $\overline{\sigma}(V)$ and $\overline{Y}(V)$ on $D_2(V)$ properly drops out [22]. For computational reasons it is easier to employ equations (3.42a) and (3.42b) with any convenient value for $D_2(V)$. The application of this analytical solution for a mass-concentration detector has been described elsewhere

[22]. Its application to both mass-concentration and light-scattering detectors is given later in this chapter. It should be mentioned that when equations (3.42a) and (3.42b) with $p = 0$ are used with the LALLSP detector response, this is equivalent to equating $\bar{M}_w(V,Uc)$ and $\bar{M}_z(V,Uc)$ in their development rather than $\bar{M}_n(V,Uc)$ and $\bar{M}_w(V,Uc)$. This subject is dealt with more fully in Section 3.9.

3.7.2. General Spreading Function

The analytical solution of the integral equation for the detector response can be generalized by decomposing the original integral equation with a non-Gaussian spreading function into a sum of integrals, each one of which has a nonuniform Gaussian kernel. This procedure is illustrated with a simple skewed spreading function as follows:

$$F(V) = \frac{1}{\sqrt{2\pi\sigma(V)^2}} \int_0^\infty W(Y)[1 + f(Y)(V - Y)] \exp\left[-\frac{(V - Y)^2}{2\sigma(V)^2}\right] dY \tag{3.43}$$

$$\begin{aligned} F(V) = {} & \frac{1}{\sqrt{2\pi\sigma(V)^2}} \int_0^\infty W(Y) \exp\left[-\frac{(V - Y)^2}{2\sigma(V)^2}\right] dY \\ & + \frac{V}{\sqrt{2\pi\sigma(V)^2}} \int_0^\infty W_1(Y) \exp\left[-\frac{(V - Y)^2}{2\sigma(V)^2}\right] dY \\ & + \frac{1}{\sqrt{2\pi\sigma(V)^2}} \int_0^\infty W_2(Y) \exp\left[-\frac{(V - Y)^2}{2\sigma(V)^2}\right] dY \end{aligned} \tag{3.43a}$$

where $W_1(Y) = f(Y)W(Y)$ and $W_2(Y) = f(Y)YW(Y)$ and

$$F(V) = F_1 + VF_2 + F_3 \tag{3.43b}$$

Using the analytical solution, one can relate $W(V)$ to F_1, F_2, and F_3. This then permits one to solve for F_1, F_2, and F_3 and then, of course, $W(V)$.

3.8. INSTRUMENTAL CORRECTIONS FOR PEAK BROADENING AND DETERMINATION OF THE MOLECULAR WEIGHT CALIBRATION CURVE: ABSOLUTE DETECTOR SYSTEMS

Detector systems that can measure MWD or molecular weight averages of the detector cell contents can be used to measure peak broadening parameters and the molecular weight calibration curve. The MWD and molecular weight averages can thus be obtained for a polymer sample directly without the need for calibration with a number of polymer standards.

The basis for the methods to be discussed is the function $W(V,Y)$ given by equation (3.38) with the detector systems, providing a means to measure $\bar{\sigma}(V)$ and $\bar{Y}(V)$. We will consider measurements of $\bar{M}_n(V,Uc)$ and $\bar{M}_w(V,Uc)$ and use the following equations to determine $\bar{\sigma}(V)$ and $\bar{Y}(V)$:

$$\bar{M}_n(V,Uc) = \exp\left[-\frac{(1+2p)[D_2(V)\bar{\sigma}(V)]^2}{2}\right] D_1(V) \exp[-D_2(V)\bar{Y}(V)] \tag{3.44a}$$

$$\bar{M}_w(V,Uc) = \exp\left[\frac{(1-2p)[D_2(V)\bar{\sigma}(V)]^2}{2}\right] D_1(V) \exp[-D_2(V)\bar{Y}(V)] \tag{3.44b}$$

Once $\bar{\sigma}(V)$ and $\bar{Y}(V)$ are known, $W(V)$ is given by equation (3.37). It should be noted that both $\bar{\sigma}(V)$ and $\bar{Y}(V)$ change with detector type.

3.8.1. Low-Angle Laser Light-Scattering Photometer

A LALLSP in series with a concentration detector such as a DRI gives a measure of $\bar{M}_w(V,Uc)$ across the detector responses for a polymer sample. This is true for both simple and complex polymers. $\bar{M}_w(V,Uc)$ is related to the detector responses by the equation

$$\frac{1}{\bar{M}_w(V,Uc)} = \frac{K_\theta^* C(V)}{\bar{R}_\theta(V)} - 2A_2 C(V) \tag{3.45}$$

where $\bar{R}_\theta$, the excess Rayleigh factor, is obtained from the LALLSP response and $C(V)$ from the mass-concentration detector response.

A_2 is the second virial coefficient and K_θ^* is the polymer optical constant. Details of detector response interpretation may be found in Chapter 5 and in Refs. 33 and 34. The $\overline{M}_w(c)$ for the whole polymer can then be found by direct integration using equation (3.8), hence the term *absolute detector system.*

Simple Polymers

A knowledge of the molecular weight calibration curve M(V) permits one to apply the correction equations for a nonuniform Gaussian spreading function [equation (3.40)] to determine $\sigma(V)$ across the detector response. W(V), the corrected detector response is then given by the analytical solution, equations (3.37), (3.42a), and (3.42b).

Complex Polymers

$\overline{M}_w(c)$ for the whole polymer can be found by direct integration using equation (3.8). A knowledge of $\sigma(V)$ permits one to use the analytical solution for W(V). It should be noted that accurate values for $D_2(V)$ are not required to determine $\overline{\sigma}(V)$ and $\overline{Y}(V)$.

3.8.2. Viscometer Detector

A viscometer detector in series with a mass-concentration detector provides a measure of $[\eta](V,Uc)$ across the detector response for both simple and complex polymers. The whole polymer intrinsic viscosity is given by

$$[\eta](c) = \frac{\int_0^\infty F(V)[\eta](V,Uc)\, dV}{\int_0^\infty F(V)\, dV} \tag{3.46}$$

where F(V) is the mass-concentration detector response. Details of the interpretation of the detector response for a viscometer detector are described in Chapter 5 and may also be found elsewhere [35]. A knowledge of $\sigma(V)$ permits one to use the analytical solution to solve for W(V) for each detector response. The use of the two corrected detector responses gives $[\eta](V)$ and then application of the universal molecular weight calibration curve provides M(V) for simple polymers and $\overline{M}_n(V)$ for complex polymers [see equation (3.17)]. $\overline{M}_n(c)$ for the whole polymer is then found by first converting $\bar{M}_n(V)$ to $\bar{M}_n(V,Uc)$ and then direct integration. A combined detector system that gives both $\bar{M}_n(V,Uc)$ and $\bar{M}_w(V,Uc)$ would permit one to determine $\bar{\sigma}(V)$ and $\bar{Y}(V)$ with equations (3.44a) and (3.44b) and thus $W(V,Y)$ and $W(V)$.

3.8.3. Quasielastic Light-Scattering Detector

In principle an on-line quasielastic light-scattering detector [36] could give a measure of the mean and variance of $W(V,Y)$ and provide a direct measure of $\bar{\sigma}(V)$ and $\bar{Y}(V)$. Direct integration using equation (3.21) gives the corrected detector response $W(V)$.

Whole polymer MWDs are too broad for precise interpretation of quasielastic light-scattering data obtained off-line. The MWD in the detector cell should be sufficiently narrow to permit accurate measurements.

3.9. EXPERIMENTAL MOLECULAR WEIGHT AND PEAK BROADENING CALIBRATION

In principle, the availability of truly monodispersed polymer standards would permit the measurement of peak broadening parameters with a once-through technique. The detector response could be integrated to provide the mean retention volume, the variance about the mean, and other shape parameters, such as those that give a measure of skewing. Unfortunately, even the relatively narrow MWD polystyrene standards are too polydispersed and significant broadening due to molecular size separation results with a once-through technique. Methods devised to overcome this lack of monodispersed standards are now discussed.

3.9.1. Narrow Molecular Weight Distribution Standards

Reverse-Flow Technique

This technique, originally developed by Tung et al. [37], assumes that molecular size separation in SEC is reversible and therefore that when the narrow MWD polystyrene standard is injected, allowed to pass through half of the column combination, and then the flow is reversed, the contribution is canceled out, leaving the contribution to axial dispersion in the columns. This reverse-flow technique has been successfully applied [24] for macropacking but is not suitable for micropacking [38].

Once-Through Technique

Balke and Hamielec [8] have proposed the use of semiempirical correction equations (3.34a) and (3.34b) based on Tung's integral equation to account for MWD separation. The success of this technique depends on the accuracy of the $\overline{M}_n$ and $\overline{M}_w$ values for the polystyrene standards. It has been pointed out that narrow MWD polystyrene standards of high molecular weight prepared by anionic polymerization may contain small amounts of low molecular weight polymer. In measurement of $\overline{M}_n$ of such standards by osmometry, this low molecular weight polymer may pass through the membrane and not be detected. SEC would detect a low molecular weight tail and provide a measure of the skewing parameters, SK, which is larger than it should be.

3.9.2. Broad Molecular Weight Distribution Standards

When the concept of universal peak broadening is not acceptable, one must resort to the use of broad MWD standards for which $\overline{M}_n$, $\overline{M}_w$, and $[\eta]$ values are known for the whole polymers. Techniques based on the use of the universal molecular weight calibration curve with broad standards are now discussed. The universal molecular weight calibration can be expressed as

$$[\eta](V)M(V) = \phi(V) \tag{3.47}$$

and

$$M(V) = X\phi(V)^Y \tag{3.47a}$$

where $X = K^{-Y}$, $Y = 1/(1 + a)$, and K and a are Mark-Houwink constants for a simple polymer.

One Broad Molecular Weight Distribution Standard

A knowledge of $\sigma(V)$ permits calculation of $\bar{\sigma}(V)$, $\bar{Y}(V)$, and $W(V)$ using the analytical solution, equations (3.37), (3.42a), and (3.42b). The following equations permit a single-variable search for Y and direct calculation of X:

$$\bar{M}_n(c) = X \left\{ \int_0^\infty \phi^{-Y}(V) F(V) \frac{\sigma(V)}{\bar{\sigma}(V)} \exp\left[-\frac{[V - \bar{Y}(V)]^2}{2\bar{\sigma}(V)^2} \right] dV \right\}^{-1} \quad (3.48a)$$

$$\bar{M}_w(c) = X \left\{ \int_0^\infty \phi^{Y}(V) F(V) \frac{\sigma(V)}{\bar{\sigma}(V)} \exp\left[-\frac{[V - \bar{Y}(V)]^2}{2\bar{\sigma}(V)^2} \right] dV \right\} \quad (3.48b)$$

$$[\eta](c) = KX^a \left\{ \int_0^\infty \phi^{aY}(V) F(V) \frac{\sigma(V)}{\bar{\sigma}(V)} \exp\left[-\frac{[V - \bar{Y}(V)]^2}{2\bar{\sigma}(V)^2} \right] dV \right\} \quad (3.48c)$$

It should be emphasized that if the Mark-Houwink constants for polystyrene used to construct the universal molecular weight calibration curve are accurate, then in principle X and Y values may be used to find accurate Mark-Houwink constants for the polymer in question. If, however, inaccurate polystyrene Mark-Houwink constants were used, the molecular weight calibration curve M(V) for the polymer will still be correct in principle even though the K and a values are now incorrect.

Two Broad Molecular Weight Distribution Standards

When $\sigma(V)$ values and M(V) are not known, one can employ two standards and the analytical solution of the integral equation for a uniform Gaussian spreading function [39]. One can solve for X, Y and σ_1, σ_2, peak broadening parameters, one for each standard.

3.9.3. Absolute Detector Systems

In principle, an absolute detector system can be used to determine the molecular weight calibration curve and peak broadening parameters simultaneously. These detector systems are generally more difficult to operate and hence should be used for calibration. Once the molecular weight calibration curve and peak broadening parameters are known, the simpler mass-concentration detectors such as DRI should be used alone for routine SEC molecular weight characterization.

Low-Angle Laser Light-Scattering Photometer

Recent publications on the use of aqueous SEC with a DRI/LALLSP detector system for the molecular weight characterization of dextrans and nonionic polyacrylamides are now discussed in some detail [40, 41]. It is shown how the molecular weight calibration curve and the peak broadening parameters $\sigma(V)$ can be measured over a wide range of molecular weights or retention volumes for dextrans and nonionic polyacrylamides, both of which are treated as simple polymers. For these specific polymers for which narrow MWD standards are not available, it is recommended that the DRI/LALLSP detector system be used to calibrate the SEC and determine both molecular weight calibration curve and peak broadening parameters $\sigma(V)$. Upon completion of calibration, however, the easier-to-employ DRI detector should be used alone to characterize other dextran and nonionic polyacrylamides. To obtain acceptable signal-to-noise ratios for the LALLSP detector with aqueous SEC, one must employ the smallest in-line filter possible without filtering out high molecular weight material in the polymer sample. With dextrans, which are small compact molecules, a 0.22-μm Millipore filter was used, and with the much larger nonionic polyacrylamides a 0.45-μm Millipore filter was used. Typical DRI/LALLSP detector responses for dextran and nonionic polyacrylamide are shown in Figures 3.6 and 3.7. With the larger in-line filter a much noisier LALLSP detector response is obtained. This can become a major problem when dealing with very high molecular weight polymers. An examination of the correction equation for $\overline{M}_w(V,Uc)$ for a mass-concentration detector [see equation (3.40b)] reveals that one can minimize the local correction for peak broadening by blending broad MWD standards to produce a very broad MWD standard with a specific detector response shape. On the high molecular weight end (low retention volume end) of the detector response, $F(V - D_2(V)\sigma(V)^2/F(V)$ is less than unity, whereas $\exp((D_2(V)\sigma(V))^2/2)$ is always greater than unity. To minimize peak broadening correction, the product of these two terms should be close to unity. The purpose of blending available broad MWD standards is to extend the high molecular weight end, giving a gradual slope to the low retention volume end of the detector response. When the local correction factor for peak broadening is close to unity over a wide range of molecular weights, measurements of $\overline{M}_w(V,Uc)$ give points on the molecular weight calibration

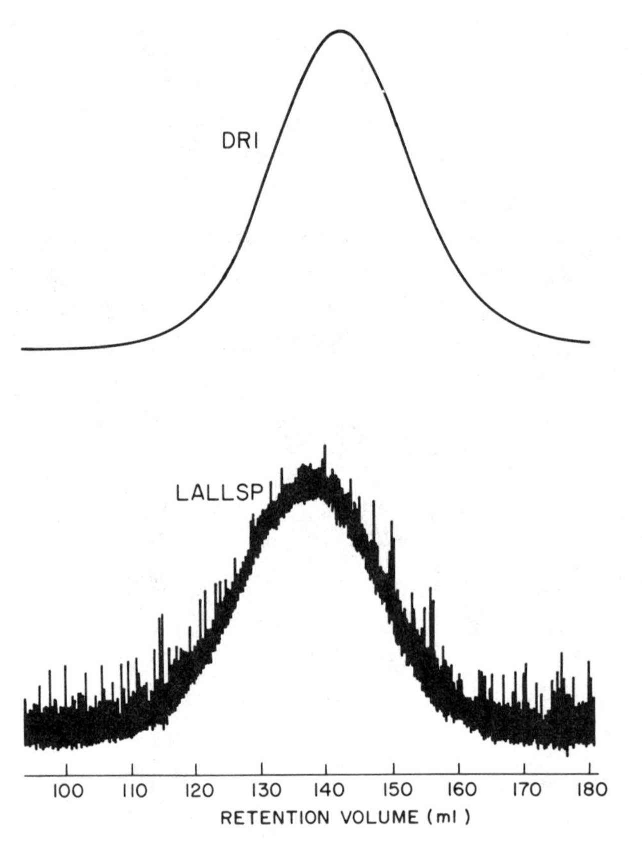

Figure 3.6. Typical DRI/LALLSP detector responses for a dextran using aqueous SEC with a 0.22-μm in-line Millipore filter. (From Ref. 40.)

curve. The need to know σ(V) and correct for broadening locally is thus eliminated. Such a standard, now called standard X, was made by blending Pharmacia dextran standards in the following proportions: T250 (53 wt %), T40 (29 wt %), and T10 (18 wt %). DRI/LALLSP detector responses for standard X are shown in Figure 3.8. The effect of the size of σ(V) on the peak broadening correction $\overline{M}_w(V,Uc)/M(V)$ for standard X is given in Table 3.4. It is clear that standard X provides a means of calibrating for molecular weight calibration curve over a wide range of retention volume without the need to apply a peak broadening correction. It is therefore assumed that M(V) is known and now the next step is to determine σ(V). Reference to

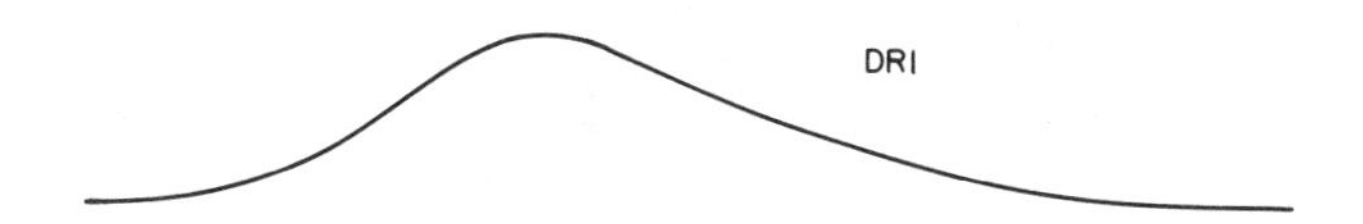

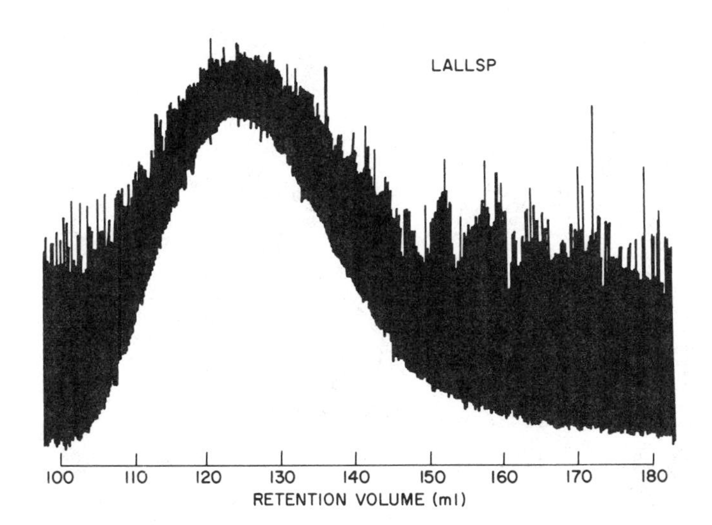

Figure 3.7. Typical DRI/LALLSP detector responses for a nonionic polyacrylamide using aqueous SEC with a 0.45-μm in-line Millipore filter. (From Ref. 41.)

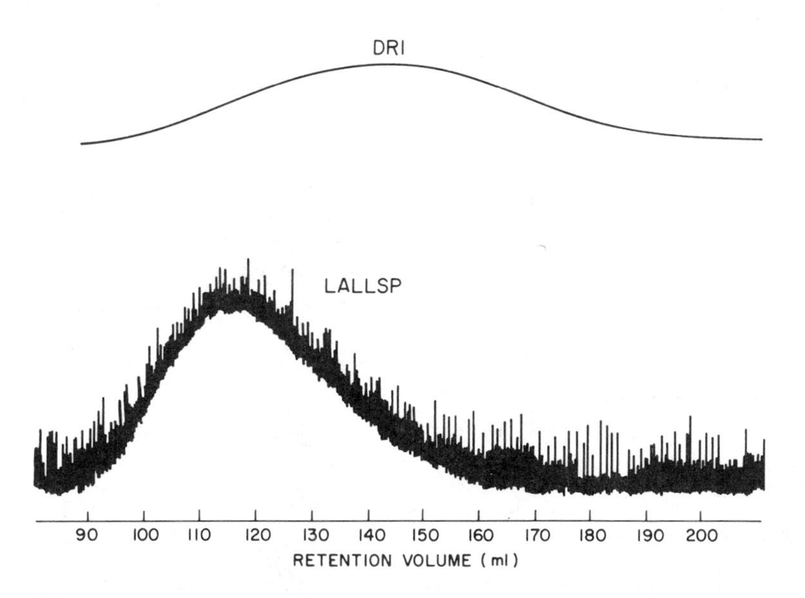

Figure 3.8. DRI/LALLSP detector responses for a dextran standard X [a blend of Pharmacia dextrans: T250 (53 wt %), T40 (29 wt %), and T10 (18 wt %)] used for molecular weight calibration.

Table 3.4. Standard X: Sensitivity to Peak Broadening Parameter, $\sigma(V)$

Retention volume (ml)	$\bar{M}_w(V,Uc)/M(V)$				
	$\sigma(V)$ actual	$0.8\sigma(V)$	$0.5\sigma(V)$	$1.2\sigma(V)$ [a]	$1.5\sigma(V)$ [a]
110.6	0.922	0.951	0.981	--	--
115.5	0.915	0.952	0.983	--	--
120.6	0.972	0.983	0.994	0.961	--
125.6	1.011	1.006	1.002	1.019	1.036
130.6	1.036	1.025	1.010	1.052	1.076
135.6	1.028	1.018	1.007	1.032	1.057
140.6	1.032	1.021	1.008	1.045	1.067
145.6	1.032	1.020	1.008	1.045	1.072
150.6	1.026	1.016	1.006	1.038	1.060
155.6	1.018	1.011	1.004	1.026	1.041
160.6	1.017	1.011	1.004	1.024	1.038
165.6	1.018	1.012	1.005	1.026	1.040
170.6	1.019	1.012	1.005	1.027	1.042
175.6	1.025	1.016	1.006	1.036	1.059
180.6	1.015	1.010	1.004	1.021	1.031

[a] Calculations at low retention volumes were not possible because of excessively large $\sigma(V)$.

equation (3.40b) reveals that given $\bar{M}_w(V,Uc)/M(V)$, $D_2(V)$ one can use single-variable search to find $\sigma(V)$. Peak broadening parameters obtained in this manner with a dextran standard [a blend containing T250 (71 wt %) and T40 (29 wt %)] are shown together with the detector responses in Figure 3.9. Whole polymer $\bar{M}_n(c)$ and $\bar{M}_w(c)$ for a number of Pharmacia standards were calculated using two calculational paths. One path (method 1) involved the integration of $\bar{M}_n(V,Uc)$ and $\bar{M}_w(V,Uc)$ using the DRI detector response according to equations (3.7) and (3.8), and the other calculational path (method 2) involved first finding corrected detector responses using the analytical solution, equations (3.37), (3.42a), and (3.42b) for DRI. A corrected detector response for Pharmacia standard T70 is shown in Figure 3.10. These corrected

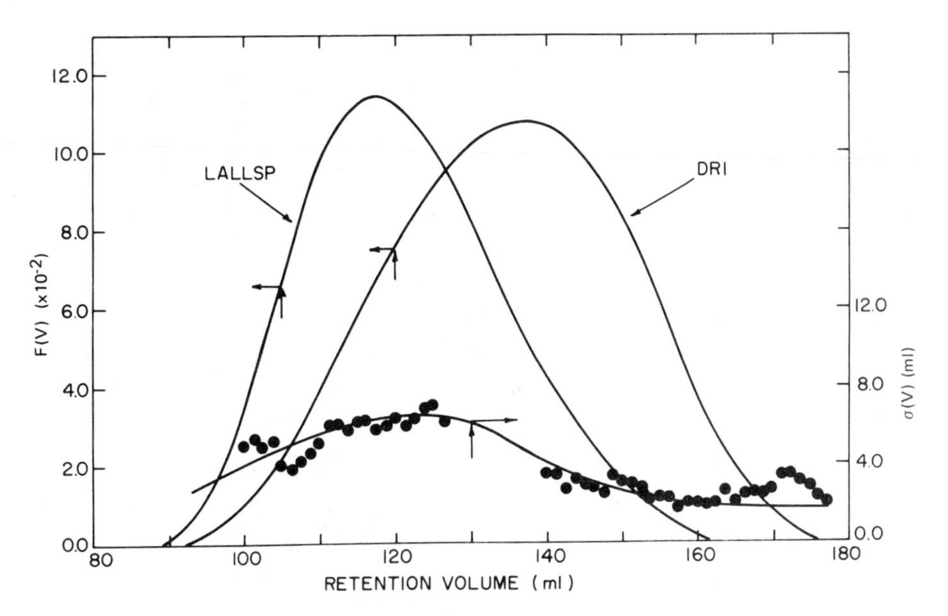

Figure 3.9. DRI/LALLSP detector responses for a dextran standard [a blend of Pharmacia dextrans: T250 (71 wt %) and T40 (29 wt %)] used for peak broadening calibration. Measured broadening parameter σ(V) is also shown.

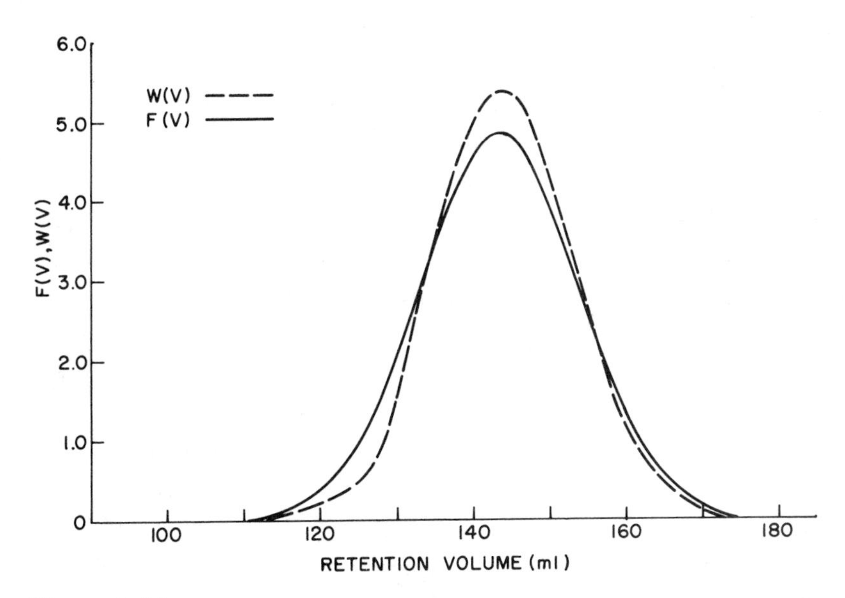

Figure 3.10. Typical raw and corrected DRI detector response for a dextran using the analytical solution of Tung's integral equation [equations (3.37), (3.42a), and (3.42b)] with p = 0.

Table 3.5. Molecular Weight Characterization (x 10^{-3}) of Pharmacia Dextran Standards by Aqueous SEC Using Two Calculational Paths

	Pharmacia		Method 1		Method 2	
Standard	$\bar{M}_n$	$\bar{M}_w$	$\bar{M}_n(c)$	$\bar{M}_w(c)$	$\bar{M}_n(c)$	$\bar{M}_w(c)$
T250	112.5	231.0	100.3	226.0	90.0	256.0
T150	86.0	154.0	76.9	141.0	69.6	139.1
T110	76.0	106.0	79.2	100.5	72.3	100.3
T70	42.5	70.0	43.0	70.4	40.1	69.4
T40	28.9	44.4	25.6	42.7	25.7	42.2
T20	15.0	22.3	16.9	22.7	14.7	22.7

DRI responses are integrated in the normal manner to find $\bar{M}_n(c)$ and $\bar{M}_w(c)$ according to equation (3.4) with p = 0 and k = 1, 2. The whole polymer $M_n(c)$ and $M_w(c)$ values found by the two calculational paths are compared with values reported by Pharmacia in Table 3.5. The agreement is within experimental error except for some of the $\bar{M}_n(c)$ values found using the latter calculational path. The reason for this discrepancy is now known.

A third calculational path was used with the nonionic polyacrylamides to determine $\bar{M}_n(c)$ and $\bar{M}_w(c)$. Both detector responses were corrected for peak broadening using the analytical solution [equations (3.37), (3.37a), and (3.37b)]. Application of equations (3.37a) and (3.37b) to the DRI detector response is equivalent to the satisfaction of $\bar{M}_n(V,Uc)$ and $\bar{M}_w(V,Uc)$ across the response. Application of equations (3.37a) and (3.37b) to the LALLSP detector response to determine $\bar{\sigma}^2(V)$ and $\bar{Y}(V)$ is equivalent to satisfaction of $\bar{M}_w(V,Uc)$ and $\bar{M}_z(V,Uc)$ across the response. The corrected detector responses and the parameters $\bar{\sigma}^2(V)$ and $\bar{Y}(V)$ are shown in Figures 3.11 and 3.12. The excellent agreement for $\bar{\sigma}^2(V)$ and $\bar{Y}(V)$ values found for detector responses which greatly differ in shape was satisfying and suggests that a single set of equations (3.37a) and (3.37b)

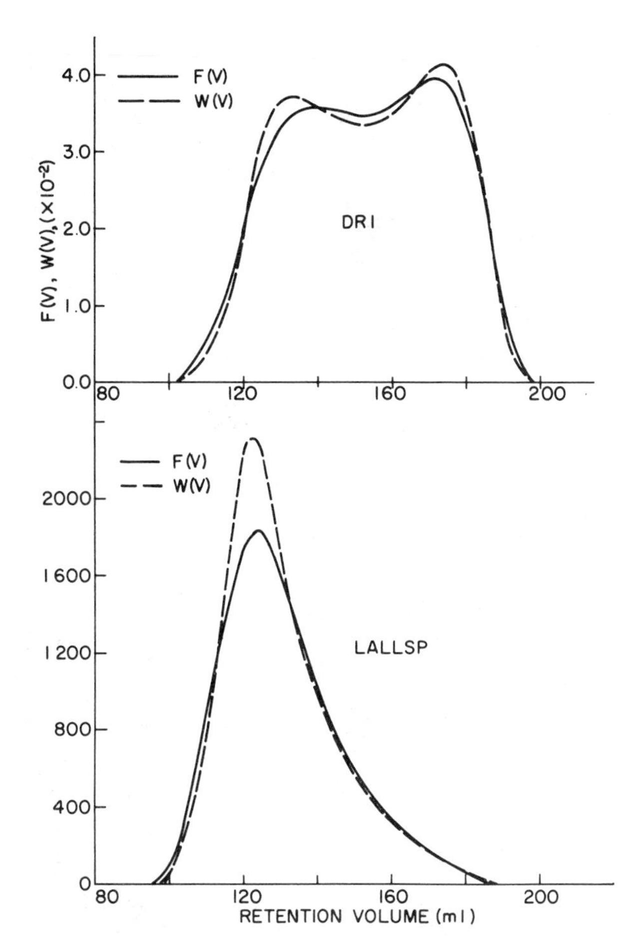

Figure 3.11. Typical raw and corrected DRI/LALLSP detector responses for a nonionic polyacrylamide using the analytical solution of Tung's integral equation [equations (3.37), (3.37a), and (3.37b)] with p = 0 for both responses.

should be used with both DRI and LALLSP detectors for practical reasons. The combined parameters from both detectors should provide better estimates for $\bar{\sigma}^2(V)$ and $\bar{Y}(V)$ over a wider range of retention volumes. Further experimentation is required to confirm this possibility.

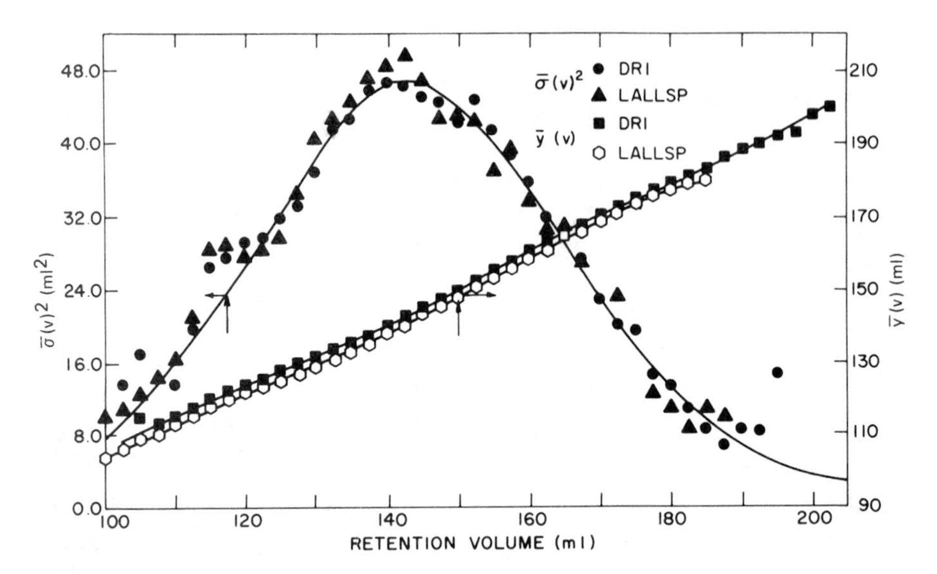

Figure 3.12. Dispersion in the detector cell: parameters of W(V,Y) [$\bar{\sigma}^2(V)$ and $\bar{Y}(V)$] measured using equations (3.37a) and (3.37b) and the DRI and LALLSP detector responses.

For nonionic polyacrylamide, Standard Y [a blend of Polysciences nonionic polyacrylamide standards: PAM500 (65 wt %), PAM2000 (13 wt %), and PAM74 (22 wt %)] was the test polymer. The two detector responses corrected for peak broadening were used to find the molecular weight

Table 3.6. Molecular Weight Characterization (x 10^{-3}) of Polysciences Nonionic Polyacrylamide Standards by Aqueous SEC Using Two Calculational Paths

	Polysciences	Method 2		Method 3	
Standard	$\bar{M}_w$	$\bar{M}_n(c)$	$\bar{M}_w(c)$	$\bar{M}_n(c)$	$\bar{M}_w(c)$
PAM2000	2,180	959	1,800	811	1,590
PAM1000	1,000	551	988	515	956
PAM500	532	288	571	280	611

calibration curve. No further peak broadening corrections were then necessary. The molecular weight calibration curve was then used with the DRI response to calculate $\overline{M}_n(c)$ and $\overline{M}_w(c)$ in the usual manner (see Table 3.6, method 2). In general, the W(V) for the DRI detector should give a more accurate $\overline{M}_n(c)$, as it has more sensitivity than LALLSP at the low molecular weight end. The $\overline{M}_w(c)$ calculated from W(V) given by LALLSP should be the more accurate. The results of these calculations are given in Table 3.6 under method 3.

The recommended procedure for $\overline{M}_n(c)$ and $\overline{M}_w(c)$ determination is to use W(V) from the mass-concentration detector to calculate $\overline{M}_n(c)$ and W(V) from the LALLSP detector to calculate $\overline{M}_w(c)$ and higher molecular weight averages.

REFERENCES

1. Yau, W. W., and Bly, D. D., Size Exclusion Chromatography (GPC) (T. Provder, ed.), ACS Symp. Ser., *138*, 197, 1980.
2. Giddings, J. C., Kucera, E., Russel, C. P., and Myers, M. N., J. Phys. Chem., *72*, 4397, 1968.
3. Ackers, G. K., Biochemistry, *3*, 723, 1964.
4. Cassassa, E. F., J. Phys. Chem., *75*, 3929, 1971.
5. Yau, W. W., Kirkland, J. J., and Bly, D. D., *Modern Size-Exclusion Liquid Chromatography*, Wiley-Interscience, New York, 1979.
6. Friis, N., and Hamielec, A. E., Adv. Chromatogr., *13*, 41, 1975.
7. Ouano, A. C., and Barker, J. A., Sep. Sci., *8*, 673, 1973.
8. Balke, S. T., and Hamielec, A. E., J. Appl. Polym. Sci., *13*, 1381, 1969.
9. Provder, T., and Rosen, E. M., Sep. Sci., *5*, 437, 485, 1970.
10. Cantow, M. J. R., Porter, R. S., and Johnson, J. F., J. Polym. Sci., B, *4*, 707, 1966.
11. Mahabadi, H. K., and Rudin, A., Polym. J., *11*, 123, 1979.
12. Mori, S., J. Appl. Polym. Sci., *21*, 1921, 1977.
13. Yau, W. W., Kirkland, J. J., Bly, D. D., and Stoklosa, H. J., J. Chromatogr., *125*, 219, 1976.
14. Grubisic, Z., Rempp, P., and Benoit, H., J. Polym. Sci., B, *5*, 753, 1967.
15. Hamielec, A. E., and Ouano, A. C., J. Liq. Chromatogr., *1*, 111, 1978.
16. Tung, L. H., and Runyon, F. R., J. Appl. Polym. Sci., *13*, 2397, 1969.
17. Tung, L. H., J. Appl. Polym. Sci., *13*, 775, 1969.
18. Hamielec, A. E., and Ray, W. H., J. Appl. Polym. Sci., *13*, 1319, 1969.

19. Grushka, E., Anal. Chem., *44*, 1733, 1972.
20. Yau, W. W., Stoklosa, H. J., and Bly, D. D., J. Appl. Polym. Sci., *21*, 1911, 1977.
21. Hamielec, A. E., J. Liq. Chromatogr., *3*, 381, 1980.
22. Hamielec, A. E., Ederer, H. J., and Ebert, K. H., J. Liq. Chromatogr., *4*, 1697, 1981.
23. Hamielec, A. E., J. Appl. Polym Sci., *14*, 1519, 1970.
24. Duerksen, J. H., and Hamielec, A. E., J. Polym. Sci., C, *21*, 83, 1968.
25. Cantow, M. J. R., Johnson, J. F., and Porter, R. S., J. Polym. Sci., B, *4*, 707, 1966.
26. Duerksen, J. H., and Hamielec, A. E., J. Polym. Sci., C, *21*, 83, 1968.
27. Duerksen, J. H., and Hamielec, A. E., J. Appl. Polym. Sci., *12*, 2225, 1968.
28. Chang, K. S., and Huang, R. Y., J. Appl. Polym. Sci., *16*, 329, 1972.
29. Ishige, T., Lee, S. I., and Hamielec, A. E., J. Appl. Polym. Sci., *15*, 1607, 1971.
30. Vozka, S., and Kubin, M., J. Chromatogr., *139*, 225, 1977.
31. Kato, Y., and Hashimoto, T., Kobunshi Kagaku, *30*, 409, 1973.
32. Silebi, C. A., and McHugh, A. J., J. Appl. Polmy. Sci., *23*, 1699, 1979.
33. Ouano, A. C., and Kaye, W., J. Polym. Sci., Polym. Chem. Ed., *12*, 1151, 1974.
34. Hamielec, A. E., Ouano, A. C., and Nebenzahl, L. L., J. Liq. Chromatogr., *1*, 527, 1978.
35. Ouano, A. C., J. Polym. Sci., A1, *10*, 2169, 1972.
36. Nose, T., and Chu, B., Macromolecules, *12*, 590, 1979.
37. Tung, L. H., Moore, J. C., and Knight, G. W., J. Appl. Polym. Sci., *10*, 1261, 1966.
38. Vozka, S., Kubin, M., and Samay, G., J. Polym. Sci., Polym. Symp., *68*, 199, 1980.
39. Hamielec, A. E., and Omorodion, S. N. E., Am. Chem. Soc. Ser., *138*, 183, 1979.
40. Kim, C. J., Hamielec, A. E., and Benedek, A., J. Liq. Chromatogr., *5*, 425, 1982.
41. Kim, C. J., Hamielec, A. E., and Benedek, A., J. Liq. Chromatogr., *5*, 1277, 1982.

—4—

EFFECT OF EXPERIMENTAL CONDITIONS

SADAO MORI / *Mie University, Tsu, Mie, Japan*

4.1. INTRODUCTION

As shown in Chapters 1 and 2, the separation mechanism in SEC is primarily steric exclusion; thus retention volume is a function of the effective size of solute molecules in solution. This relation should not depend on the nature of the mobile phase, or on the nature and structure of the solute. However, abnormality in retention has been observed when polymers are eluted in relatively poor solvents or when column packing materials are active: solutes are separated by a mixed mechanism involving steric exclusion and polymer-gel (column packing) interactions. This abnormality causes difficulty in calculating the molecular weight of polymers using universal calibration. Experimental precautions must be taken.

The observation that the position of the hydrodynamic volume plot for polymers is dependent on the type of mobile phase makes the application of a universal calibration method difficult. It is therefore desirable to understand the nonexclusion interactions occurring during separation in order to select an eluant system for which secondary separation mechanisms are minimized. Nonexclusion interactions involve solute-gel, solute-solvent, and solute-gel-solvent (or generally solute-packing-solvent) interactions, and are explained by partial adsorption and partition, incompatibility, solvation, and ionic exclusion and inclusion. The first review on nonexclusion effects in SEC was that by Audebert [1].

Several other experimental variables that affect the results of SEC are sample concentration, column temperature, injection volume, and flow rate. Column arrangement and column length will affect separation efficiency. Errors caused by temperature and flow rate variations are discussed in Chapter 7. Optimization of experimental conditions is important to eliminate all the interfering factors that affect V_R and to obtain precise and accurate SEC results. The values of the theoretical plate number increase in general with decreasing flow rate, injection volume, and sample concentration. However, by using smaller injection volume and sample concentration, we also sacrifice detector sensitivity. The smaller flow rate consumes analytical time, but it is preferable from the viewpoint of the lifetime of the column. Consequently, a compromise should be made among these variables. A review of the influence of operational variables on SEC results has been made by Janča [2].

4.2. NONEXCLUSION PHENOMENA

4.2.1. SEC with Polystyrene Gel

Validity of Hydrodynamic Volume Concept

As shown in Chapter 2, hydrodynamic volume $[\eta]M$ is a universal calibration parameter in SEC for homopolymers and copolymers having linear and branched structure. When a packing material is cross-linked polystyrene gel and the mobile phase is compatible with the polystyrene gel (i.e., a good solvent for uncross-linked polystyrene), solutes do not display preferential affinity for the mobile or the stationary phase. Hence partition and adsorption mechanisms do not influence the solute size separation. For example, o-dichlorobenzene is a good solvent ($a = 0.70$ at 135°C and $a = 0.73$ at 87°C) for polystyrene and a poor solvent ($a = 0.57$ at 138°C) or a theta solvent ($a = 0.5$ at 87°C) for poly(dimethyl siloxane). Here a is exponent of Mark-Houwink equation (2.29). A universal calibration is found to be valid for polystyrene and poly(dimethyl siloxane) at both temperatures [3]. Good solvents for some polymers are listed in Tables 2.1 and 2.2.

Several deviations from universal calibration are observed when polymers are eluted in poor or theta solvents for polystyrene (and polystyrene gel). Chloroform is a good solvent for polystyrene, poly(dimethyl siloxane), and polyisoprene, which are separated solely by steric exclusion, since they follow the same plot of log $[\eta]M$ versus V_R [4]. Cyclohexane at 35°C is a theta solvent for polystyrene ($a = 0.5$), but a good solvent for the other two polymers. The plot of log $[\eta]M$ versus V_R for polystyrene in cyclohexane is displaced to the high V_R values with respect to those for poly(dimethyl siloxane) and polyisoprene as shown in Figure 4.1, and for polybutadiene [5].

Trans-decaline is a theta solvent for polystyrene ($a = 0.52$) close to ambient temperature and a good solvent for poly(dimethyl siloxane) ($a = 0.72 - 0.76$). At 25°C, the plot of log $[\eta]M$ versus V_R for polystyrene is displaced to high V_R values with respect to the plot for poly(dimethyl siloxane), the displacement increasing as

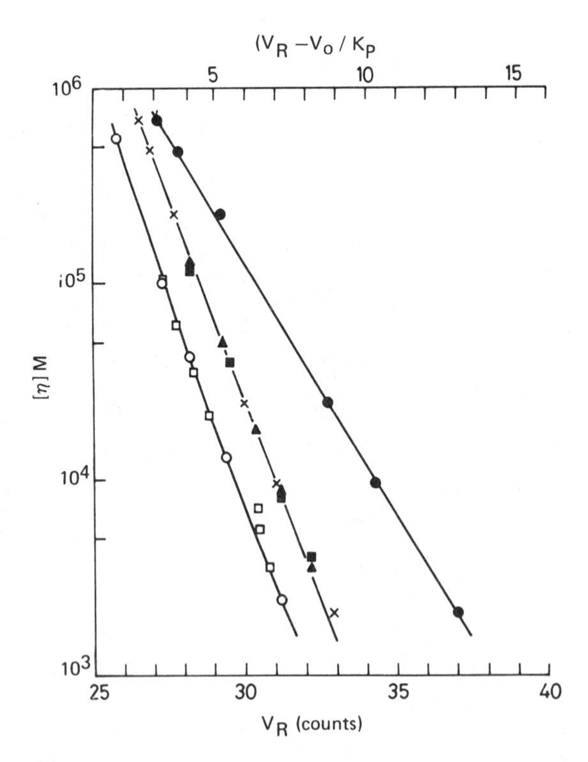

Figure 4.1. Discrepancy of hydrodynamic volume versus V_R calibration curves (lower abscissa), and a universal calibration curve constructed according to equation (4.14) (upper abscissa). Polystyrene (○) and (●), poly(dimethyl siloxane) (□) and (■), and polyisobutyrene (▲) in chloroform (open symbols) and in cyclohexane (filled symbols) at 35°C. Polystyrene (●), poly(dimethyl siloxane) (■) and polyisobutyrene (▲) with $K_P = 1.00$, and polystyrene (x) with $K_P = 1.45$ (see Section 4.3.2). (Lower abscissa from Ref. 4; upper abscissa from Ref. 36.)

V_R increases [6]. The deviation between the two plots decreases as the temperature increases, becoming almost negligible at 100°C. It was demonstrated that trans-decaline becomes a better solvent for polystyrene as the temperature is raised (a = 0.64 at 60°C and a = 0.67 at 100 and 138°C) and that the displacement of the plot of hydrodynamic volume for the polystyrene standards decreases as the temperature is raised. In a calibration curve of log $[\eta]$M versus V_R with polystyrene in trichloroethylene, the points of poly(vinyl acetate) fractions are shifted toward smaller retention volumes [7]. As a general rule, the mobile phase for use with cross-linked polystyrene gel should have the Mark-Houwink exponent for uncross-linked polystyrene in the range 0.67-0.80.

N,N-Dimethylformamide (DMF) is a poor solvent for linear polystyrene (a = 0.60 at 20°C) and a universal calibration curve for polystyrene is displaced to high V_R values with respect to those for polyacrylonitrile [8] (see Figure 2.10) and for polypropylene glycol [9]. Polyacrylonitrile and copolymers of acrylonitrile and styrene have the same universal calibration curve in DMF. These copolymers also share the same universal calibration curve with polystyrene in chloroform [8]. The retention volumes for poly(p-nitrostyrene), polyvinylpyrrolidone, poly(vinyl acrylate), and polystyrene at the same hydrodynamic volume were increased in that order [10]. Deviation of the universal calibration curves for some polymer solutes is phenomenally the retardation of elution of some solutes. SEC for polystyrene and polyethylene glycol in DMF and in tetrahydrofuran (THF) was performed on similar column sets. The retardation of elution and the peak broadening for polystyrene were observed to be predominant in DMF, whereas polyethylene glycols eluted at almost identical retention volumes in both eluants [11]. Similar results have also been observed by Dubin et al. [10].

Otocka and Hellman observed a progressive displacement of the plot of log $[\eta]M$ versus V_R for polystyrene to high retention volumes as the mobile phase is changed from chloroform (a = 0.79), to THF (a = 0.72), to dioxane (a = 0.69), and finally to methyl ethyl ketone (MEK) (a = 0.60) [12]. The displacement increases as polymer-solvent interaction decreases, as shown in Figure 4.2. The slopes of the calibration curves change in proportion to *a* of the mobile phase, while the polymer peak widths remain constant, leading to variations in apparent polydispersity.

Universal calibration curves for polystyrene, poly(2-vinylpyridine), and poly(styrene-2-vinylpyridine) block and graft copolymers in THF and DMF were constructed [13]. In THF, the curve for poly(2-vinylpyridine) deviated to higher retention volume over the value

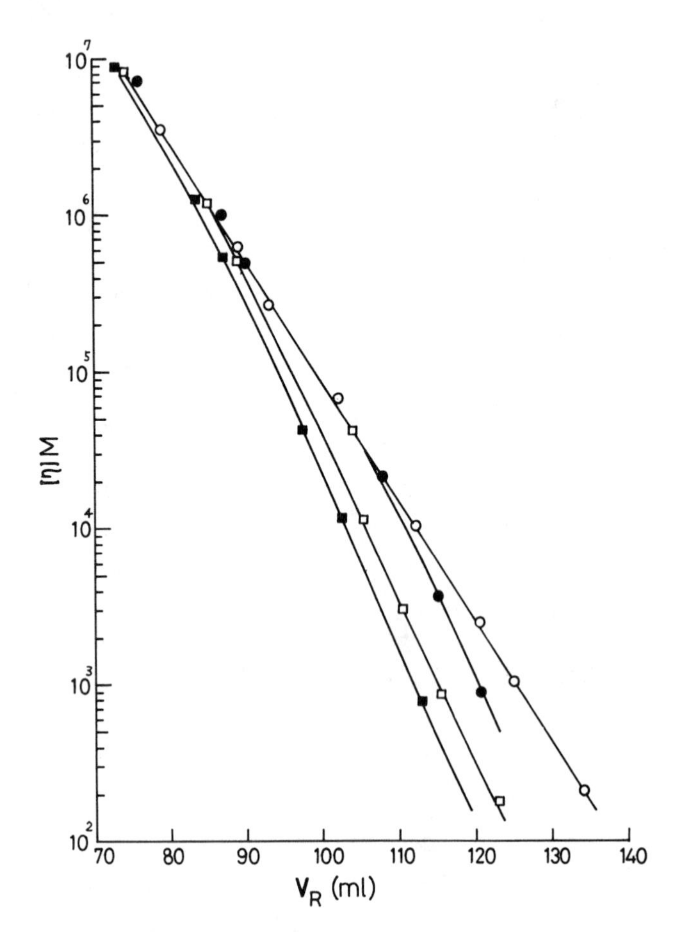

Figure 4.2. Calibration curve [η]M versus V_R for polystyrene in several eluants. ○, MEK; ●, dioxane; □, THF; ■, chloroform. Column: styrene gel. (From Ref. 12.)

$[\eta]M > 10^6$. In DMF, the situation of the curves for two homopolymers was the same as those for polystyrene and polyacrylonitrile. The experimental points for the copolymer were located between two curves for homopolymers, the value of V_R for a given [η]M value being a function of the chemical composition of the copolymer.

Multiple peaks were observed for acrylonitrile copolymers in DMF on polystyrene gel [14-16]. Polyester-based polyurethane and quaternized polyurethane were analyzed by SEC in DMF and multiple

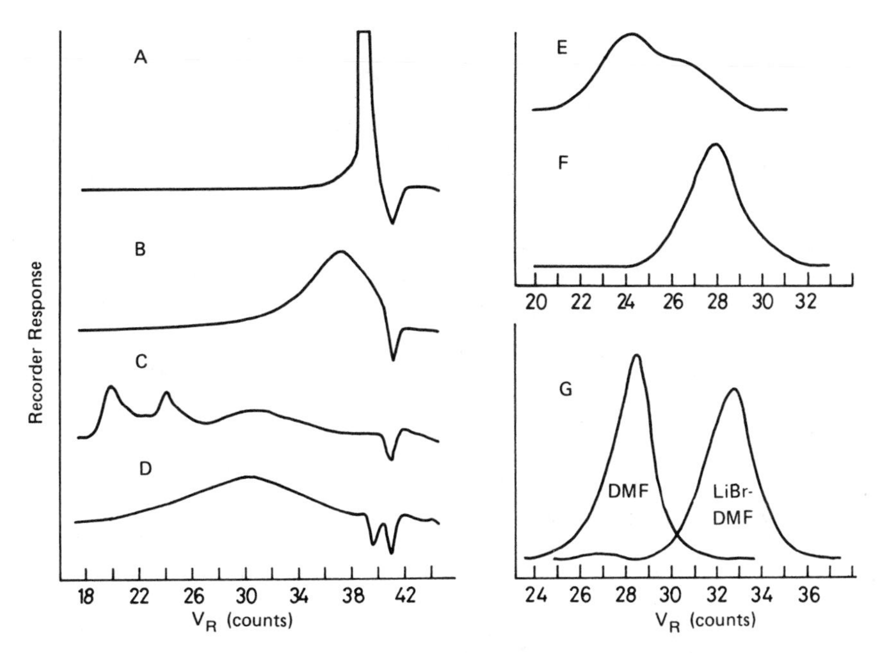

Figure 4.3. Curiosities of chromatograms and peak retardation in DMF and in LiBr-DMF. A, Quaternized polyurethane in DMF [17] and B, in 0.05 M LiBr-DMF [17]; C, polyester-based polyurethane in DMF [17] and D, in 0.05 M LiBr-DMF [17]; E, acrylonitrile-vinyl ether copolymer in DMF [16] and F, in 0.05 M LiBr-DMF [16]; G, acrylonitrile-methyl acrylate copolymer in DMF and in 0.1 M LiBr-DMF [15]. (From Refs. 15, 16, and 17, © John Wiley & Sons, Inc.)

peaks at earlier retention volume or elution at the extremely small molecular weight region were observed [17]. These polymers should have yielded normal Gaussian distribution curves based on the polymerization mechanism. The addition of LiBr to the DMF mobile phase eliminated the multiple peaks or retardation observed in size exclusion chromatograms for these polymers. Examples are shown in Figure 4.3. When LiBr was added to DMF, the SEC curves for both polyacrylonitrile and polystyrene were shifted to higher retention volumes: calibration curves for polystyrene of log M versus V_R [14,16] and log $[\eta]M$ versus V_R [15,17] in LiBr-DMF shifted to higher retention volume with respect to those in DMF (see Figure 4.4).

A particular solvent-temperature combination, although satisfactory for one polymer, may be inadequate for another. If the

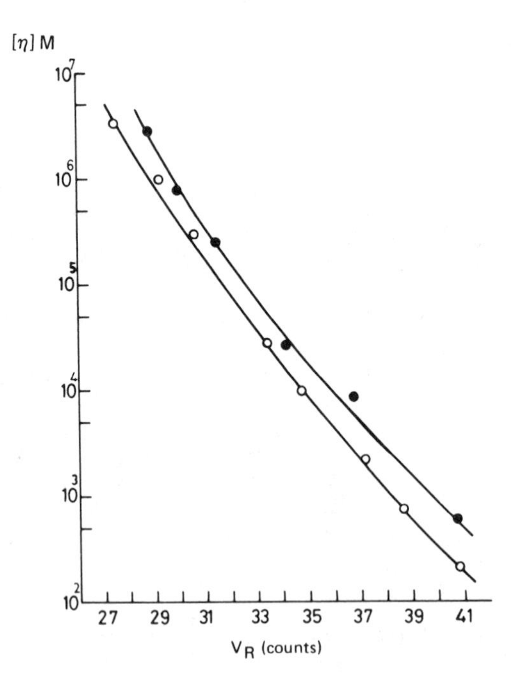

Figure 4.4. Universal calibration curves for polystyrene. ○, In DMF at 80°C; ●, in 0.1 M LiBr-DMF. (From Ref. 15, © John Wiley & Sons, Inc.)

choice of the mobile phase then requires a poor solvent for polystyrene, universal calibration based on polystyrene standards will be inaccurate. Therefore, universal calibration methods should be used cautiously for polar polymers separating on polystyrene gels.

Secondary Effects for Oligomers

Oligomer-solvent interactions are more remarkable than polymer-solvent interactions in separation by SEC. Elution often changes due to hydrogen bonding to the solvent. Properties of low molecular weight compounds other than molecular size appear to have the greatest effects on the elution characteristics of solutes [18]. The effects of polarity and stereochemistry of oxygen-containing low molecular weight substances on SEC were examined [19]. Molecules that possess an appreciable permanent dipole moment emerge from a SEC column earlier than would be predicted on the basis of molar-volume estimates [20].

The retention of several oligomers in THF and chloroform has been examined, choosing n-hydrocarbons as reference standards, which are assumed to elute without the solute-solvent association or adsorption on the gel [21]. Oligostyrene, epoxy resin, and polyethylene glycol (PEG), which eluted at the same retention volume as the n-hydrocarbon in THF, eluted earlier in chloroform than the n-hydrocarbon. p-Cresol novolak resin in chloroform eluted later than the n-hydrocarbon, which eluted at the same retention volumes as the p-cresol novolak resin in THF. Epoxy resin and PEG, which eluted at the same retention volume as oligostyrene in THF, eluted earlier than the oligostyrene in chloroform.

4.2.2. SEC with Inorganic Porous Packings

The SEC behavior of polystyrene in good and theta solvents has been compared using polar inorganic packings (porous silica) [22,23]. Figure 4.5a shows the universal calibration curves of polystyrene in benzene, chloroform, and the theta mixtures benzene-methanol and chloroform-methanol. All the experimental points fall on the same calibration curve in the cases of benzene and chloroform, which are good solvents for polystyrene. However, in different theta mixtures different curves were obtained. The calibration curve in a mixture of benzene and methanol (90:10 vol %) falls between those in benzene and in a theta mixture benzene-methanol (77.8:22.2). Figure 4.5b shows the universal calibration curves of polystyrene in MEK and in mixtures of MEK and n-heptane. Retentions of polystyrene in benzene-methanol or benzene-chloroform are obviously different from that in MEK/n-heptane. The increase of methanol content in benzene or chloroform results in the acceleration of elution of polystyrene, in contrast to the increase of n-heptane content in MEK, which retards the elution of polystyrene.

Similarly, in pure THF and in its mixtures with water up to 8.9% on porous glass columns, V_R of polystyrene decreased with increasing

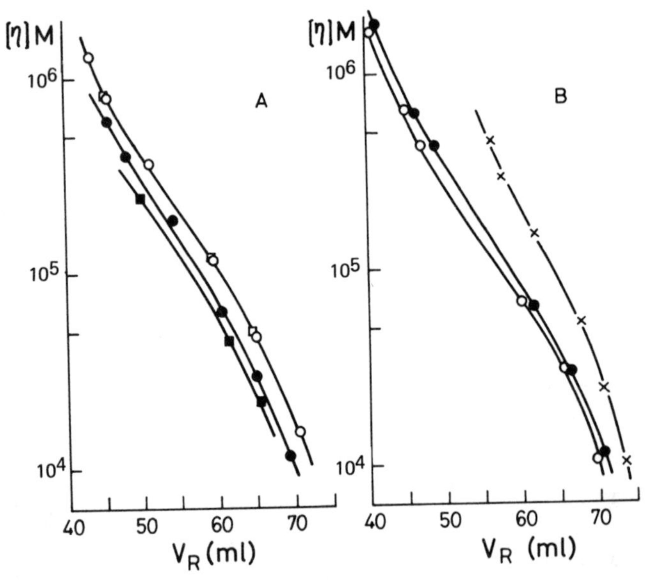

Figure 4.5. Universal calibration curves for polystyrene in several eluants: (a) ○, benzene; □, chloroform; ●, theta mixture benzene-methanol (77.8:22.8, vol %); ■, theta mixture chloroform-methanol (74.7:25.3, vol %); (b) ○, MEK; ●, MEK/n-heptane (68.4:31.6, vol %); ×, MEK/n-heptane (50:50, vol %). (The values of V_R were extrapolated to zero concentration.) (Reproduced from Ref. 22, by permission of the publishers, Hüthig & Wepf Verlag, Basel; and from Ref. 23, © John Wiley & Sons, Inc.)

water content [24]. Campos et al. [25] examined retention of polystyrene on porous silica columns in several good solvents and their mixtures with methanol or n-heptane and obtained similar results to those of Berek et al. [22,23]; V_R of polystyrene decreased with increasing methanol content in benzene, chloroform, and carbon tetrachloride, and increased with increasing n-heptane content in MEK and benzene. V_R of poly(dimethyl siloxane) in benzene and mixtures of benzene-methanol were coincident with those of polystyrene, having the same hydrodynamic volumes [26]. However, poly(methyl methacrylate) (PMMA) eluted later than polystyrene in these solvents as well as in chloroform, chloroform-methanol, dioxane, MEK, and ethyl acetate. Universal calibration curves for PMMA in these solvents shifted to higher V_R. Otocka and Hellman examined the hydrodynamic volume plot for polystyrene in four separate eluants using two different packings [12]. In contrast to the case of polystyrene gel, the porous glass packing is not affected by solvent changes and gives reasonable confirmation of the universal calibration concept for the same polymer in different solvents.

Discrepancy of universal calibration curves for polystyrene, polyvinylpyrrolidone, and other polymers in DMF on silanized porous glass was observed as well as on polystyrene gel [10], although somewhat diminished polymer-substrate interactions were observed. However, on untreated porous glass no discrepancy was observed for these polymers, except polyvinylpyrrolidone, which failed completely to elute. Poly(N-vinyl acetamide) was characterized by SEC on silanized porous glass columns in DMF and in 0.01 M LiBr in DMF [27]. Chromatograms obtained in pure DMF were very different in appearance from those obtained using 0.01 M LiBr in DMF. The effect of LiBr concentration in the mobile phase on chromatogram shape was also observed.

Various commercially available porous packings were tested for aqueous SEC of polyacrylamides, poly(acrylamide-sodium acrylate), and poly(sodium styrene-sulfonate) [28]. Test runs with porous silica packing, silica treated with sililated reagent, and polystyrene gel in salt solutions or in pure water proved unsatisfactory because these polymers were adsorbed to some extent or largely excluded. Polyacrylamides were completely excluded with porous glass in pure water. When the mobile phase was changed to a 0.1 M Na_2SO_4 solution, separation of polyacrylamides by SEC mechanism was possible. Polyacrylamides were separated by SEC mechanism on porous glass in a KCl-formamide solution [29]. Hydrolyzed polyacrylamides were found to retard and to elute at the total permeation region in this system.

4.2.3. SEC with Other Column Packings

Discrepancies in universal calibration were also observed on dextran gel columns in 0.3% sodium chloride-water solution [30]. The plots of log $[\eta]M$ versus V_R for dextran and polyvinylpyrrolidone differed greatly from those for poly(ethylene oxide) and poly(vinyl alcohol), the former groups having larger V_R than those of the latter groups of the same hydrodynamic volume.

Several adsorption phenomena on dextran gel for low molecular weight compounds have been observed: phenols, anilines, benzoic acids [31], aromatic nitrocompounds [32,33], aromatic compounds [34], and hydrocarbons and their hydroxyl derivatives [35]. The distribution coefficients K_{SEC} for these compounds have higher values than unity.

4.3. INTERPRETATIONS FOR NONEXCLUSION INTERACTIONS

4.3.1. Thermodynamic Treatments

K_P Values

Dawkins defined K_{SEC} for a polymer in SEC by

$$K_{SEC} = K_D K_P \tag{4.1}$$

where K_D is the distribution coefficient for steric exclusion and K_P is the distribution coefficient for solute-gel (porous packing) interaction effects [36]. Network-limited partition and adsorption mechanisms have been assumed for separating polystyrene in poor and theta solvents. These separations involve solute-gel interactions which must be weak and reversible, so that polystyrene is not completely retained in the stationary phase. As reviewed in Chapter 1, for polymers separating solely by steric exclusion, K_P is unity and K_D will fall between zero and unity. K_P will be greater than unity if some nonexclusion interactions are observed. According to the definition, the term K_P may not be identical to the distribution coefficient for solute partition between the stationary and mobile phases in liquid-liquid partition chromatography.

The steric exclusion mechanism at equilibrium is explained by the loss in conformational entropy when a polymer molecule transfers from the mobile phase to a pore within the gel [37]. Then the distribution coefficient K_D is defined as the ratio of accessible conformations for a polymer within the gel to those in the mobile phase. Solute-gel interaction effects will be determined by the standard enthalpy change on solute transfer to the pore surface in the gel. Finally, K_p is defined by equation (1.37) [38], which suggests that K_p should decrease on raising the temperature. In addition, the magnitude of $\Delta H°$ will change as the solute-gel interactions increase or decrease. A value of $\Delta H°$ greater than zero is explained as the incompatibility and one smaller than zero as adsorption and partition.

Solubility Parameters

Solute retention may be explained on the basis of the solubility parameters [39]. A solvent having the solubility parameter δ close to that for the polystyrene gel is compatible with the gel. The partial molar energy of mixing of the solvent into the polystyrene gel is given by

$$\Delta E_{solvent} = V_{solvent}(1 - \phi_1)^2(\delta_{gel} - \delta_{solvent})^2 \qquad (4.2)$$

Similarly, those of the solute into the gel, ΔE_{solute}, and of the solute into the solvent, $\Delta E'_{solute}$, are expressed as

$$\Delta E_{solute} = V_{solute}(1 - \phi_2)^2(\delta_{gel} - \delta_{solute})^2 \qquad (4.3)$$

and

$$\Delta E'_{solute} = V_{solute}(1 - \phi_3)^2(\delta_{solvent} - \delta_{solute})^2 \qquad (4.4)$$

where $V_{solvent}$ and V_{solute} are molar volumes of the solvent and the solute, and ϕ_1, ϕ_2, and ϕ_3 are the volume fractions of solvent in gel, solute in gel, and solute in solvent, respectively.

This partial molar energy of mixing could substitute for the energy of the adsorption and the desorption of the solute on the gel. Since the retention (the adsorption) of the solute onto the polystyrene gel is a kind of solvation, the solute, which is more miscible with the gel, adsorbs strongly onto the gel surface, so that ΔE_{solute} is the measure of the magnitude of the solute adsorption. The smaller the value of ΔE_{solute}, the stronger will be the adsorption of the solute onto the gel surface. The desorption of the solute from the gel surface is accelerated if $\delta_{solvent}$ approaches δ_{gel}. The solute, which is more soluble in the solvent, tends to desorb from the gel surface. Therefore, the magnitude of the desorption of the solute is expressed by $\Delta E_{solvent} + \Delta E'_{solute}$.

Hence the net energy involved in the adsorption of the solute on the gel surface is

$$\Delta E = \Delta E_{solute} - (\Delta E_{solvent} + \Delta E'_{solute}) \tag{4.5}$$

When the value of ΔE approaches zero, the adsorption and desorption equilibrium of a solute onto the gel surface balances. The value of ΔE is comparable to the difference in the chemical potential, which is a function of the distribution coefficient [see equation (1.34)]. Assuming that $V_{solute} = V_{solvent}$ and $(1 - \phi_1) \cong (1 - \phi_2) \cong (1 - \phi_3) \cong 1$, the next equation can be derived:

$$\log K_I = \frac{2V_{solute}}{2.3RT} [(\delta_{solvent} - \delta_{gel})(\delta_{solvent} - \delta_{solute})] \tag{4.6}$$

K_I is a distribution coefficient similar to K_p, but not identical by definition. The term K_I may be identical to the distribution coefficient in liquid-liquid partition chromatography. Similar treatment to the equation (4.6) has been proposed for partitioning of metal complexes on polystyrene gel [40,41].

The equation (4.6) explains that when the solubility parameter of the solvent is equal to that of the gel or that of the solute, the distribution coefficient is unity ($K_I = 1$), and the separation is performed by steric exclusion. In this instance, the separation of solute depends only on the pore size of the gel. Network-limited partition and adsorption theory is still valid. When $\delta_{solvent} = \delta_{gel}$, the solvent molecules adsorb strongly on the gel surface, preventing access of the solute molecules to the gel surface, minimizing the interactions between the solute molecules and the gel surface. The strong solvent layer formed is retained around the gel surface and will not migrate with the mobile phase. Preferential solvation of polystyrene [$\delta = 18.6(J^{1/2}/cm^{3/2})$] with benzene ($\delta = 18.8$) is observed [42]. The increase of the difference between $\delta_{solvent}$ and δ_{gel} contributes to weakening of the binding of the solvent layer to the gel surface (i.e., the adsorption force of the solvent molecules to the gel surface decreases), and then the solute molecules and the mobile-phase molecules other than the solvent layer become accessible to the gel surface, so that the adsorption and desorption processes of the solute molecules and/or the mobile phase molecules to the gel surface become notable. The separation mode will be normal-phase adsorption chromatography if solubility parameter of the gel is greater than that of the solvent, and reverse-phase adsorption if it is smaller than that of the solvent.

Interaction Parameters

A theory of polymer partition in organic polymer gels has been proposed by several workers. Lecourtier et al. dealt with the partition mechanism of solutes by selective dissolution in swollen gel and derived the equation for the distribution coefficient K_{ave} of the solute partition between the gel and the mobile phases in a simplified form by taking the value of r_s as unity [43].

$$\ln K_{ave} = \frac{n_p}{n_s} [\ln (1 - \phi_g) + \phi_g(\chi_{gs} + \chi_{ps} - \chi_{pg})] \quad (4.7)$$

where n_p and n_s are the numbers of segments composing solute and solvent molecules, ϕ_g is the volume fraction of the gel in the swollen gel, and χ_{gs}, χ_{ps}, and χ_{pg} are the Flory-Huggins interaction parameters related to segment interactions of gel (g)-solvent (s), solute (p)-s, and p-g. The value n_p/n_s can be expressed as the polymerization degree of the solute by taking r_s as unity. K_{ave} is defined by equation (1.5). The steric exclusion and the partition (dissolution) effects occur simultaneously and lead to a double equilibrium between the solute in the mobile phase, the gel phase, and the gel pores. Consequently, the expression for V_R can take the form [44]

$$V_R = V_o + K_{ave}(V_p + V_m) + K_D K_{ave} V_p \quad (4.8)$$

Belenkii et al. tried to calculate the magnitude of compatibility between gel and solute polymer [30]. In swollen gels the value of the gel volume accessible to solute polymers, V^{acc}, and therefore the value of V_R depends on the total pore volume. The ability of polymer chains to diffuse into dense regions of the swollen gel is closely related to the thermodynamic compatibility of polymers with the gel matrix and the accessible volume of the gel increases by a certain value, ΔV^{acc}. If solute polymers are incompatible with the gel matrix, the retention volume for these polymers is expressed as

$$V_R^{incomp} = V_o + V_{SEC}^{acc} \tag{4.9}$$

where V_{SEC}^{acc} is the gel volume that is accessible to polymers according to steric exclusion mechanism. For a polymer compatible with the gel, the retention volume is

$$V_R^{comp} = V_o + V_{SEC}^{acc} + \Delta V^{acc} = V^{incomp} + \Delta V^{acc} \tag{4.10}$$

The value ΔV^{acc} can be calculated as a function of χ parameters.

Due to preferential solvation of the gel by one of the components of the mixed eluant, a difference in the thermodynamic quality of the mobile and stationary phases, and consequently, additional partition of the solute, may occur [23]. The contribution to V_R caused by partition of the solute into the stationary phase ΔV_R^P is expressed as

$$V_R^P = \frac{M_m}{M_s \rho_m} \exp\,(\chi_{ps}^{\infty} - \chi_{pg}^{\infty}) \tag{4.11}$$

where M_m and M_s are molecular weights of the mobile and the stationary phases, ρ_m the density of the mobile phase, and χ^{∞} the Flory-Huggins interaction parameter at infinite dilution for solute-solvent and solute-gel. The χ parameter may be expressed by the solubility parameters; for example,

$$\chi_{ps} = \frac{V_{solute}}{RT} (\delta_{solvent} - \delta_{solute})^2 \tag{4.12}$$

4.3.2. Evaluation of Thermodynamic Treatments

K_P Values

The term K_D in equation (4.1) is expressed as a function of hydrodynamic volume:

$$K_D = -A \log\,[\eta]M + B \tag{4.13}$$

where A and B are constants [36]. Substitution of equations (4.1) and (4.13) into equation (1.2) and rearrangement give

$$\frac{V_R - V_0}{K_p} = V_p(-A \log [\eta]M + B) \tag{4.14}$$

Therefore, a plot of log $[\eta]M$ versus the left-hand side of equation (4.14) will give a universal calibration when solutes are separated by steric exclusion alone ($K_p = 1$), or when solutes are separated both by steric exclusion and partition ($K_p > 1$).

In Figure 4.1, assuming that $K_p = 1.45$ for polystyrene and $K_p = 1.00$ for poly(dimethyl siloxane) and polyisoprene in cyclohexane at 35°C, a universal calibration curve, a plot of log $[\eta]M$ versus $(V_R - V_0)/K_p$, for all three polymers is obtained [36], suggesting the participation of network-limited adsorption and partition mechanisms. From the data given by Kranz et al. [8], universal calibration is obtained with $K_p = 1.37$ for polystyrene and with $K_p = 1.00$ for polyacrylonitrile in DMF. The dependence of K_p on gel porosity is found in the gel of narrow pore sizes. Since polymer-solvent interaction increases with raising temperature, solute-gel interaction effects should become less prevalent. However, although the divergence of the universal calibration curves for polystyrene and poly(dimethyl siloxane) in cyclohexane is less at 45°C than at 35°C, it is still sufficient to invalidate the hydrodynamic volume procedure for this case. Universal calibrations for polystyrene and poly(dimethyl siloxane) in trans-decaline are obtained with values of K_p for polystyrene of 1.25, 1.20, and 1.10 at 25, 60, and 100°C, respectively, and with $K_p = 1$ for poly(dimethyl siloxane) at all temperatures [6].

The shift of V_R of poly(vinyl acetate) is considered to be the effect of incompatibility between sample and gel arising from repulsive interaction [7]. Assuming that polystyrene is separated on polystyrene gel by steric exclusion alone (i.e., $K_p = 1.0$), K_p values for poly(vinyl acetate) have been calculated [45] from the literature data [7]. The values of K_p found for poly(vinyl acetate) depend on the molecular weight, decreasing with increasing molecular weight, which is at variance with the representation of the most experimental

Table 4.1. K_P Values for Polystyrene on Silica Gel from Figure 4.5a

Mobile phase	ε°	[η]M	K_P
Chloroform-methanol (74.7:25.3, v/v)	0.87	2.2×10^5	1.00
		1.0×10^5	1.00
		0.4×10^5	1.00
Benzene-methanol (77.8:22.2, v/v)	0.85	2.2×10^5	1.07
		1.0×10^5	1.04
		0.4×10^5	1.03
Benzene	0.32	2.2×10^5	1.22
Chloroform	0.40	1.0×10^5	1.15
		0.4×10^5	1.16

Source: Ref. 46.

data over a wide molecular weight range by a single value of K_P (K_P values were 0.38, 0.55, and 0.86 with [η]M 3.74×10^4, 1.94×10^4, and 9.85×10^3, respectively). In theta mixtures of benzene-methanol and chloroform-methanol, polystyrenes are supposed to be separated solely by steric exclusion [22]. Using this assumption, the data of Berek and coworkers [22] have been plotted according to equation (4.14), giving the values of K_P in Table 4.1. The data in the literature [19] were also plotted, and K_P values for monocarboxylic acids and glycols were calculated, assuming the hydrocarbons to be separated by steric exclusion alone (K_P = 1.0) [47]. There were K_P = 1.09 for monocarboxylic acids and K_P = 1.15 for glycols.

The method of determining values of K_P using equation (4.14) can be criticized on the grounds that polymer-gel interactions may still be present even for the system of chloroform-methanol in Table 4.1. Campos and Figueruelo [48] proposed a correction K_P as

$$K_P^{(1)} = K_P^{(0)} f \tag{4.15}$$

where $K_P^{(0)}$ is obtained in a reference system and f is a coefficient showing the deviation in the K_P value of the new system with respect

Table 4.2. Polystyrene Relative Distribution Coefficients f on Silica Gel in Different Mobile Phases

Mobile phase (v/v)	f	Mobile phase (v/v)	f
Carbon tetrachloride (TC)	Very large	TC-methanol (90:10)	0.84
Benzene/n-Heptane (70:30)	Large	Benzene-methanol (90:10)	0.78
MEK/n-Heptane (40:60)	1.49	Chloroform-methanol (90:10)	0.75
MEK/n-Heptane (45:55)	1.28	TC-methanol (85:15)	0.75
MEK/n-Heptane (50:50)	1.11	Benzene-methanol (84:16)	0.70
Benzene/n-Heptane (92:8)	1.05	Chloroform-methanol (84:16)	0.70
MEK/n-Heptane (60:40)	1.01	TC-methanol (78:22)	0.70
MEK/n-Heptane (75:25)	1.00	Chloroform-methanol (75:25)	0.71
Reference system {Chloroform, THF, Benzene, MEK}	1.00	Benzene-methanol (75:25)	0.65

Source: Ref. 25.

to that of the reference system. The values of f were calculated for several retention data of polystyrene on porous silica columns in several mobile phases [25] and are shown in Table 4.2. Systems for good solvents are taken as reference. A problem may arise when we attribute a physical meaning to the $K_p < 1$ values, but a coefficient f should allow one to evaluate the secondary effects in some systems relative to that in a reference system.

The equation (1.37) suggests that the value K_p may be calculated if one knows the enthalpy change $-\Delta H°$. Mori and Suzuki obtained $\Delta H°$ values in the course of a study for the effect of column temperatures on the retention volume of polystyrene in a THF-polystyrene gel system and calculated K_p values for several polystyrenes, oligostyrenes, and epoxy resin oligomers [49]. The K_p values thus obtained are shown with K_D values in Table 4.3. The K_p values increased with increasing molecular weight for polymers and with decreasing molecular weight for oligomers. The enthalpy changes for epoxy resin oligomers at 15°C were positive, and the K_p values were lower than unity, suggesting partial exclusion by incompatibility.

Table 4.3. Two Distribution Coefficients, K_P and K_D, in a THF-Polystyrene Gel System at 25°C[a]

	K_P		K_D
	A80M system		
Benzene	1.64		0.59
PS 2,100	1.71		0.39
PS 20,400	1.77		0.31
PS 97,200	1.53		0.26
PS 180,000	1.86		0.18
PS 411,000	1.82		0.14
PS 670,000	2.32		0.08
PS 1,800,000	3.44		0.025
	A802 system		
PS			
n = 2	1.44	(1.10)	0.46
n = 3	1.33	(1.08)	0.45
n = 4	1.28	(1.00)	0.42
n = 5	1.24	(1.00)	0.40
EPIKOTE			
n = 0	1.23	(0.97)	0.51
n = 1	1.16	(0.92)	0.36
n = 2	1.09	(0.81)	0.29
n = 3	1.08	(0.71)	0.23
n = 4	1.05	(0.55)	0.19

[a]*Notes*: (1) A80M is a polystyrene gel column used for polymer fractionation. (2) PS refers to polystyrene and the number after PS refers to molecular weight. EPIKOTE refers to epoxy resins. (3) n is the number of the repeating unit. (4) The values in parentheses were obtained at 15°C.

Source: Reprinted with permission from Ref. 49. Copyright 1982 American Chemical Society.

Solubility Parameters
Mori's theory has been experimentally verified by using low molecular weight compounds and large pore gels to make K_{SEC} unity [39]. The examples are shown in Figure 4.6. The elution order and V_R of solutes are governed by the solubility parameter of the mobile phase. As the application of Mori's theory, SEC and normal-phase partition chromatography (NPPC) have been performed on the same column packed with polystyrene gels used for SEC of small molecules. Chloroform was used as the mobile phase in SEC and a mixture of chloroform and n-hexane was used for NPPC [50].

The solubility parameters δ are 18.6 and 16.7 for polystyrene and cyclohexane, respectively, suggesting that the K_I value in equation (4.6) is larger than unity and polystyrene will prefer the

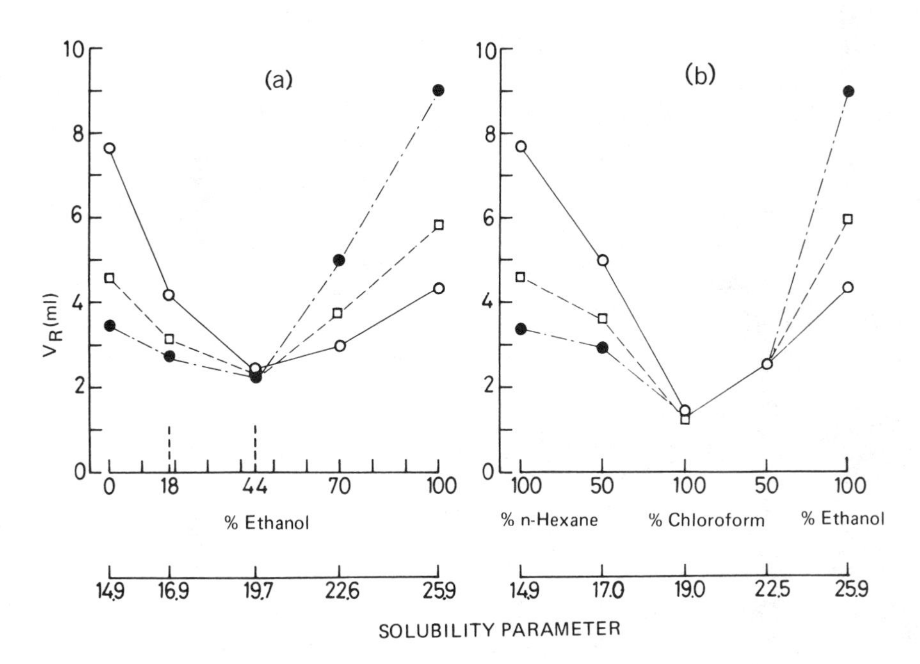

Figure 4.6. Retention volume of solutes versus solubility parameter. (a) In n-hexane, ethanol, and the binary mixtures: ○, dimethyl phthalate; □, dibutyl phthalate; ●, dioctyl phthalate. (b) In ethanol-chloroform and n-hexane/chloroform systems. Symbols are the same as in (a). (Reprinted with permission from Ref. 39. Copyright 1978 American Chemical Society.)

polystyrene-like gel environment to the mobile phase. Because the solubility parameters for poly(dimethyl siloxane) and polyisoprene are 15.5 and 16.5, respectively, K_I values for these polymers in cyclohexane are near unity and these polymers are separated solely by steric exclusion (see Figure 4.1).

The solubility parameters for DMF and THF are 24.8 and 18.6, respectively, and those for polyacrylonitrile (δ = 25.3) and PEG are close to that for DMF. Substituting these values in equation (4.6), the retention behavior of polyacrylonitrile and polystyrene in DMF [8] and that of PEG and polystyrene in DMF and in THF [11] can be elucidated, as well as the other data in DMF [9,10,13].

Altgelt and Moore pointed out that adsorption can occur when solutes are more polar than the mobile phase and that in SEC the mobile phase should have a similar solubility parameter to that for the gel [51]. Several guidelines may be followed in order to minimize polymer-gel interactions [52]. When $\delta_{solute} < \delta_{gel}$, K_P will be 1.0 with $\delta_{solvent} > \delta_{solute}$ provided that $\delta_{solvent} \cong \delta_{gel}$. When $\delta_{solute} > \delta_{gel}$, K_P should be 1.0 with $\delta_{solvent} > \delta_{solute}$. Thus poly(vinyl pyridine) (δ = 21.4) is irreversibly adsorbed on polystyrene gel (δ = 18.6) with chloroform (δ = 19.0), but separated by steric exclusion [K_I in equation (4.6) is unity] with N,N-dimethylacetamide (δ = 22.1) [53].

Interaction Parameters

For a given solvent-gel system, equation (4.7) may be simplified as

$$\ln K_{ave} = [-A + B(\chi_{ps} - \chi_{pg})]n_P \quad (4.16)$$

where A and B are coefficients that depend on the gel structure [54]. For a benzene (solvent)-polystyrene or alkane (solute)-triazinic gel system, $(\chi_{ps} - \chi_{pg})$ is a negative term. Coefficients A and B are positive for this gel, so $\ln K_{ave}$ is always a negative number. When $(\chi_{ps} - \chi_{pg})$ is considered constant, $\ln K_{ave}$ decreases with increasing values of n_P. These theoretical predictions were in good agreement with experimental results.

In the system polystyrene-MEK/n-heptane on silica gel [23], MEK is preferentially adsorbed on the gel. As MEK is a thermodynamically better solvent for polystyrene, the mobile phase then becomes thermodynamically poorer than the quasistationary phase adsorbed on the gel and thermodynamical partition of the solute between the two phases may occur. Since the stationary phase is enriched by MEK, the value $(\delta_{\text{solvent}} - \delta_{\text{solute}})^2$ is higher than that $(\delta_{\text{gel}} - \delta_{\text{solute}})^2$ or $\chi_{ps}^{\infty} > \chi_{pg}^{\infty}$. According to equations (4.11) and (4.12), it ensures that partition may contribute to some extent in SEC (for further discussion, see Section 4.3.4).

4.3.3. Incompatibility Between Solute and Gel

Incompatibility between solute and gel arises from repulsive interaction between them, taking place in good as well as in poor solvents [7]. It is caused by the low entropy of mixing, which cannot overcome even relatively weak positive heats of mixing and which thus leads to positive free energy of mixing. The incompatibility is dependent on the structure, size, and concentration of solute. The early elution of poly(vinyl acetate) ($K_p < 1$) suggests partial exclusion by polymer incompatibility between the solute and the gel. Molecular weight dependence of K_p values confirms this suggestion. Therefore, poly (vinyl acetate) may be separated by both steric exclusion and partial exclusion by incompatibility between the solute and the gel [45].

Incompatibility can be explained by using Flory-Huggins interaction parameters. Equation (4.7) is simplified as [45]

$$\ln K_p = -n_p\phi_g(1 + \chi_{pg}) \tag{4.17}$$

Strong polymer-gel interaction in partition effect results in a negative value of χ_{pg}, whereas the incompatibility effect leads to a positive value. Therefore, K_p in equation (4.17) decreases with increasing molecular weight, that is, greater exclusion as polymer incompatibility increases.

Discrepancies in universal calibration on dextran gel are suggested to be caused by different degrees of thermodynamic compatibility of the eluted polymers with the gel matrix [30]. The amount of compatibility between gel and polymer has been calculated.

4.3.4. Adsorption and Partition Mechanisms

Except for incompatibility, ionic exclusion, and inclusion, all other interactions for solute-gel, solute-solvent, and solvent-gel are assigned to the categories of adsorption and partition mechanisms. In this section we discuss the interactions from the points of adsorption and partition.

Inorganic Packings

Adsorption and partition mechanisms may result from solute-gel interaction effects and will both give K_P (or K_I) > 1.0. Discrimination between the two mechanisms is difficult and sometimes meaningless. Adsorption or partition effects which will be evaluated by calculating K_P or K_I values may occur whenever the polarity of the mobile phase is very different from that of the gel (porous packing) and/or the polymer solute.

The experimental results shown in Figure 4.5 can be explained by adsorption or partition effects and by the selective sorption of mixed solvent components on polymer solute and/or porous silica [22, 23]. The plots of log (hydrodynamic volume) for polystyrene in a theta solvent formed from a binary liquid mixture are always displaced to low V_R with respect to plots for polystyrene in single liquids which are good solvents for polystyrene. Polystyrene remains adsorbed in the silica gel column when carbon tetrachloride or a mixture of benzene and n-heptane (1:1) is used as the mobile phase. The extent of adsorption may be controlled by the polarity of the solvent used. In benzene, chloroform, and MEK, the adsorption sites on gel are assumed to be solvated by these solvent molecules. On passing the polystyrene-containing zone through the column, a dynamic competitive equilibrium is established between the molecules of the mobile phase and the solute on the adsorption sites of the gel.

Consequently, adsorption contributes partly to the increase of V_R of the solute.

In mixtures of benzene-methanol or chloroform-methanol, methanol is selectively sorbed in the pores of silica gel. This result may be interpreted in terms of polarity considerations, as given by the solvent strength parameter $\varepsilon°$. Increasing the level of methanol ($\varepsilon° = 0.95$) reduces the adsorption of polystyrene on the silica surface even when the binary liquid mixture is a theta solvent for polystyrene. The sorbed molecules of methanol may reduce the adsorption of polystyrene on the silica gel by blocking the active sites on its surface, which will result in a decrease in the corresponding V_R. The retardation of the polystyrene solute should, as a result of adsorption, be smaller in methanol mixtures than in pure solvents. A partition separation mechanism between the mobile phase and the stationary phase may arise, because this preferential adsorption of methanol gives a solvent environment at the pore surface which is poorer than the mobile phase for polystyrene, which prefers the better solvent environment.

A different situation arises for the mixture MEK/n-heptane, where MEK ($\varepsilon° = 0.51$) is preferentially adsorbed on the gel. The adsorption of MEK will tend to compete with polystyrene adsorption. As the fraction of MEK is reduced, polystyrene adsorption will increase. Furthermore, the preferential adsorption of MEK gives a solvent environment at the pore surface, so that polystyrene will undergo a partition separation where the solvent in the pore surface is the preferred phase. Thus polystyrene has greater affinity to the gel than the mobile phase, which is rich with nonpolar component n-heptane ($\varepsilon° = 0.01$). This leads to retardation of polystyrene in mixtures of MEK with n-heptane. For the retention of polystyrene in pure THF and in mixtures with water on porous glass columns [24], an explanation similar to that for the systems benzene-methanol and chloroform-methanol can be applied. Water exhibits the same role as methanol in the binary mixtures described above, so that an increase in water content reduces the V_R. These results suggest that polymer-gel interaction for inorganic porous packings decreases with increasing polarity of the mobile phase because of the higher affinity of the mobile phase for the surface of the packing. Consequently, solvents of low polarity should be avoided because irreversible adsorption of polymer may occur in the packing, for example, for polystyrene in carbon tetrachloride ($\varepsilon° = 0.18$). Moreover, to avoid partition effects, a single solvent that is more polar than the polymer solutes should be used as the mobile phase, so that solvent is preferentially adsorbed on surface sites.

The preferential solvation of polymers in binary solvent mixtures has been studied and the preferential solvation parameter has been determined by SEC using silica gel columns [55]. In the system benzene-methanol-poly(dimethyl siloxane), the polymer is preferentially solvated by benzene. In the system benzene-methanol-PMMA, methanol is preferentially adsorbed by the polymer in the mixtures of low methanol content and benzene is preferentially adsorbed by the polymer in methanol-rich mixtures. The difference of solvation between two solutes of closed bulkiness may affect the separation, especially in oligomer analysis [56]. The shift to higher V_R of PMMA with respect to polystyrene is attributed to stronger polarity with respect to polystyrene leading to an increase in its polymer-silica gel interactions and therefore in its V_R [26] (see Section 4.2.2). The abnormal retention of PMMA by silica gel in chloroform is assumed to be due to the existence of strong interactions via hydrogen bonding between carbonyl groups of PMMA and silanol sites of the gel. The addition to chloroform of solvent containing hydroxy groups normalizes the PMMA retention behavior. These results are shown in Figure 4.7. Addition of decaline to THF retards the elution of polystyrene with higher molecular weights, and in the range of THF content smaller than 25%, polystyrene elutes with higher V_R for higher molecular weights [57].

The role of adsorption in SEC can be evaluated quantitatively by using the solvent strength parameter or the solubility parameter [23]. The V_R of a solute at the same hydrodynamic volume $[\eta]M$ in different mobile phases decreases with an increase in these parameters

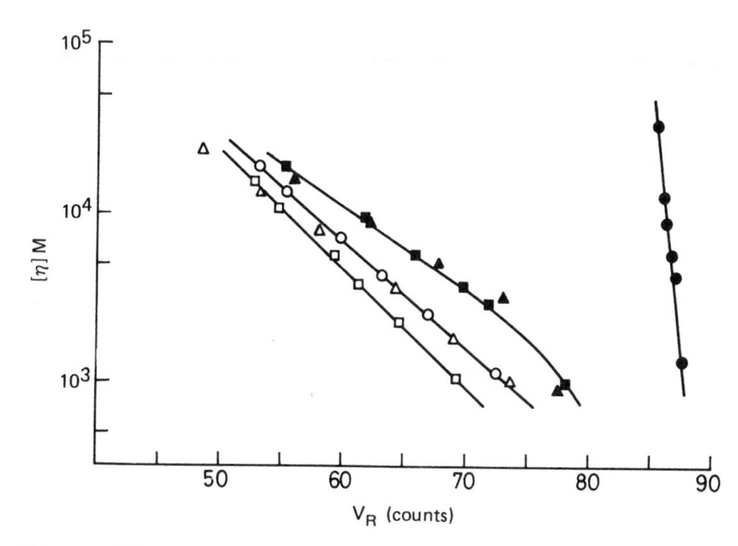

Figure 4.7. Abnormal retention behavior of PMMA with respect to polystyrene on silica gel. In chloroform (○) and (●), chloroform-methanol (90:10) (□) and (■), and dioxane (△) and (▲), for polystyrene (open symbols) and PMMA (filled symbols). (From Ref. 26.)

of the mobile phases. The retention of polystyrene on porous silica gel in several mobile phases has been related to the solvent strength parameter $\varepsilon°$ and the *a* exponent of the Mark-Houwink equation (2.29) [25]. An equation that relates V_R to distribution coefficient K_P and average pore radius $\bar{r}$ has also been proposed. For $\varepsilon° > 0.45$, a linear dependence with negative slope between $\bar{r}$ and $\varepsilon°$ exists, whereas for $\varepsilon° < 0.45$, $\bar{r}$ values seem to display a joint dependence $\varepsilon°$ and a. The product $a\varepsilon° = 0.17$ is the limit for obtaining the experimental polystyrene elution curve; below that value the polymer is not recovered. At high $\varepsilon°$ values partition effects will be mainly responsible. For high $\varepsilon°$ and $f < 1$ values [see equation (4.15) for f], the thick quasi-stationary layer of the mobile phase strongly interacting with the gel prevents the polymer to approach the gel. At low $\varepsilon°$ and $f > 1$ values, the layer thickness decreases, allowing the polymer to approach the gel, and in addition to partition effects, adsorption of solute onto the gel starts to play a role.

The results in Table 4.1 suggest that polymer adsorption onto the SEC packing decreases as eluant polarity increases. For eluants with $\varepsilon°$ in the range 0.40-0.56, deviations from universal hydrodynamic volume plot are interpreted in terms of a separation operating by steric exclusion and adsorption mechanisms in which the degree of polystyrene adsorption becomes more prevalent as polymer-solvent interactions decrease [52].

Belenkii and his coworkers [58] described the dependence of adsorption of polymers on silica gel using the interaction energy $-\varepsilon^* = \Delta H/N_a RT$ (N_a is the number of adsorbed segments of the polymer molecule), which is the difference between the interaction energies of the polymer segment, $-\varepsilon^*_{pa}$, and of the solvent, $-\varepsilon^*_{sa}$, with the surface of the adsorbent:

$$-\varepsilon^* = -\varepsilon^*_{pa} - (-\varepsilon^*_{sa}) \tag{4.18}$$

The value $-\Delta G°/RT$ increases with increasing values of $-\varepsilon^*$, and at $-\varepsilon^* = -\varepsilon^*_{cr}$, the value $-\Delta G°$ becomes zero and changes its sign to positive. At $-\varepsilon^* < -\varepsilon^*_{cr}$, the entry of polymers into the gel pores is governed primarily by the steric exclusion effect; at $-\varepsilon^* > -\varepsilon^*_{cr}$ the adsorption effect begins to predominate. At the critical energy $-\varepsilon^*_{cr}$, these two effects mutually compensate and the distribution coefficient K_{SEC} is unity for polymers of any molecular weight. Since the steric exclusion and the adsorption effects are interrelated, even at $-\varepsilon^* < -\varepsilon^*_{cr}$ the adsorption effect participates in the separation of polymers. This theory was compared with the experimental results in thin layer chromatography. Refinements will probably be necessary because the calculation of $-\varepsilon^*$ should take into account polymer-solvent interaction, which plays a part in the adsorption phenomenon [1].

Polymer-substrate interactions can be divided into adsorption and partition [59]. The former effect involves interactions of functional groups of the solute and substrate and the latter-type affinity may arise when the polymer and the substrate are weakly solvated [10]. The retention of polymers on polystyrene gel in DMF may be examples of the partitioning effect.

In view of the difficulties in eliminating secondary mechanisms with inorganic packings, some attempts have been made to reduce surface adsorption by silanization of the active sites on the packing. However, the structure of the resulting bonded phase must be selected carefully. The example was reported by Dubin et al. [10], who examined porous glass treated with trimethylsilyl reagents. It was found that the universal calibration curves for polystyrene, poly(methyl acrylate), poly(ethylene oxide), and several other polymers in DMF did not superimpose, in contrast to the results for untreated porous glass, on which the universal calibration curves for these polymers fell on the same line.

Organic Gels

Equation (4.7) indicates that there is no partition effect when a gel is not swollen ($\phi_g = 1$ and $K_{ave} = 0$). For poor swelling the value K_{ave} increases with decreasing molecular weight of the solute. When the affinity of the solute toward the solvent is weak ($\chi_{ps} > 0$) or when its affinity toward the gel is strong ($\chi_{pg} < 0$), solutes are eluted in the order of increasing molecular weight. An example was the case of a sequence of n-alcohols on dextran gel in water. The solubility of n-alcohols in water decreases with increasing molecular size, and V_R increases with increasing molecular weight of the solute. Polyhydroxy compounds have a strong affinity for water and are eluted in the order of decreasing molecular weight in the system dextran-water [43].

In SEC by dextran gel, adsorption and partition mechanisms are considered by several workers. The adsorption of aromatic molecules from methanol and isopropanol solution by Sephadex LH-20 has been considered to be a function of π bonding for hydrocarbons and a combination of π and hydrogen bonding for substituted compounds and heterocycles [34]. The retention of some solutes in dioxane-water mixtures on Sephadex G type and LH type has been studied and the results were used to formulate a method of design of SEC systems using mixed solvents which will be free of adsorption and partition effects [60]. Hydrophobic interactions in the adsorption of solutes onto dextran gel and the reduced affinity due to the polarity of solutes are discussed [35]. The partition of homologous series of low molecular weight polar and nonpolar compounds has been studied on cellulose gel in DMF [61]. The solute partitioning is markedly affected by the nature of the functional groups present in the gel matrix, solute, and solvent. Effects of solvents on the separation of oligomer polyethers on Sephadex LH-20 have been discussed [62].

Adsorption effects have been observed when using cross-linked poly(vinyl acetate) gels for separation of epoxy resins in THF [63]. When the resin was treated so as to block the hydroxyl groups chemically by trimethylsilylation or by reaction with butyl vinyl ether, normal SEC has been obtained.

4.3.5. Ionic Exclusion and Inclusion

The retention of polyacrylonitrile and acrylonitrile copolymers with and without ionic groups is strongly influenced by polymer-solvent and polymer-polymer interactions. Polyelectrolyte problems will be eliminated by the use of LiBr with DMF [14-17]. Ionic groups on the copolymer chain may repel each other (intramolecular electrostatic interaction) and thus produce a species with a hydrodynamic volume somewhat larger than its uncharged counterpart of the same molecular weight (polyelectrolyte expansion) [14]. The electrolyte (LiBr) may neutralize the ionic groups and allow the molecules to shrink on a smaller hydrodynamic volume, detected as an increase in V_R (contraction of molecular configuration). The question that required answering is how LiBr could affect the polymer-solvent interaction of polyacrylonitrile and polystyrene, which are not considered ionic (see Figure 4.4).

An acrylonitrile-vinyl ether copolymer fraction shown in Figure 4.3E is supposed to be a neutral polymer, so that the bimodal effect must arise from other than ionic charges [16]. When the LiBr-DMF system is used, a single modal chromatogram is observed, which is caused by the shielding of the dipoles with the Li ions. The cause of the bimodal distribution would be related to association rather than the polyelectrolyte effect. In Figure 4.3C, polyester-based polyurethanes are highly polar and retain characteristics of polyelectrolytes in DMF solutions [17]. The charge nature of these polyurethanes results in the formation of super macromolecules or aggregates [15]. In the case of the quaternized polyester-based polyurethanes, polymer adsorption onto the gel surface appeared to be predominant. Ionic exclusion (repulsion) and ionic inclusion (adsorption related to charge) associated with polymer-gel or solvent-gel interactions should also be considered for these separation mechanisms. The viscosity data for polyacrylic acid showed that this polymer behaves as a polyelectrolyte in DMF, an effect suppressed by the addition of LiBr [10].

Retention of polyelectrolytes (copolymers of acrylonitrile and methallylsulfonate sodium salt) in DMF or DMF with $NaNO_3$ on silica gel has been examined and the salt rejection mechanism proposed [64]. When a polyelectrolyte is dissolved and eluted in a solvent containing an added salt, the SEC chromatogram presents two peaks, A and B. Peak B is attributed to the salt directly related to the quantity of injected polyelectrolyte. Peak A corresponds to the polymer deficient of salt equal to the quantity under peak B. The salt peak excluded for high V_R (peak B) was interpreted by taking into account a Donnan equilibrium. For low salt content, the responses of the differential refractive index and the conductivity difference can be used to deduce the charge distribution of the copolymer as a function of V_R. SEC measurements must be performed with the mobile phase containing a neutral salt at a concentration of at least 5×10^{-2} M to prevent electrostatic exclusion.

A solution of poly(N-vinylacetamide) in pure DMF was supposed to form stable microcrystalline aggregates leading to the appearance of exclusion limit peaks for the higher molecular weight samples [27]. The effect of the addition of LiBr in DMF is attributed to strong ion-dipole interactions between the polymer and LiBr, preventing the formation of such aggregates and resulting in normal chromatographic behavior.

Klein and Westerkamp explained salt effect in a polyelectrolyte solution [28]. In a low-ionic-strength solvent, the polyelectrolyte effect results in concentration-dependent uncoiling of the polyelectrolyte molecule. This uncoiling decreases the precision of SEC measurements appreciably because the concentration of a polymer species decreases during its elution, depending also on polydispersity of the polymer. By addition of low molecular weight electrolytes to a SEC solvent, the uncoiling of polyelectrolytes can largely be prevented. Ion inclusion is prominent on the porous support, which acts like a semipermeable membrane for the polydisperse polyelectrolytes, because the equilibrium of electrical charges on the column is disturbed by the steric exclusion of part of the polymer from the pore. The Donnan effect resulting from the unequilibrium of electrical charges on the column will cause an additional permeation of the low molecular weight part of the polymer or salt ions into the gel pore beyond the steric distribution equilibrium. This results in the retardation of the low molecular weight part of the polymer. Ion exclusion is attributed to a repulsion of equal charges on the surface of the gel and polymer coils at low ionic strength. In this way the polymer will be eluted earlier than expected from the steric equilibrium alone. Ionic inclusion and exclusion become less important when the ionic strength of the solvent is raised. The behavior of nonionic polymers cannot be explained by a polyelectrolyte effect or by ion exclusion. LiBr may produce a salting-out effect with coil contraction [15]. For ionic polymers, a supermolecular structure is assumed and salt addition may destroy all superstructures.

4.4. EXPERIMENTAL VARIABLES

4.4.1. Concentration Effects

Concentration Dependence of Retention Volume

It is a well-known phenomenon that in the SEC of polymers, sample concentration affects the molecular weight-peak retention volume relationship [65-67]. The peak retention volume of a polymer increases with increasing concentration of its injected solution. A shift in peak maximum of the chromatogram with sample concentration is more pronounced for polymer with higher molecular weights. This shift is almost eliminated for the sample with molecular weight lower than 10^4. Figure 4.8 shows examples of concentration dependence of V_R.

This phenomenon is called *concentration effect* or *overload effect*. However, an appreciable shift in peak retention volume with sample size is also observed in the low concentration range (e.g., 0.01-0.05%), so that the term "concentration effect" is preferable to "overload effect" for this phenomenon. The shift of V_R caused by the flow nonuniformities in the system at higher concentrations

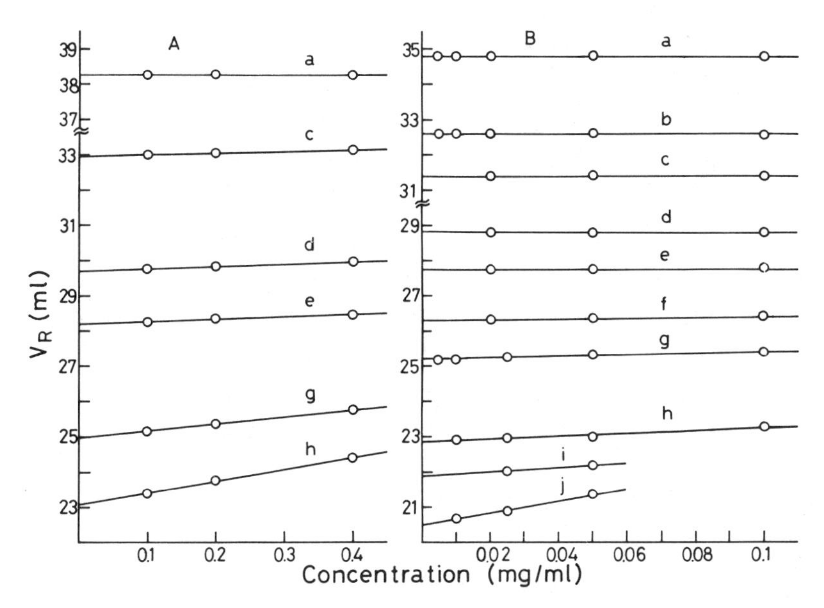

Figure 4.8. Concentration dependence of V_R for polystyrene in good solvents on polystyrene gel columns. (A) In toluene on micro-Styragel columns (3/8 in. x 1 ft x 4) (10^6, 10^5, 10^4, and 10^3 Å nominal porosity) at flow rate 2 ml/min and injected volume 0.25 ml [68]. (B) In THF on Shodex A 80M columns (8 mm x 50 cm x 2) (mixed polystyrene gels of several nominal porosities) at flow rate 1.5 ml/min and injected volume 0.25 ml. [unpublished data]. Molecular weight of polystyrene standards: a, 2100; b, 10,000; c, 20,400; d, 97,200; e, 180,000; f, 411,000; g, 670,000; h, 1,800,000; i, 3,800,000; j, 8,500,000. (From Ref. 68 and unpublished data.)

(unusual use or preparative scale) may be attributed to column overloading [69].

Distribution coefficients of chloroform and benzoic acid in a carbon tetrachloride-polystyrene gel system increase with decreasing solute concentration [70], which is assumed to be based on solute-gel adsorption interactions. This phenomenon is in contrast to the concentration dependence of V_R of polymers. Concentration effects will therefore have other causes.

In theta solvents, V_R is essentially independent of polymer concentration [71]. In a good solvent (benzene) the values of V_R changed with the concentration of the injected polymer more than in a poor (theta) solvent. The concentration dependence of V_R of polystyrene in benzene, chloroform, and theta mixtures of benzene-methanol (77.8:22.2, vol %) and chloroform-methanol (74.7:25.3, vol %) are presented in Figure 4.9. A mixture of benzene-methanol (77.8:22.2) is a theta solvent for polystyrene at 25°C, but at 60°C the thermodynamic quality of the mixed solvent is improved substantially and the concentration dependence becomes relatively distinct [72]. In the system polystyrene (solute)-THF (eluant)-silica gel, the slope of V_R (concentration dependence) which is a straight line, decreases with increasing amounts of water in THF (i.e., with decreasing thermodynamic quality of the mixed solvent for polystyrene solutes). The slope is zero when the mixture of THF and water is 91.1:8.9 (v/v), which is a poorer solvent compared with the theta solvent.

As a polymer is separated in SEC according to its hydrodynamic volume rather than its molecular weight, the relation between the hydrodynamic volume and V_R must be known. Rudin showed that the hydrodynamic volume of a given species is inversely related to the concentration [73]. The concentration effects are greater the higher the molecular weight. The hydrodynamic volume V_h of an unswollen polymer molecule is estimated from

$$V_h = \frac{M}{\rho N_A} \tag{4.19}$$

where ρ is the density of the amorphous part of the polymer, M is the

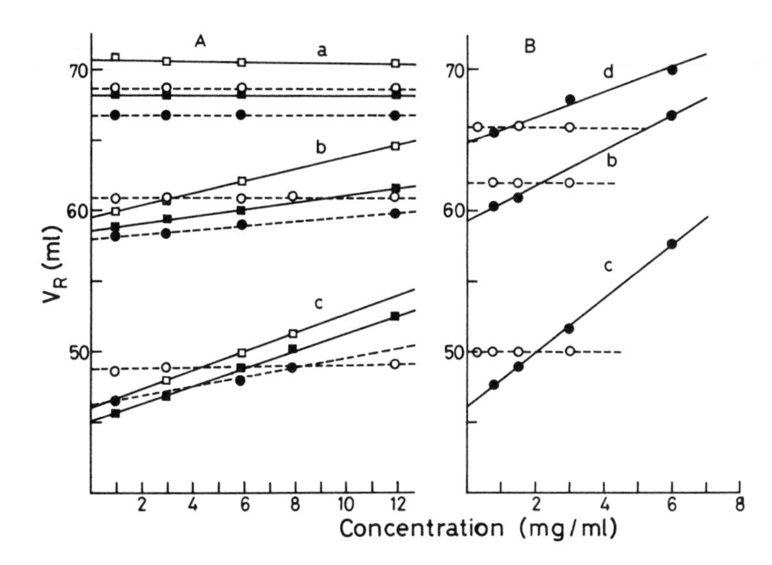

Figure 4.9. Concentration dependence of V_R for polystyrene in good and theta solvents on silica gel columns. (A) In benzene (□) and (■), and in benzene-methanol (77.8:22.8) (○) and (●), at 25°C (open symbols), and at 60°C (filled symbols). a, PS MW 51,000; b, PS MW 160,000; c, PS MW 498,000. (Reproduced from Ref. 72.) (B) In chloroform (●) and in chloroform methanol (74.7:25.3) (○). (From Ref. 22, by permission of the publishers, Hüthig & Wepf Verlag, Basel.)

molecular weight, and N_A is Avogadro's number. The hydrodynamic volume of a swollen polymer [the size of a dissolved (solvated) polymer] at a finite concentration C is the product of V_h and ξ, where ξ is an effective swelling factor (dimensionless) given by [74]

$$\frac{1}{\xi} = \frac{1}{\xi_0} + \frac{C(\xi_0 - \xi_x)}{0.507\rho\xi_0} \qquad (4.20)$$

In equation (4.20), ξ_0 and ξ_x are defined as an effective swelling factor at infinite dilution and the critical volume factor expressed as

$$\xi_0 = \frac{KM^a\rho}{2.5} \qquad (4.21)$$

and

$$\xi_x = 2.60 + \frac{(0.34 \times 10^{-3})M}{M_0} \quad (4.22)$$

where M_0 is the formula weight of the repeating unit. Hydrodynamic size for polystyrene has been calculated by differential (or vacancy) SEC using equations (4.19) to (4.22) [75].

As equation (4.20) is an empirical formula, it does not fit experimental observations in theta solvents. A revised equation, which is less empirical and fits the retention behavior of polymers in all solvents in SEC, has been proposed [76] as

$$\frac{1}{\xi} = \frac{1}{\xi_0} + \frac{C(\xi_0 - 1)}{C_x \xi_0} \quad (4.20a)$$

where C_x is the concentration at critical theta dimensions of polymer coil, and consequently,

$$\xi = \frac{9.3 \times 10^{24}[\eta]}{[\eta]_\theta[(9.3 \times 10^{24}) + 4\pi C N_A([\eta] - [\eta]_\theta)]} \quad (4.23)$$

Therefore, the size of dissolved polymer (the hydrodynamic volume of the solvated polymer) at finite concentration, $V_h\xi$, can be estimated from the equation

$$V_h\xi = \frac{4\pi[\eta]M}{9.3 \times 10^{24} + 4\pi(6.022 \times 10^{23})C([\eta] - [\eta]_\theta)} \quad (4.24)$$

This equation means that the hydrodynamic volume decreases from a maximum value at infinite dilution (C = 0) and it is unchanged at any concentration in theta solvents.

The dimensions of flexible coil polymers decrease with increasing polymer concentration in solution. Consequently, an increase in polymer concentration is accompanied by a decrease in hydrodynamic volume and in an increase in V_R. The interpretation of the concentration effect as a consequence of the decreasing effective hydrodynamic volume of polymer coil with increasing concentration has been proposed by the estimation of thermodynamic quality of solvents [72]. In the range of very low concentrations, deviations from the linear dependence of V_R on concentration (V_R increases with decreasing concentration) can be expected as a result of both the adsorption effect and the thermodynamic partition of the solute. Adsorption phenomena in the system of theta solvent may play an important part [22]. Because of interactions between polymers and the pore wall, the variation of the polymer activity with concentration is different inside the pore to that in bulk solution. This difference causes a concentration dependence of the distribution coefficient, which is explained by using a model for a concentration-dependent distribution which explicitly accounts for a coupling between pore-polymer and polymer-polymer interactions [77].

Janča (and his coworkers) investigated concentration effects both theoretically and experimentally as complex processes from various viewpoints and in various experimental arrangements [78-87], and he proposed a complete and unifying theory of concentration effects under conditions where the gel structure remains unchanged and interactions such as adsorption and incompatibility do not take place [83]. In his theoretical model, three basic contributions are taken into account: a change in the effective size of permeating molecules, viscosity phenomena of the polymer solution in the interstitial volume, and secondary exclusion due to occupancy of a pore by another polymer molecule. The first two contributions lead to an increase in V_R, while the last causes reduction in V_R with increasing concentration. The viscosity phenomenon is a contribution due to the lower mobility of the viscous polymer solution compared with the pure solvent of the mobile phase. Secondary exclusion is a contribution due to the reduction of the accessible pore volume of the column packing by the proper volume of the separated macromolecules, which increases with increasing concentration. The factors taken into account in the theoretical calculation are that the changes in the concentration of the injected polymer solution occur immediately at the beginning of the column because of the distribution between the mobile and the stationary phases, and during the elution, because of the longitudinal spreading accompanied by the movement of the sample zone along the column.

Assuming an instantaneous change in the effective hydrodynamic volume 1/h with a change in the concentration, the retention volume (V_S) can be calculated as [78,81,83]

$$V_S = P + Q\left[\ln V_h + \frac{B^2C_I^2\sigma_I^2}{(\sigma_T^2 - \sigma_I^2)A^2} \ln \frac{\sigma_T A + BC_I\sigma_I}{\sigma_I A + BC_I\sigma_I} - \frac{BC_I\sigma_I}{(\sigma_T + \sigma_I)A} + \frac{\sigma_I^2}{\sigma_T^2 - \sigma_I^2} \ln (A + BC_I) - \frac{\sigma_T^2}{\sigma_T^2 - \sigma_I^2} \ln \frac{\sigma_T A + BC_I\sigma_I}{\sigma_T}\right] \quad (4.25)$$

where P and Q are constants of the calibration curve

$$V_R = P + Q \ln (V_h\xi) \quad (4.26)$$

Here V_S is the calculated retention volume as a function of the hydrodynamic volume, concentration, and efficiency of the column. ξ is the swelling factor at a given and ξ_0 at zero concentration. Also, C_I is the concentration of the injected sample solution, σ_T the overall standard deviation of the elution curve for a monodisperse polymer at the column end, and σ_I the standard deviation of contribution of the injected volume. A and B are defined as

$$A = \frac{1}{\xi_0} \quad \text{and} \quad B = \frac{\xi_0 - 1}{C_x\xi_0}$$

The contribution to the retention volume due to viscosity effect in the interstitial volume is proportional to the difference between the viscosities of the polymer solution and the solvent, and is calculated as

$$V_V = k'\left(\frac{2[\eta]C_I\sigma_I}{\sigma_T + \sigma_I} + \frac{2k_H[\eta]^2C_I^2\sigma_I^2}{\sigma_T^2 - \sigma_I^2} \ln \frac{\sigma_T}{\sigma_I}\right) \quad (4.27)$$

where k' is the proportionality constant calculated using this equation and by substituting the retention volumes at different concentrations of excluded molecules. k_H is the Huggins constant for the respective polymer-solvent system, expressed from Huggins equation

$$\eta_{sp} = [\eta]C + k_H[\eta]^2C^2 \tag{4.28}$$

where η_{sp} is the specific viscosity of polymer solution related to pure solvent.

The retention volume increment due to secondary exclusion is calculated using the equation

$$V_E = -\frac{2(\overline{V}_R - V_0)\phi_x\sigma_I C_I}{\sigma_T^2 - \sigma_I^2}\left(\frac{\sigma_T - \sigma_I}{A} - \frac{BC_I\sigma_I}{A^2}\ln\left|\frac{A\sigma_t + BC_I\sigma_I}{A\sigma_I + BC_I\sigma_I}\right|\right) \tag{4.29}$$

where $\overline{V}_R$ is the averaged retention volume [81,83] and ϕ_x the critical volume fraction of the polymer in solution at which the volume of macromolecules is the same as under theta conditions. Then the retention volume experimentally obtained in the chromatogram caused by a change in concentration starting from zero concentration is expressed as

$$V_R = V_S + V_V + V_E \tag{4.30}$$

The equations (4.25), (4.27), and (4.29) show that the changes of V_R caused by concentration effects become higher with increased volume of the polymer solution injected and with increased column efficiency. The conclusions are in agreement with experimental results [68,78,80]. The contribution of viscosity phenomena in the interstitial volume is approximately four times that due to the changes in the hydrodynamic volume of the polymer coils eluating in the central part of the calibration curves; that is, the contribution of the concentration dependence of the hydrodynamic volume does not exceed about 20% of the total change in V_R [83]. This contribution increases with increasing molecular weight. The viscosity phenomena can be almost compensated for by the effect of secondary exclusion in the total permeation region and the former effect prevails over the latter from the middle to the exclusion region [83]. The secondary exclusion contribution does not appear to be important for the polymer concentrations that are generally used in calibration experiments [86].

The effect of the viscosity of the injected polymer solution is predominant if the specific viscosity of the injected solution exceeds unity, showing a more complicated chromatogram [79,87]. This phenomenon is attributed to viscous fingering [7,69]. The average retention volume of totally excluded polymers was proportional to the specific viscosity of the injected solution and increased with increasing viscosity [79].

The absolute viscosity of the mobile phase is of minor importance as far as viscosity effects in SEC are concerned [82], which has been experimentally verified [49]. Similar results for viscosity effects were obtained by Goetze et al., who injected sample solutions of different viscosities [88]. Viscosities of sample solutions were changed by adding polymer or changing solvent. According to them, the viscosity effects caused a change in V_R, but absolute viscosity alone was not wholly responsible for the change in V_R.

Under equilibrium stationary conditions, where the phases remain immobile, to eliminate viscosity effects the two other factors were investigated [86]. Experimental results showed that at least under stationary conditions, secondary exclusion is operative to a less important degree. The dependence of HETP on η_{sp} of the injected solutions of totally excluded polymers is linear and increases with increasing η_{sp} [82]. HETP of permeating polymers decreased with increasing concentration of the injected solution [68].

Reduction of Errors by Concentration Effects

As the effects of sample concentration on V_R are based on the essential nature of a sample solution in SEC, it is necessary to select the experimental conditions so as to reduce the errors produced by the concentration effects. In the calculation of molecular weight averages from SEC chromatograms, it is assumed that V_R of the individual species is not affected by the sample concentration and by the presence of the other components in the sample. This assumption leads to the result that the molecular weight averages change with changing sample concentration, even when calibration concentrations are the same as sample concentrations. Some examples are shown in

Table 4.4. Molecular Weight Averages of Polystyrene NBS 706 Measured at Various Concentrations

Case	Sample concentration (%)	Calibration concentration (%)	$\bar{M}_w \times 10^{-5}$	$\bar{M}_n \times 10^{-5}$	$\bar{M}_w/\bar{M}_n$
1	0.4	0.4	3.35	1.47	2.27
	0.2	0.2	3.01	1.43	2.11
	0.1	0.1	3.18	1.39	2.29
2	0.4	0	2.30	1.23	1.87
	0.2	0	2.44	1.25	1.94
	0.1	0	2.84	1.26	2.26
3	0.4	0.1	2.55	1.35	1.90
	0.2	0.1	2.71	1.38	1.97
4	0.4	Concentration of each species at elution points	2.57	1.35	1.90
	0.2		2.55	1.31	1.95

Source: Ref. 68.

Table 4.4. Some attempts have been made to minimize the variation of calculated molecular weight averages from SEC chromatograms due to the concentration effects.

Cantow and his coworkers have proposed employing some extrapolation procedure for treating results [65]. They plotted reciprocal apparent average molecular weights of a polymer in various concentrations as a function of concentration and found that the value obtained by plotting to infinite dilution was very close to that measured independently by other methods. Boni et al. measured molecular weight averages with the use of a calibration curve extrapolated to infinite dilution (at zero concentration) [66]. However, as shown in Table 4.4, molecular weight averages calculated by Boni's method still depend on sample concentrations. The concentration of the injected polymer solution starts to change immediately at the beginning of the column because of the molecular weight distribution of the polymer and also because of the peak spreading effects: that is, the concentration of polymer in the eluant at the column outlet is different from that in the injected solution. Hence the concentration of each species in the column is not that of the original sample solution or the infinite dilution. The variation of polymer dimensions with polymer concentration in bulk solution may not apply to a polymer molecule in a pore. Moreover, other phenomena may contribute to the shift in V_R with polymer concentration.

One possible method to reduce errors caused by concentration effects has been proposed [89]. Calibration curves at several finite and zero concentrations are first constructed. The SEC chromatogram of a sample was divided into several points and the concentration at each retention point is obtained from a concentration-peak height calibration curve. Molecular weight at the point is obtained from a molecular weight-V_R calibration curve corresponding to a concentration of species at the point, and molecular weight averages are calculated by using the usual method. The results are shown in Table 4.4, with the values obtained by other methods for comparison. A similar correction procedure was proposed by Nakano and Goto [90].

James and Ouano have calculated molecular weight averages from the SEC chromatograms of polystyrenes obtained from columns in their normal ordering (high- to low-permeability limit), reverse ordering, and random ordering, and they have found that the latter two ordering systems are less sensitive to concentration effects and to errors caused by misuse of calibration curves [91].

The $[\eta]M$ versus V_R calibration curve is based on the assumption that the hydrodynamic volume of a solvated polymer species in SEC columns is that which pertains at infinite dilution. This is not true for most cases, as already mentioned. The equations to estimate hydrodynamic volumes of polymers at finite concentrations [equations (4.20), (4.23), and (4.24)] presented by Rudin et al. provide a universal calibration at any concentration [74,76]. Equations (4.25), (4.27), and (4.29) may also be interpreted as a calibration function. Using these equations, one can construct a calibration curve that explicitly compensates the concentration effects. Consequently, there may be no need to extrapolate to zero concentration or to use other correction methods [2,83].

4.4.2. Effects of Other Experimental Conditions

Column Temperature

When polyisobutylene fractions were eluted in the system polystyrene gel/1,2,4-trichlorobenzene at four different temperatures, 150, 110, 70, and 35°C, a pronounced shift with increasing column temperature toward lower V_R was observed [92]. Calibration curves in the system polystyrene gel-THF at column temperatures 15, 25, 35, and 45°C are shown in Figure 4.10.

Two main factors that cause V_R variations are assumed to be an expansion or a contraction of the mobile phase in the column and the adsorption effect of a solute to the gel phase [49]. When column temperature is 10°C higher than room temperature, the mobile phase will expand about 1% when it enters the columns. The magnitude of the retention volume dependence on the solvent expansion is about one-half of the total change in the retention volume. The residual change in V_R after correcting the change in V_R due to the difference between column temperature and solvent temperature is assumed to be due to gel-solute interactions (adsorption). This magnitude increases with decreasing molecular weight. However, the enthalpy change itself increases with increasing molecular weight, resulting in an increase in K_P value (see Table 4.3), and precise control of column temperature is one of the important factors for operation in SEC of high polymers.

The other factors affecting V_R variations are assumed to be viscosity, the expansion of gel pores, and the effective size of a solute molecule [92]. In our experiments, however, the mobile-phase viscosity did not affect V_R at all [49]. The hydrodynamic volume of a solute increases with increasing temperature, but the estimated change in V_R caused by the increase in hydrodynamic volume is small. The expansion of gel pores and a decrease of V_R with increasing in diffusion coefficient are improbable. The effect of temperature on partitioning may

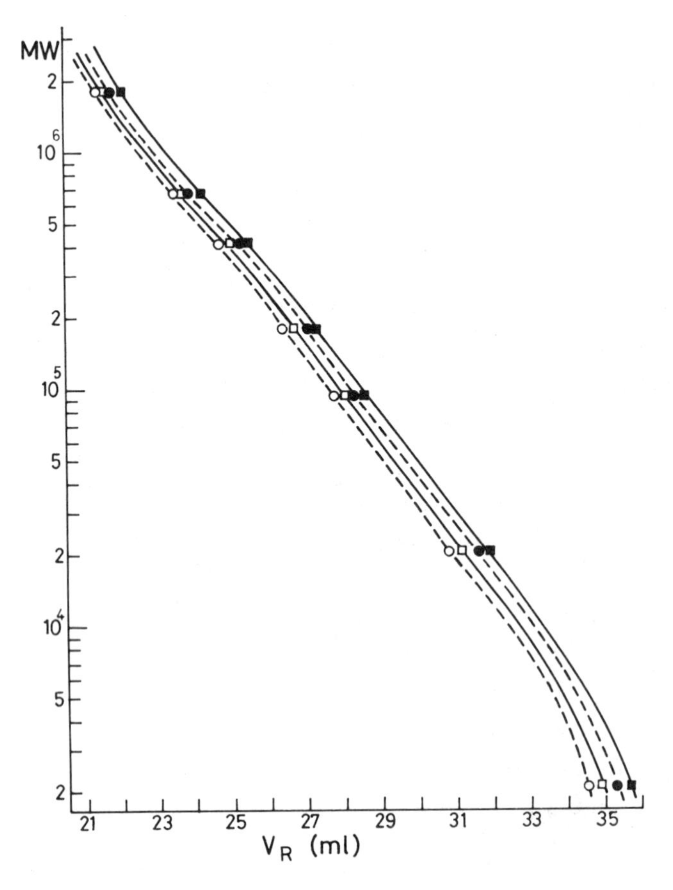

Figure 4.10. Dependence of V_R on column temperature for polystyrene. Column temperature (°C): ○, 45; □, 35; ●, 25; ■, 15. (Reprinted with permission from Ref. 49. Copyright 1980 American Chemical Society.)

play an important role in some systems [59,93]. In a system of oligosaccharides on dextran gel in water, K_{SEC} increased with increasing column temperature [59]. However, when DMF was used instead of water, K_{SEC} decreased with increasing temperature [93].

The theoretical plate number increases with increasing column temperature. From the viewpoint of peak separation, operation at higher temperature is preferred. At higher column temperatures, increased column efficiency and therefore better resolution can be expected because of decreased viscosity, resulting in better mass transfer [94]. A conventional SEC apparatus with a column oven can be operated at higher temperatures. Several factors at higher column temperatures, such as inferior reproducibility and repeatability for the measurement of molecular weight averages, lower column lifetime, and the problem of solvent expansion, force us to control the column oven at near room temperature.

Flow Rate

Conflicting observations on the flow rate dependence of V_R were reported. Boni et al. observed that the V_R values of narrow distribution polystyrenes passed through a maximum as the flow rate was increased [66]. By correcting errors arising from flow rate monitoring with a siphon, they found that V_R was independent of flow rate. Little et al. [94,95] observed the same independence. Yau et al., on the other hand, found that V_R decreased with increasing flow rate [96]. The flow rate dependence increased greatly with molecular weight. As the molecular weight of polystyrene increases, the diffusion coefficient decreases, so that they concluded that nonequilibrium distribution of the polymer molecules between the mobile and stationary phases should account for the change in V_R with flow rate. However, this explanation is not well accepted.

Mori also presented the flow rate dependence of V_R [68]. V_R decreased in the range below 1.0 ml/min and increased in the range over 1.0 ml/min with increasing flow rate. The increase of V_R with increasing flow rate was also observed by other workers [97,98]. These observations have been attributed to nonequilibrium effects [96,97], because polymer diffusion between intrapore and extrapore of gels is sufficiently slow that equilibrium cannot be attained at each point in the column. An explanation based on a flow-rate-dependent equilibrium distribution coefficient has been offered [98]. Nonuniform spatial concentration profiles in the column may arise from interactions between the fluid mechanics and the macromolecular shape during flow through the column. This nonhomogeneous flow effect may produce the change of distribution coefficient with flow rate. With increasing flow rate the efficiency decreased, and the resolution was increased as flow rate was decreased [95]. Bimodal distribution of a polystyrene standard (NBS 705) with narrow molecular weight distribution was clearly observed at slower flow rates [99]. For excluded polymers, the HETP remained constant at flow rates in the range 0.03-3.0 ml/min [82].

Injected Volume

The effect of injected volume on V_R has been observed, and V_R increased with increasing injected volume [66,68]. The increment of V_R between 0.1 and 0.25 ml of injected volume on a given column set was about three times larger than that between 0.25 and 1.0 ml, suggesting that precise or constant injection is required if the injected volume is smaller [68]. Subtraction of one-half the injected volume from the observed V_R produced a constant value in the case of a larger injected volume as in conventional SEC. However, in a small injected volume, as in high-performance SEC, increments of V_R were larger than those of injected volumes, and use of the same injected volume for the sample of interest as when constructing the calibration curve is required. The peak width increased with increasing injected volume and the effect of injected volume depended on sample concentration [68]. There is a limiting sample size in SEC below which the column efficiency is independent of sample size [100]. Theoretically, the effect of injected volume is described by equations (4.25), (4.27), (4.29), and (4.30).

When selecting optimal experimental conditions, it must be considered whether it is more advantageous to inject a smaller volume of solution of higher concentration, or vice versa [68,101]. It may be inferred from the theoretical calculation that a rise in concentration compensated by a simultaneous decrease in the injected volume of polymer solution (with the same total injected weight amount of the polymer) leads to an increase in V_R. However, this change in most of the practical applications is almost comparable with experimental errors. Thus the injection conditions have to be chosen so that the total amount of the injected polymer is constant and the contribution to the total width of the elution curve due to injected volume is negligible within the limits of experimental error. The theoretical calculation was confirmed by the experiment that V_R remained unchanged with decreasing concentration and increasing injected volume and that the ratio of concentration to the injected solution volume may be chosen within broad limits without affecting the results of separation [101]. The changes of the efficiency at equal sample load tell us that lower concentration and larger injected volume are more desirable than the opposite [68].

4.4.3. Column Arrangements

In SEC, several columns packed with gel beads of different pore size are arranged to permit a broad range of molecular weight species to be separated in a single run. The usual column arrangement is from a high-permeability limit column down to a low-permeability limit column (normal ordering). The effects of column arrangements (normal, reverse, and random ordering) on SEC separation process have been compared with the experimental results and computer simulations [91]. The data show significantly different results. Random ordering is preferred because of better resolution and less concentration effects than in the other two.

To obtain an adequate molecular weight separation, the molecular weights of the separated species must be properly matched to the pore size of the gel beads; in other words, column sets must be tailored to the specific separation. Using a "gapped" column set (i.e., a discontinuous transition of gel pores) proves to be detrimental to the resolution of molecular species, especially when analyzing narrow molecular weight distribution (MWD) samples [99]. Accuracy of calibration curves can be improved by developing linear calibration curves.

Average molecular weights and the integral MWD curves for standard polystyrenes NBS 705 and NBS 706 measured using different column combinations have been compared [102]. The average molecular weights calculated were not affected by the column combination, but the integral MWD curves showed definite differences among several column sets.

The most probable integral distribution curve will be obtained from a column packed with mixed gels of different porosities. A column combination without a gap in porosity (i.e., with many different gel porosities) is preferable. The column set should have an exclusion limit of about 10 times the weight-average molecular weight for the polymer sample.

REFERENCES

1. Audebert, R., Polymer, *20*, 1561, 1979.
2. Janča, J., Adv. Chromatogr., *19*, 37, 1981.
3. Dawkins, J. V., and Hemming, M., Makromol. Chem., *155*, 75, 1972.
4. Dawkins, J. V., and Hemming, M., Makromol. Chem., *176*, 1777, 1975.
5. Iwama, M., and Homma, T., Kogyo Kagaku Zasshi, *74*, 277, 1971.
6. Dawkins, J. V., and Hemming, M., Makromol. Chem., *176*, 1815, 1975.
7. Altgelt, K. H., Sep. Sci., *5*, 777, 1970.
8. Kranz, D., Pohl, H. U., and Baumann, H., Angew. Makromol. Chem., *26*, 67, 1972.
9. Zinbo, M., and Parsons, J. L., J. Chromatogr., *55*, 55, 1971.
10. Dubin, P. L., Koontz, S., and Wright, K. L., III, J. Polym. Sci., Polym. Chem. Ed., *15*, 2047, 1977.
11. Matsuzaki, T., Inoue, Y., Ookubo, T., Tomita, B., and Mori, S., J. Liq. Chromatogr., *3*, 353, 1980.
12. Otocka, E. P., and Hellman, M. Y., J. Polym. Sci., Polym. Lett. Ed., *12*, 331, 1974.
13. Mencer, H. J., and Grubisic-Gallot, Z., J. Liq. Chromatogr., *2*, 649, 1979.
14. Cha, C. Y., J. Polym. Sci., B, *7*, 343, 1969.
15. Coppola, G., Fabbri, P., Pallesi, B., and Bianchi, U., J. Appl. Polym. Sci., *16*, 2829, 1972.
16. Kenyon, A. S., and Mottus, E. H., Appl. Polym. Symp., *25*, 57, 1974.
17. Hann, N. D., J. Polym. Sci., Polym. Chem. Ed., *15*, 1331, 1977.
18. Edwards, G. D., and Ng, Q. Y., J. Polym. Sci., C, *21*, 105, 1968.
19. Cazes, J., and Gaskill, D. R., Sep. Sci., *2*, 421, 1967.
20. Cazes, J., and Gaskill, D. R., Sep. Sci., *4*, 15, 1969.
21. Mori, S., and Yamakawa, A., J. Liq. Chromatogr., *3*, 329, 1980.
22. Berek, D., Bakoš, D., Bleha, T., and Šoltés, L., Makromol. Chem., *176*, 391, 1975.
23. Bakoš, D., Bleha, T., Ozima, A., and Berek, D., J. Appl. Polym. Sci., *23*, 2233, 1979.
24. Spychaj, T., and Berek, D., Polymer, *20*, 1108, 1979.
25. Campos, A., Soria, V., and Figueruelo, J. E., Makromol. Chem., *180*, 1961, 1979.
26. Figueruelo, J. E., Soria, V., and Campos, A., Makromol. Chem., *182*, 1525, 1981.

27. Dubin, P. L., J. Liq. Chromatogr., *3*, 623, 1980.
28. Klein, J., and Westerkamp, A., J. Polym. Sci., Polym. Chem. Ed., *19*, 707, 1981.
29. Onda, N., Furusawa, K., Yamaguchi, N., and Komuro, S., J. Appl. Polym. Sci., *23*, 3631, 1979.
30. Belenkii, B. G., Vilenchik, L. Z., Nesterov, V. V., Kolegov, V. J., and Frenkel, S. Ya., J. Chromatogr., *109*, 233, 1975.
31. Brook, A. J. W., and Munday, K. C., J. Chromatogr., *47*, 1, 1970.
32. Mori, S., and Takeuchi, T., J. Chromatogr., *50*, 419, 1970.
33. Mori, S., and Takeuchi, T., J. Chromatogr., *95*, 159, 1974.
34. Streuli, C. A., J. Chromatogr., *56*, 219, 1971.
35. Haglund, A. C., and Marsden, N. V. B., J. Polym. Sci., Polym. Lett. Ed., *18*, 271, 1980.
36. Dawkins, J. V., and Hemming, M., Makromol. Chem., *176*, 1795, 1975.
37. Casassa, E. F., and Tagami, Y., Macromolecules, *2*, 14, 1969.
38. Dawkins, J. V., J. Polym. Sci., Polym. Phys. Ed., *14*, 569, 1976.
39. Mori, S., Anal. Chem., *50*, 745, 1978.
40. Yamamoto, M., and Yamamoto, Y., Anal. Chim. Acta, *87*, 375, 1976.
41. Saitoh, K., and Suzuki, N., Bull. Chem. Soc. Jpn., *51*, 116, 1978.
42. Berek, D., Bleha, T., and Pevna, Z., J. Polym. Sci., Polym. Lett. Ed., *14*, 323, 1976.
43. Lecourtier, J., Audebert, R., and Quivoron, C., J. Chromatogr., *121*, 173, 1976.
44. Lecourtier, J., Audebert, R., and Quivoron, C., Pure Appl. Chem., *51*, 1483, 1979.
45. Dawkins, J. V., Polymer, *19*, 705, 1978.
46. Dawkins, J. V., Pure Appl. Chem., *51*, 1473, 1979.
47. Dawkins, J. V., J. Chromatogr., *135*, 470, 1977.
48. Campos, A., and Figueruelo, J. E., Makromol. Chem., *178*, 3249, 1977.
49. Mori, S., and Suzuki, T., Anal. Chem., *52*, 1625, 1980.
50. Mori, S., and Yamakawa, A., Anal. Chem., *51*, 382, 1979.
51. Altgelt, K. H., and Moore, J. C., in *Polymer Fractionation* (M. J. R. Cantow, ed.), Academic Press, New York, 1967, Chap. B4.
52. Dawkins, J. V., J. Liq. Chromatogr., *1*, 279, 1978.
53. Dawkins, J. V., and Hemming, M., Polymer, *16*, 554, 1975.
54. Lecourtier, J., Audebert, R., and Quivoron, C., in *Chromatography of Synthetic and Biological Polymers*, Vol. 1 (R. Epton, ed.), Ellis Horwood, Chichester, England, 1978, p. 156.
55. Campos, A., Borque, L., and Figueruelo, J. E., J. Chromatogr., *140*, 219, 1977.
56. Cazes, J., and Herron, S., Sep. Sci., *8*, 395, 1973.
57. Klein, J., and Leidigkeit, G., Makromol. Chem., *180*, 2753, 1979.
58. Belenkii, B. G., Gankina, E. S., Tennikov, M. B., and Vilenchik, L. Z., J. Chromatogr., *147*, 99, 1978.
59. Brown, W., J. Chromatogr., *59*, 335, 1971.
60. Bush, B., Jones, T. E. L., and Burns, D. T., J. Chromatogr., *49*, 448, 1970.

61. Chitumbo, K., and Brown, W., J. Chromatogr., *80*, 187, 1973.
62. Berek, D., and Bakoš, D., J. Chromatogr., *91*, 237, 1974.
63. Hope, P., Angew. Makromol. Chem., *33*, 191, 1973.
64. Domard, A., Rinaudo, M., and Rochas, C., J. Polym. Sci., Polym. Phys. Ed., *17*, 673, 1979.
65. Cantow, M. J. R., Porter, R. S., and Johnson, J. F., J. Polym. Sci., B, *4*, 707, 1966.
66. Boni, K. A., Sliemers, F. A., and Stickney, P. B., J. Polym. Sci., A2, *6*, 1567, 1968.
67. Ouano, A. C., J. Polym. Sci., Al, *9*, 2179, 1971.
68. Mori, S., J. Appl. Polym. Sci., *21*, 1921, 1977.
69. Moore, J. C., Sep. Sci., *5*, 723, 1970.
70. Freeman, D. H., and Angeles, R. M., J. Chromatogr. Sci., *12*, 730, 1974.
71. Berek, D., Bakoš, D., Šoltés, L., and Bleha, T., J. Polym. Sci., Polym. Lett. Ed., *12*, 277, 1974.
72. Bleha, T., Bakoš, D., and Berek, D., Polymer, *18*, 897, 1977.
73. Rudin, A., J. Polym. Sci., Al, *9*, 2587, 1971.
74. Rudin, A., and Hoegy, H. L. W., J. Polym. Sci., Al, *10*, 217, 1972.
75. Bartick, E. G., and Johnson, J. F., Polymer, *17*, 455, 1976.
76. Rudin, A., and Wagner, R. A., J. Appl. Polym. Sci., *20*, 1483, 1976.
77. Anderson, J. L., and Brannon, J. H., J. Polym. Sci., Polym. Phys. Ed., *19*, 405, 1981.
78. Janča, J., J. Chromatogr., *134*, 263, 1977.
79. Janča, J., and Pokorný, S., J. Chromatogr., *148*, 31, 1978.
80. Janča, J., and Pokorný, S., J. Chromatogr., *156*, 27, 1978.
81. Janča, J., J. Chromatogr., *170*, 309, 1979.
82. Janča, J., and Pokorný, S., J. Chromatogr., *170*, 319, 1979.
83. Janča, J., Anal. Chem.,*51* , 637, 1979.
84. Janča, J., J. Chromatogr., *187*, 21, 1980.
85. Janča, J., Polym. J., *12*, 405, 1980.
86. Janča, J., Pokorný, S., Bleha, M., and Chiantore, O., J. Liq. Chromatogr., *3*, 953, 1980.
87. Janča, J., Pokorný, S., Vilenchik, L. Z., and Belenkii, B. G., J. Chromatogr., *211*, 39, 1981.
88. Goetz, K. P., Porter, R. S., and Johnson, J. F., J. Polym. Sci., A2, *9*, 2255, 1971.
89. Mori, S., J. Appl. Polym. Sci., *20*, 2157, 1976.
90. Nakano, S., and Goto, Y., J. Appl. Polym. Sci., *19*, 2655, 1975.
91. James, P. M., and Ouano, A. C., J. Appl. Polym. Sci., *17*, 1455, 1973.
92. Cantow, M. J. R., Porter, R. S., and Johnson, J. F., J. Polym. Sci., Al, *5*, 987, 1967.
93. Chitumbo, K., and Brown, W., J. Chromatogr., *87*, 17, 1973.
94. Little, J. N., Waters, J. L., Bombaugh, K. J., and Pauplis, W., J. Chromatogr. Sci., *9*, 341, 1971.
95. Little, J. N., Waters, J. L., Bombaugh, K. J., and Pauplis, W. J., J. Polym. Sci., A2, *7*, 1775, 1969.

96. Yau, W. W., Suchan, H. L., and Malone, C. P., J. Polym. Sci., A2, *6*, 1349, 1968.
97. Gudzinowicz, B. J., and Alden, K., J. Chromatogr. Sci., *9*, 65, 1971.
98. Aubert, J. H., and Tirrell, M., Sep. Sci. Technol., *15*, 123, 1980.
99. Ambler, M. R., Fetters, L. J., and Kesten, Y., J. Appl. Polym. Sci., *21*, 2439, 1977.
100. Chuang, J.-Y., and Johnson, J. F., Sep. Sci., *10*, 161, 1975.
101. Janča, J., J. Liq. Chromatogr., *4*, 181, 1981.
102. Mori, S., J. Chromatogr., *174*, 23, 1979.

—5—

USE FOR POLYMER ANALYSIS

CLAUDE QUIVORON / *École Supérieure de Physique et de Chimie Industrielle, Université Pierre et Marie Curie, Paris, France*

5.1. INTRODUCTION

Since the introduction of SEC by Moore [1], this physicochemical method of polymer characterization has progressed at an extraordinary rate, and no laboratory working in the polymer field can ignore this versatile technique and its widespread applications. Extensive progress has been made in calibration, data correction, and data computerization, but the most important and basic improvements of SEC have undeniably been acquired by technological efforts toward the availability of very efficient micropackings allowing high-speed SEC, and by the use of specific detectors, described in this chapter, which have considerably enlarged the SEC capacity for solving numerous polymer problems.

Polymer analysis by SEC has been reviewed by Cobler [2,3], Janča [4], and more recently by Hagnauer [5]. Readers can also find a discussion of some aspects of SEC applications in the book by Yau et al. [6].

Traditionally, SEC has been widely used for the determination of average molecular weights and molecular weight distributions of linear polymers (Section 5.2), but also provides more and more accurate information on both the molecular weight and chemical composition of branched polymers (Section 5.3) and copolymers (Section 5.4), through the use of concentration- and molecular-weight-sensitive detectors. Moreover, it is an unrivaled method to follow polymerization and chain degradation processes versus time (Section 5.5). Nonconventional applications, such as the study of aggregation, molecular interactions, latex size, or pore size in porous materials, presented in Sections 5.6 and 5.7, lead us to think that SEC may be extended, in the near future, to the resolution of other problems for which molecular size is one of the dominant parameters. The numerous examples used in this chapter have been chosen to illustrate the multiple facets of SEC methodology.

5.2. MOLECULAR WEIGHT AVERAGES AND DISTRIBUTIONS

Molecular weight (MW) average and molecular weight distribution (MWD) determinations actually include all the calculations that can be made from experimental chromatograms, leading to the structural characteristics of polymer chains: (a) average molecular weights, (b) molecular weight distribution curves, and (c) Mark-Houwink viscosity laws.

By contrast, with methods such as osmometry and light scattering, the main limitation of SEC is that it is a nonabsolute method for determining polymer structural characteristics, since it requires preliminary calibration. The various calibration methods (primary, universal, and empirical ones), recently reviewed [7,8], are discussed in Chapter 2.

First, we will describe the general procedures employed in handling data to convert SEC chromatograms into MW and MWD quantities, with the help of any calibration curve. A large part of this section will be devoted to the coupling of continuous detections, especially concentration- and molecular-weight-sensitive detectors. Microgel packings (5-10 μm) are used more and more in SEC. Conventional siphon counters for monitoring the solvent flow are not convenient in this case, because of the small amount of solvent delivered. Finally, as the major difficulty involved in SEC data treatment is to get the most accurate MW and MWD values, we have deemed it advisable to emphasize here the main sources of errors often encountered in SEC data handling, whose correction is discussed in Chapters 3 and 7.

5.2.1. General Procedure for Data Treatment

As often as not, SEC apparatus contains only one concentration detector, commonly a differential refractive index (DRI) device, and the general procedure is to convert chromatographic raw data into MW and MWD values through the response of that kind of detector. Procedures for data treatment depend on the nature of the calibration curve used.

Primary Calibration

With a primary calibration curve (i.e., molecular weight versus retention volume), monodisperse samples are analyzed using the peak position method; the retention volume of the chromatographic peak is precisely measured and the molecular weight is deduced from the calibration curve.

For polydisperse samples, the quantitative procedure consists in digitizing chromatograms (i.e., refractometric peaks) by drawing vertical lines at equally spaced retention volumes, as illustrated in Figure 5.1. Such artificial fractions must be characterized by their heights h_i, proportional to the corresponding solute concentrations C_i, and by their areas A_i, which allow one to calculate cumulative polymer weight values according to

$$I(V) = \frac{1}{A_T} \sum_i A_i \tag{5.1}$$

where A_T is the total peak area and $\sum_i A_i$ is the sum relative to fractions having retention volumes greater than the considered one V_i.

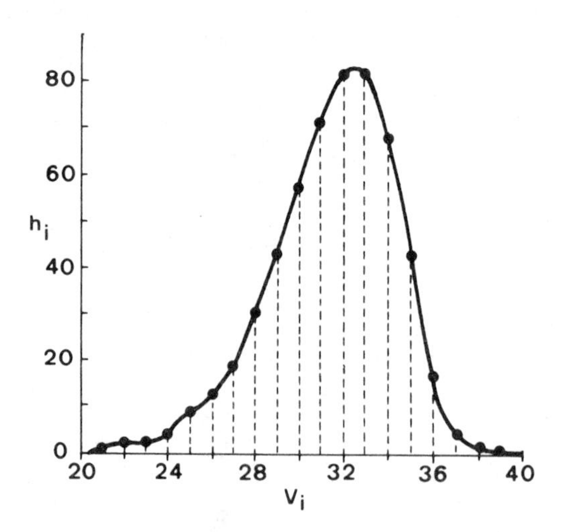

Figure 5.1. Example of digitized chromatogram obtained with a concentration detector (i.e., h_i proportional to solute concentration C_i). (From Ref. 9.)

After conversion of retention volumes V_i into molecular weights M_i through a primary calibration curve, the various (number-, weight-, z-, and viscosity-) average molecular weight values, respectively, are deduced from the following relationships:

$$\bar{M}_n = \frac{\sum_i h_i}{\sum_i h_i/M_i} \qquad \bar{M}_w = \frac{\sum_i h_i M_i}{\sum_i h_i} \qquad \bar{M}_z = \frac{\sum_i h_i M_i^2}{\sum_i h_i M_i} \qquad \bar{M}_v = \left[\frac{\sum_i h_i M_i^a}{\sum_i h_i}\right]^{1/a} \tag{5.2}$$

The ultimate MWD is then obtained by calculating the weight fraction W(M) from equation (2.1). If needed, the differential number fraction molecular weight distribution can be deduced by dividing W(M) values by M:

$$N(M) = \frac{W(M)}{M} = \frac{h_i}{\sum_i h_i} \frac{\Delta V_i}{\Delta M_i} \tag{5.3}$$

Actually, this general procedure, performed by digitizing a chromatogram into equal fractions, is less obvious than it appears [10], since a chromatogram is a very distorted representation of MWD because of the logarithmic nature of the retention. If W(M) and F(V) are, respectively, the MWD and the experimental chromatogram, the total weight of solute can be expressed by

$$\sum_i W^i(M)\ \Delta M_i = \sum_i F^i(V)\ \Delta V_i \tag{5.4}$$

which becomes, considering the calibration equation V = g(M) and its derivative $g'(M) = \Delta V_i/\Delta M_i$,

$$\sum_i \frac{W^i(M)}{g_i'(M)} \Delta V_i = \sum_i F^i(V)\ \Delta V_i \tag{5.5}$$

and, consequently,

$$F^i(V) = \frac{W^i(M)}{g_i'(M)} \tag{5.6}$$

This relationship illustrates the MWD deformation by the SEC process, but it is fortunately balanced by the detection system, which analyzes equal volumes. For example, equation (5.6) can be written

$$\sum_i M_i W^i(M)\ \Delta M_i = \sum_i M_i g_i'(M) F^i(V)\ \Delta M_i = \sum_i M_i F^i(V)\ \Delta V_i \qquad (5.7)$$

and from equation (5.4), the weight-average molecular weight can be represented by

$$\overline{M}_w = \frac{\sum_i M_i W^i(M)\ \Delta M_i}{\sum_i W^i(M)\ \Delta M_i} \equiv \frac{\sum_i M_i F^i(V)\ \Delta V_i}{\sum_i F^i(V)\ \Delta V_i} \qquad (5.8)$$

The same holds true for the other MW averages. As a result, MW averages can be calculated either by digitizing in ΔM_i the MWD distribution W(M) or by exploiting directly the chromatogram F(V) digitized in ΔV_i [10]. These two ways are valid whatever the calibration curve may be (primary or universal).

With primary calibration, sample analysis must be carried out under identical experimental conditions as those used for the elution of the standards. Moreover, standards and sample must be strictly of the same chemical and structural (linear) characteristics; otherwise, MW and MWD calculated values must be considered as either apparent or relative. Empirical procedures have been proposed to deduce MW and MWD of unknown samples from polystyrene primary calibration. Among those, the Q factor method is not realistic, as demonstrated for instance by Bly [11] and in Chapter 2. A better method is to consider that the coefficient C_1 in equation (2.3) is constant for both standards and sample to be analyzed and to evaluate the A value for the sample from its intrinsic viscosity and an exponent a [12].

The general exploitation of SEC chromatograms can be achieved either manually or automatically. A number of on-line minicomputer systems have been developed for collecting and processing SEC data. In general, the data systems are designed for handling SEC data, computer hardware for data acquisition and handling, and computer software for data reduction. SEC data are plotted, calculations are performed, and the results are reported.

Universal Calibration

Benoit et al. [13] considerably improved MW and MWD calculations from chromatograms by recommending the use of the product $[\eta]M$ as a universal calibration parameter (see Chapter 2). This concept has been widely verified and is now unanimously accepted, except for some highly branched polymers [14,15] and for questions of sample polydispersity [16].

Once the universal calibration is known, the general data treatment indicated above is strictly the same here if the correspondence between retention volumes and molecular weights can be made. As each V_i is directly related to $[\eta]_i M_i$, a first approach is to measure $[\eta]_i$ of narrow chromatographic fractions or, better, continuously. These two procedures will be dealt with later (see Section 5.2.2). A more usual way to make this conversion is to use some iterative methods [17,18] combining chromatographic data and one measured value $\bar{M}_n$, $\bar{M}_w$, or $[\eta]$. These procedures are discussed in detail in Section 2.4.3. The reliability of this iterative method [17] was checked by Samay et al. [19] by comparing for various homo- and copolymers the MW results obtained with primary and universal calibrations. As illustrated in Table 5.1 for methacrylic homopolymer fractions, there is rather good agreement for both calculated weight-average molecular weights and intrinsic viscosities. Tables 5.2 and 5.3 show that there is a good accord between intrinsic viscosities determined via SEC and those measured separately, even for copolymers that may have additional heterogeneity in chemical composition. In summary, chromatographic data treatments using only a concentration detector (i.e., refractometer) can lead to MW and MWD determinations whose precision and accuracy are dependent on both the calibration method chosen and the data procedure employed.

5.2.2. Data Treatment Using Detector Coupling

One of the most important improvements in the development of SEC during the past years is the use, in addition to the traditional refractive index detection, of two (or more) detectors in series so as to get complementary information for polymer samples.

In fact, dual detection concerns two kinds of coupling according to the nature of the detector response. The first group is related to the coupling of a usual refractometer with another concentration

Table 5.1. Comparison of the MW and [η] Data Obtained with the Weiss and Cohn-Ginsberg (WCG) Iterative Method [17] and with Primary (PC) and Universal (UC) Calibrations

$\bar{M}_w \times 10^{-3}$					[η] (ml/g)	
LS[a]	PC	UC[b]	WCG	UC[c]	WCG	Measured
			Poly(ethyl methacrylate) fractions			
937	1023	1011	1000	1026	186	186
726	795	781	767	780	154	148
562	604	590	573	579	124	120
340	409	395	386	385	94	88
270	230	218	253	249	69	80
2050	1729	1728				
1510	1474	1467				
1050	1228	1218	1059	1088	195	202
620	649	634	610	616	129	118
			Poly(butyl methacrylate) fractions			
1331	1384	1215				
572	647	655	811	688	111	118
592	520	550	658	571	99	129
501	545	577	739	634	106	98
330	355	401	494	439	83	66
207	185	227	281	262	59	52
99	75	95	117	117	34	34
			Poly(octyl methacrylate) fractions			
1560	1641	1877				
601	728	717	698	698	99	96
2090	2365	1984				
1060	1296	1195	1134	1134	134	134
815	1101	1039	887	887	115	99
480	547	561	532	533	85	84
368	377	389	390	390	71	85
194	183	199	196	196	46	45

[a] $\bar{M}_w$ from light scattering.
[b] UC with axial dispersion correction.
[c] UC without axial dispersion correction.
Source: Ref. 19.

Table 5.2. Comparison of Intrinsic Viscosities Obtained by SEC and Classical Viscometry for Polystyrene and High- and Low-Density Polyethylene Samples

Sample	Polydispersity	[η] (dl/g) True	[η] (dl/g) SEC
	Polystyrene fractions		
2×10^6	1.30	3.92	4.16
6.7×10^5	1.15	1.79	1.73
4.11×10^5	1.10	1.28	1.23
1.60×10^5	1.06	0.62	0.68
5.1×10^4	1.06	0.28	0.24
1.98×10^4	1.06	0.138	0.136
	Polystyrene polymers		
NBS #705 (179,000)	1.07	0.740	0.728
NBS #706 (258,000	2.10	0.931	1.01
			0.967
			1.03
		Average	1.00
	Linear polyethylene		
NBS 1475	2.90	1.01	0.970
			1.010
			1.012
			1.016
			1.014
	Branched polyethylene		
NBS 1476	2.12	0.9024	0.9024
	1.95	0.9024	0.895

Source: Ref. 12.

detector which provides additional data on solutes. This is mainly the case of ultraviolet and infrared detections, especially used in SEC studies of copolymers and described in Section 5.4. Another concentration detector consists of a continuous densimeter [21-23] which can be used when refractive index increment dn/dc values are near zero and when there is still an appreciable density difference between

Table 5.3. Comparison of Intrinsic Viscosities Obtained by SEC and Direct Viscometry for Diblock Styrene-Isoprene (PS-PIP) and Styrene-Methyl Methacrylate (PS-PMM) and Triblock Styrene-Isoprene-Styrene (PS-PIP-PS) Samples in Tetrahydrofuran

Sample	Percent PS UV	$\overline{M}_{w\ LS}$	$\overline{M}_{n\ OSM}$	$\overline{M}_{w\ SEC}$	$\overline{M}_{n\ SEC}$	$[\eta]_{Ubbelohde}$ (ml/g)	$[\eta]_{SEC}$
PS-PIP-1	11.0	101,000	95,800	104,000	84,000	82	75
PIP-PS-2	52.7	159,000	137,700	167,500	136,200	103	100
PS-PIP-3	53.8	1,074,000	--	1,147,000	375,200	414	371
PIP-PS-4	93.4	114,000	88,100	124,500	89,000	57	54
PS-PMM-5	49.7	404,000	345,000	442,000	268,900	101	104
PS-PIP-PS-6	75.3	119,000	107,600	132,000	105,000	64.0	60.7
PS-PIP-PS-7	77.4	69,000	61,100	74,000	59,000	42.0	39.7

Source: Ref. 20.

solvent and solute. Other concentration detectors employed in column liquid chromatography have recently been reviewed [24].

The second group of dual detections, which will be discussed here, involves the coupling between a concentration detector (i.e., refractometer) and another detector that is specifically sensitive to solute molecular weight. This dual detection is extremely helpful for MW and MWD data treatment, since it does not require any additional information. The two concentration/molecular weight detector couplings used until now are refractometer/viscometer dual detection and, more recently, refractometer/low-angle laser light-scattering photometer (LALLSP) dual detection, both developed in the on-line and continuous mode by Ouano [25,26].

DRI/Viscometer Dual Detection

In order to apply universal calibration and determine $[\eta]_i$ values directly for the chromatographic fractions, a first approach was the use of a discontinuous but on-line viscometer. This device, first proposed by Meyerhoff [27] and later improved by several authors [20, 23,28-34], consists of connecting the refractometer outlet, through a siphon counter, to one or more automatic viscometers of the Ubbelohde type. Provided that the length and the diameter of the capillary are chosen to give flow times shorter than the time required for the siphon filling, intrinsic viscosity of each of the collected fractions can be determined through one of the following relationships:

$$[\eta]_i = \frac{1}{C_i} \frac{t_i - t_0}{t_0} \tag{5.9}$$

$$[\eta]_i = \frac{1}{C_i} \log \frac{t_i}{t_0} \tag{5.10}$$

where t_i and t_0 are, respectively, the efflux times of solution and solvent. Under the usual SEC conditions, the concentrations C_i are always very small (<0.05% at the column outlet) and intrinsic viscosity is practically identical to both reduced specific [equation

(5.9)] and inherent [equation (5.10)] viscosities. The concentrations C_i are still evaluated from the respective refractometric peak heights h_i.

On-line determinations of $[\eta]_i$ values allow direct correspondence between V_i and M_i when universal calibration is used; the general procedure seen above (Section 5.2.1) leads, normally, to sample MW and MWD values. This first type of on-line viscometer is very reliable and gives good results for intrinsic viscosity and MW values compared to those obtained by classical viscometry and light scattering, as illustrated in Table 5.4. However, this "batch" technique has the great disadvantage of being dependent on a siphon counter delivery and on the ability of classical packings to lead to convenient volumes for sample elutions (about 10 siphon volumes); modern microgels render its use impossible, since complete elutions of polymer samples correspond in this case to only 2-3 siphon volumes.

Table 5.4. Comparison of Intrinsic Viscosities and Weight-Average Molecular Weights Obtained by SEC-Viscometer Coupling and by Classical Methods (Off-Line Ubbelohde Viscometry and Light-Scattering) for Linear Polystyrene (PS) and Poly(vinyl acetate) (PVAc) and Branched PVAc Samples

Sample	$[\eta]_{Ubbelohde}$ (ml/g)	$\overline{M}_{w\ LS}$	$[\eta]_{SEC}$ (ml/g)	$\overline{M}_{w\ SEC}$
PS (linear)	11.5	14,500	11.8	15,400
PS (linear)	66.4	173,000	64.3	182,000
PVAc (linear)	36.9	66,500	37.9	73,700
PVAc (linear)	51.6	109,000	53.7	109,800
PVAc (linear)	63.6	143,000	61.2	158,000
PVAc (linear)	80.4	192,000	81.7	195,000
PVAc (branched)	31.2	48,000	26.1	51,700
PVAc (branched)	33.1	60,400	32.0	59,400
PVAc (branched)	89.5	318,000	85.5	269,000
PVAc (branched	112.0	500,000	111.7	501,300

Source: Ref. 23.

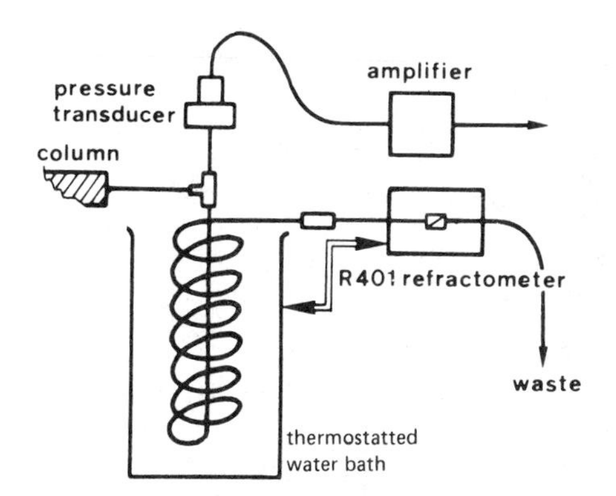

Figure 5.2. Sketch of an on-line continuous viscometer. (From Ref. 38.)

Ouano [25] greatly improved on-line viscometric detection by conceiving a clever continuous viscometer usable with modern fast columns, but curiously, few publications have appeared since then in this field [35-38] even though this detector is very simple and highly sensitive. Its principle consists in measuring, with a piezoresistive transducer, the pressure drop ΔP_i in a capillary tube (typically 0.009 in. inside diameter, 1- to 3-m length) inserted between the column set and the refractometer, as shown in Figure 5.2. If flow is laminar for both solvent and solutions, viscosity η_i is related to ΔP_i according to the Poiseuille law,

$$\eta_i = \frac{\Delta P_i \pi d^4}{128 \ell q} \tag{5.11}$$

where d and ℓ are the diameter and length of the capillary and q is the mobile-phase flow rate. If the flow rate is rigorously kept constant, η_i is strictly proportional to ΔP_i and intrinsic viscosities can be continuously determined, for example, by

$$[\eta]_i = \frac{1}{C_i} \log \frac{\eta_i}{\eta_0} = \frac{1}{C_i} \log \frac{\Delta P_i}{\Delta P_0} \tag{5.12}$$

where ΔP_0 is the measured pressure drop for pure solvent. As with the discontinuous viscometer, the concentrations C_i are deduced from the refractometric chromatogram and are assumed to be small enough to apply equation (5.10).

Figure 5.3 illustrates this continuous dual detection for both narrow and broad MWD samples. For the monodisperse sample, the chromatographic peaks are normally at almost exactly the same retention volume. To the contrary, however, the viscometer peak of a broad MWD polymer is skewed toward the lower end of the retention volume, since the detector response increases with the molecular weight.

The principle of MW and MWD calculations using universal calibration is shown in Figure 5.4. The three chromatographic factors--V_i (related to $[\eta]_i M_i$ via universal calibration), h_i (leading to C_i), and ΔP_i (from which $[\eta]_i$ is deduced)--allow us to perform automated summations giving $\overline{M}_n$, $\overline{M}_w$, $\overline{M}_z$, and $\overline{M}_v$ and to build up MWD curves.

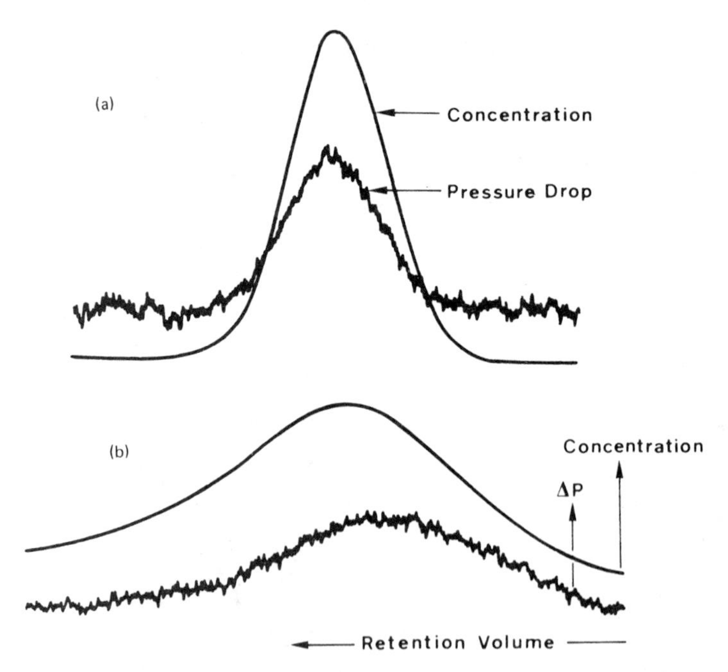

Figure 5.3. On-line viscometer and refractometer chromatograms of (a) narrow MWD polystyrene in THF; (b) broad MWD polymer sample. (From Ref. 39.)

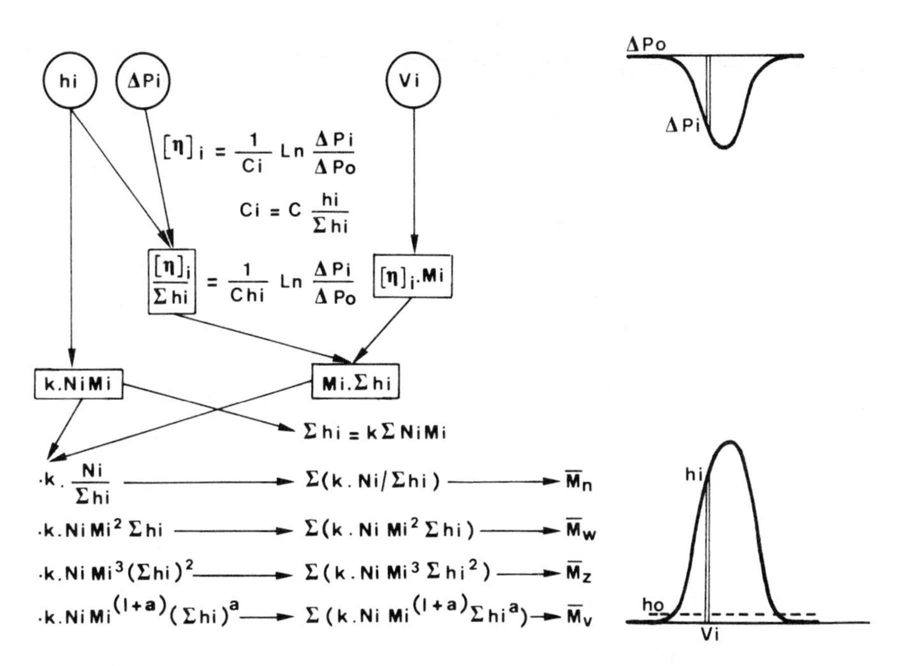

Figure 5.4. Principle of data treatment by using refractometer-viscometer continuous detection and universal calibration. (From Ref. 37.)

Moreover, the continuous viscometric signal also supplies the constants K and *a* from a single sample, provided that its polydispersity is high enough (typically > 1.3) [38]. Figure 5.5 shows an example of a complete SEC analysis (MW, MWD, and viscosity law) carried out with automatic apparatus equipped with both a refractometer and a continuous viscometer [38]. For accurate determinations, the dead volume connecting the cells of the consecutive detectors must be taken into account [40,41]. Through the work of Ouano and Lesec et al., the continuous viscometer detector has actually become a simple and realistic device, which is also a very accurate flow meter. It has been extrapolated very recently to high-temperature SEC for the study of polyolefins and their copolymers (see Section 5.3).

DRI/LALLSP Dual Detection

MW and MWD determinations through universal calibration require knowledge of the coefficients K and *a* of the polymer sample to be

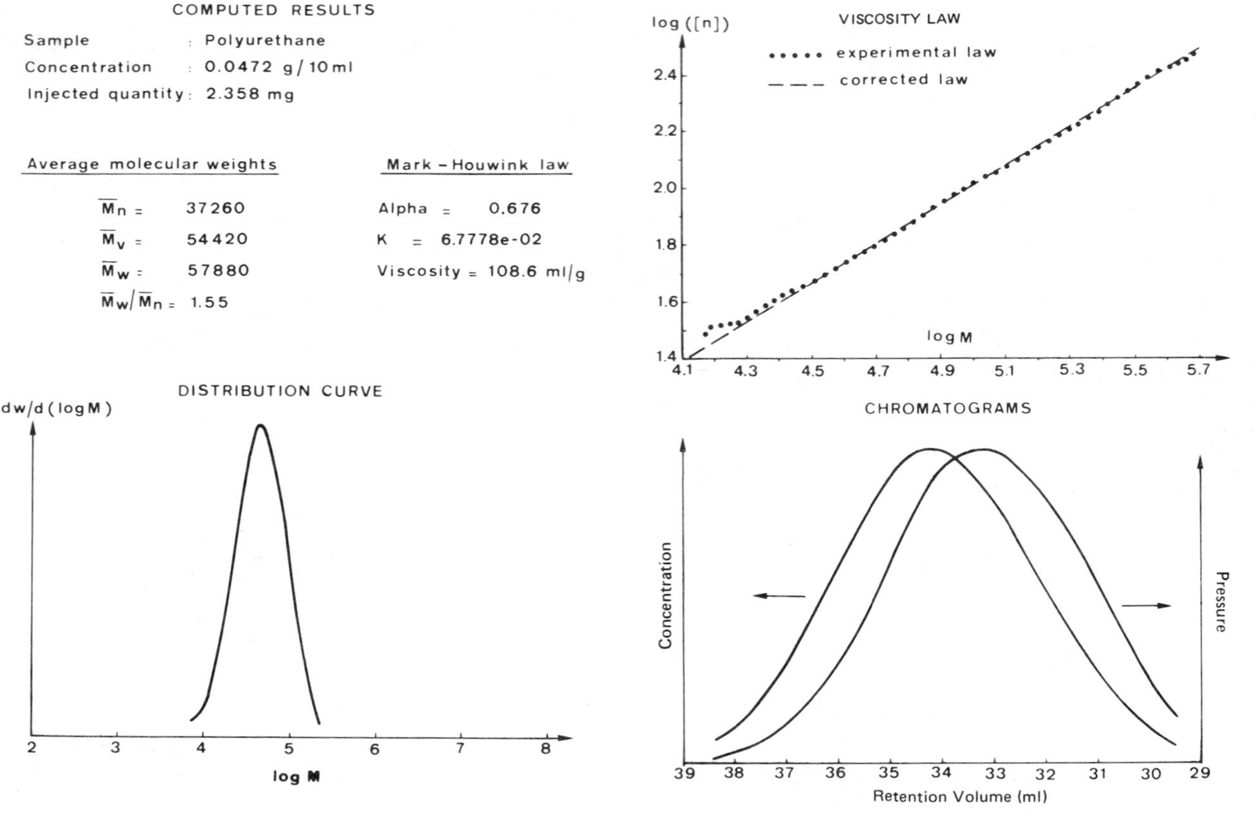

Figure 5.5. Complete SEC analysis (MW, MWD, and Mark-Houwink constant K and *a* of a polyurethane sample, using a refractometer-viscometer continuous detection and universal calibration. Solvent: THF. Points represent logarithms of viscosity data deduced from the viscometric chromatogram and the straight line is the final Mark-Houwink viscosity law, due to an axial dispersion correction.

analyzed, but these values are not always available for special and complex polymers. This great limitation can be overcome by using an on-line continuous viscometer which is convenient to establish viscosity laws as mentioned above. From an absolute point of view, the best way is to obtain the molecular weight directly at each point for a sample chromatogram and, consequently, correct MW and MWD values, without referring to any calibration curve.

This goal led Ouano et al. [26,42] to design an on-line continuous light-scattering detector coupled with a refractometric one. That second type of concentration/molecular weight detector coupling has been considerably simplified by the LALLSP version, which eliminates most of the problems generally involved in classical light scattering: extrapolation to zero concentration and angle, and use of large mixing volume cells. Since the subsequent commercialization of LALLS photometers [43], valuable papers have been devoted to SEC/LALLSP studies [43-51], showing the promising ability of this type of on-line detection.

In classical light-scattering experiments [52], the measured intensity $I(\theta)$ of the light scattered at some angle θ by a polymer solution is generally expressed through the Rayleigh ratio

$$R_\theta = r^2 \frac{I(\theta)}{I_0} \tag{5.13}$$

where r and I_0 are, respectively, the distance of observation and the intensity of the incident light. The excess Rayleigh factor (i.e., related only to solute molecules)

$$\bar{R}_\theta = R_{\theta\ \text{solution}} - R^0_{\theta\ \text{solvent}} \tag{5.14}$$

is given by the classical relationship

$$\frac{K^*_\theta C}{\bar{R}_\theta} = \frac{1}{M} + 2A_2C + 3A_3C^2 + \cdots \tag{5.15}$$

where C and M are the solute concentration and molecular weight, and A_2 and A_3 are the second and third virial coefficients, respectively.

K_θ^* is a constant, at a given angle, specific to the solute-solvent pair:

$$K_\theta^* = \frac{2\pi^2 n_0^2}{\lambda_0^4 N_A} \left(\frac{dn}{dc}\right)^2 (1 + \cos^2\theta) \tag{5.16}$$

where N_A is Avogadro's number, λ_0 the in vacuo wavelength of light, n_0 the refractive index of the solvent, and dn/dc the refractive index increment of a polymer solution.

When measurements are performed on very large polymer chains whose dimensions in solution approach that of the wavelength of the light, intramolecular light-scattering interferences must be taken into consideration through a scattering function $P(\theta)$, depending essentially on the chain radius of gyration $\langle s^2 \rangle$, as described at low angles by

$$P(\theta) = 1 - \frac{16\pi^2}{3\lambda^2} \langle s^2 \rangle \sin^2 \frac{\theta}{2} + \cdots \tag{5.17}$$

λ being the wavelength of the light in the medium.

Consequently, the general expression allowing light-scattering data treatments, at both low angle and concentration, is [52]

$$\frac{K_\theta^* C}{\overline{R}_\theta} = \frac{1}{M} + 2A_2C + \frac{16\pi^2}{3\lambda^2} \frac{\langle s^2 \rangle}{M} \sin^2 \frac{\theta}{2} + \cdots \tag{5.18}$$

and requires for solute molecular weight determination extrapolation to zero concentration and angle (Zimm plot procedure). When solutes are polydisperse, equation (5.18) leads to $\overline{M}_w$ values [52].

The LALLSP version has tremendously simplified data treatment procedures, since angles of observation are very low (typically 4-6°) and $P(\theta)$ [i.e., the third term in equation (5.18)] is close to unity even for large macromolecules. For example, with $\theta = 5°$, $\overline{M}_w = 2.5 \times 10^6$, $\langle s^2 \rangle \sim 1.3 \times 10^6$ Å^2 (polystyrene in a good solvent), and $P(\theta) = 1.003$ [39]. Thus equation (5.15) can be readily used with a LALLS photometer to evaluate $\overline{M}_w$ molecular weights.

Figures 3.6 to 3.8 illustrate results of the DRI/LALLSP detection coupling. According to equation (5.15), the LALLSP response is proportional to the product CM and, as a result, is more sensitive to high molecular weights than the refractometer response. Figure 5.6 shows this specific sensitivity toward high molecular weights, which is not apparent in the refractometer response, for a highly branched poly(vinyl acetate) sample [49].

The data treatment for obtaining MW and MWD values from DRI/LALLSP coupling is performed after a preliminary digitization of the two chromatograms; C_i is still proportional to the corresponding refractometer height h_i, and $\overline{R}_\theta$ values are deduced from the LALLSP signal height knowing the instrumental response factor. Hence each molecular weight M_i is finally given by equation (3.45), rewritten here as

$$\frac{1}{M} = \frac{K_\theta^* C_i}{\overline{R}_\theta^i} - 2A_2^i C_i \qquad (5.19)$$

and (C_i, M_i) pairs lead to MW and MWD determinations as mentioned before, but without any V_i/M_i calibration procedure.

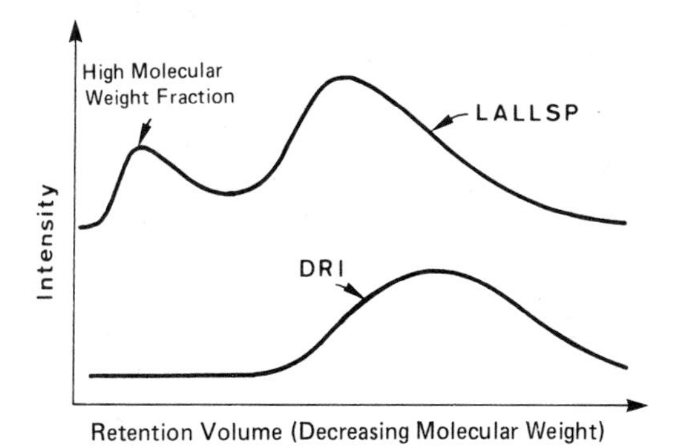

Figure 5.6. On-line LALLSP and refractometer (DRI) chromatograms of a highly branched poly(vinyl acetate) sample, showing a bimodal distribution only detected by LALLSP. (From Ref. 39.)

Strictly speaking, A_2 values depend on molecular weight according to [52]

$$A_2 = kM^{-\gamma} \tag{5.20}$$

with γ varying from 0.15 to 0.35 for many polymers in good solvents. Consequently, samples with broad MWD might require taking into account this variation for determining each A_2^i value. Practically, owing to the low value of the exponent γ, it is usually assumed that A_2^i is independent of M_i and a mean value is determined from LALLSP off-line measurements.

In most cases, satisfactory agreement is observed between $\overline{M}_w$ values obtained by SEC/LALLSP and by classical SEC using a universal calibration [42-44]. However, SEC/LALLSP coupling systematically leads to lower $\overline{M}_w/\overline{M}_n$ values than those specified by the suppliers with the help of other methods, as indicated in Table 5.5. This discrepancy is due to the polydispersity in molecular weight of the eluant in the LALLSP cell ($\overline{M}_w$ correct values but greater $\overline{M}_n$ values) and signal data must be corrected for band broadening (see Chapter 3).

Concentration/LALLSP detector coupling has already experienced growing use in MW and MWD polymer analysis, but should extend its

Table 5.5. Comparison of MW Data Obtained by SEC/LALLSP and Classical SEC Using Universal Calibration (Without Band Broadening Correction) for Some Polystyrene Standards

	$\overline{M}_w \times 10^{-5}$		$\overline{M}_n \times 10^{-5}$		$\overline{M}_w/\overline{M}_n$	
Sample	SEC	SEC-LALLSP	SEC	SEC-LALLSP	SEC	SEC-LALLSP
Duke 7.1 x 10^6	69.0	73.0	54.3	70.6	1.28	1.03
PS-1.8 x 10^6	17.4	16.0	7.94	15.4	2.19	1.04
PS-670 x 10^3	6.10	6.05	5.01	5.64	1.22	1.07
PS-411 x 10^3	4.46	4.30	3.80	4.17	1.17	1.03
PS-179 x 10^3	1.81	1.76	1.62	1.67	1.12	1.06

Source: Ref. 42.

widespread possibilities in the near future toward the resolution of more complex situations, such as polymer branching (see Section 5.3) and aggregate or microgel occurrence (see Section 5.6). On-line multidetections involving refractometers, LALLS photometers, and continuous viscometers should help to solve some of these problems.

As we have seen, SEC is a quantitative method for determining MW and MWD for both organophilic and water-soluble polymers. However, data treatment procedures include various sources of errors [53], and the principal ones have to be known in order to apply the requisite corrections (see Chapters 3, 4, and 7).

5.3. POLYMER BRANCHING

Besides MW and MWD determinations, which are the most usual application of SEC, this technique has been also used to obtain information about chain branching in polymer structures. Branching concerns another level of structure characterization, but the problems involved are much more difficult to solve, particularly from a theoretical point of view and, in all cases, are not yet definitively settled.

However, SEC has been applied extensively to branching studies of various homo- and copolymers. Most of the papers published in this field concern low-density polyethylene (LDPE), but chain branching has also been investigated by SEC for poly(vinyl acetate), polystyrene, polydienes, poly(vinyl chloride), polyamides, ethylene-vinyl acetate copolymers, and so on. Reviews of polymer branching analysis can be found in Refs. 54 and 55 and also in Refs. 4 and 5.

5.3.1. Generality of Long-Chain Branching Characterization

If a linear polydisperse structure can be simply defined by average molecular weights, the characterization of branched polymers introduces additional parameters, such as (a) the functionality of branch points, (b) the number of branches per macromolecule, and (c) the length and the distribution (star-, comb-, or random-like) of branches.

It is clear that these branching parameters are average quantities for non-tailor-made samples, even for branch point functionalities, and consequently, experimental data are not very easy to analyze. Many branching parameters have been proposed, but here we will use only two parameters, besides the functionality of branch points: m, the number of branches per polymer chain, and λ_b, the chain branching frequency (i.e., the number of branches per molecular weight unit):

$$\lambda_b = \frac{m}{M} \tag{5.21}$$

Furthermore, short-chain and long-chain branching do not lead to the same modifications of polymer properties; the first greatly alters solid-state properties, whereas the latter induces large changes in viscometric properties, both in solution and in the melt. As a result, SEC methods can provide information on long-chain branching which specifically perturbs hydrodynamic volumes of chain coils in solution.

The basic concept of the use of SEC is that the occurrence of large branches in a polymer chain reduces its hydrodynamic volume with regard to that of a linear chain of the same chemical nature and molecular weight (see Figure 5.7) [56]. The mean lower length value of chain branching affecting hydrodynamic volume depends on polymer chemical structure and the nature of the solvent. It is not really known in general, but experimental data obtained, for example, with LDPE samples in trichlorobenzene (TCB) at 135°C gave an approximate value for six-carbon branches.

To calculate the extent of long-chain branching, Stockmayer and Fixman [57] defined a branching parameter g as the ratio of the square of the radius of gyration of a branched molecule to the square of the radius of gyration of a linear chain of the same molecular weight:

$$g = \frac{\langle s^2 \rangle b}{\langle s^2 \rangle 1} \tag{5.22}$$

This ratio can be classically determined from light-scattering measurements for the branched sample if a linear sample with the same

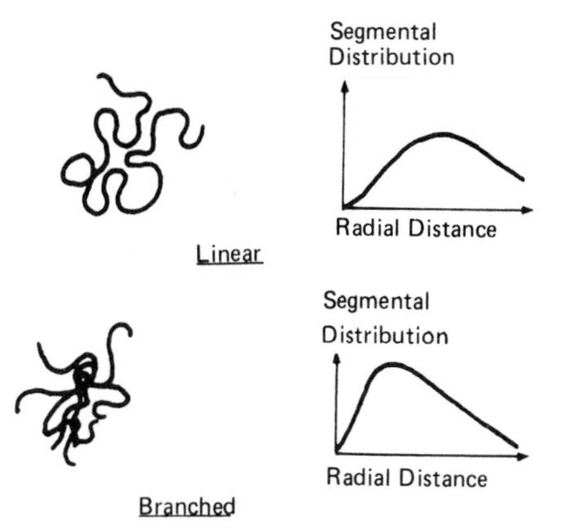

Figure 5.7. Comparison of segmental distribution in linear and branched polymers with the same molecular weight; the maximum density of polymer segments is closer to the center of gravity in the branched species, which consequently has the smaller values of radius of gyration and intrinsic viscosity. (From Ref. 56.)

molecular weight is available or if the relationship $\langle s^2 \rangle = f(M)$ is known for the linear structure.

On the other hand, g can be theoretically related to the number m of branches randomly distributed per macromolecule, assuming a given value for the branch point functionality. The most simple and commonly used relationships are for tri- and tetrafunctional branch points [58], respectively,

$$g = \left[\left(1 + \frac{m}{7}\right)^{1/2} + \frac{4m}{9\pi}\right]^{-1/2} \tag{5.23}$$

$$g = \left[\left(1 + \frac{m}{6}\right)^{1/2} + \frac{4m}{3\pi}\right]^{-1/2} \tag{5.24}$$

where m is strictly the number-average number $\bar{m}_n$. Other theoretical equations are expressed through the weight-average $\bar{m}_w$ [58].

The evaluation of m values is complicated in practice by the fact that samples are often polydisperse, and in this case, light scattering gives $\bar{M}_w$ and z-average values for $\langle s^2 \rangle$, whereas the equations above are valid for monodisperse chains in theta solvent. It can already be seen why a SEC fractionation prior to any measurement may bring helpful improvements in polymer branching studies, especially for the variation of g versus molecular weight.

As intrinsic viscosity measurements are much easier to carry out, a second branching parameter g' was introduced [59] as the ratio of the intrinsic viscosities of homologous branched and linear samples:

$$g' = \frac{[\eta]_b}{[\eta]_1} < 1 \tag{5.25}$$

If the K and *a* constants of the Mark-Houwink law are known for the linear structure,

$$[\eta] = KM_1^a \quad KM_b^a \tag{5.26}$$

and g', also written

$$g' = \frac{[\eta]_b}{KM_b^a} \tag{5.27}$$

can be determined from both intrinsic viscosity and molecular weight measurements, with the same restriction as for light-scattering experiments, when samples are polydisperse. Here also it is advisable to perform a preliminary SEC fractionation so as to apply equation (5.27) to each fraction, establish the variation $g' = f(M_i)$, and finally calculate the average branching parameter:

$$g' = \frac{\sum_i c_i [\eta]_{i,b}}{\sum_i c_i [\eta]_{i,1}} = \frac{\sum_i c_i [\eta]_{i,b}}{K \sum_i c_i M_{i,b}^a} \tag{5.28}$$

However, the evaluation of branch number values m from viscometry is not feasible, since no theoretical expression for g' has been published up to now. A first indirect way is to identify [57]

$$g' = h^3 \tag{5.29}$$

where h is the ratio of the hydrodynamic radius (equivalent Stokes radius) of the branched polymer to that of the linear polymer. This ratio can be theoretically related to m [54]. Although this procedure appears to have been more and more used recently, the most general way of interpreting branching data from viscometry is to assume an a priori relationship between g' (determined by experiment) and g (theoretically accessible). According to the well-known Fox-Flory viscosity law [equation (2.25)], it was firstly supposed that [58]

$$g' = g^{3/2} \tag{5.30}$$

but this relationship was soon questioned [57,60], especially for highly branched species. For starlike branched polymers, $g' = g^{1/2}$ was proposed [59] and experimentally confirmed [61,62], but each study so far published on long-chain branching leads to various values of the exponent

$$g' = g^{x} \tag{5.31}$$

sometimes written [63] as

$$g' = g^{0.5+\mu^*} \tag{5.32}$$

where μ^* is the skeleton polymer fraction ($\mu^* = 0$ for a star polymer).

The lack of real theoretical support for the dependence between g' and g explains the scattering of the experimental data and, for example, the values of x from 0.5 to 1.5 that can be found in papers published in the field of LDPE [54]. However, on the basis of recent studies on LDPE samples, it seems that there is a clear tendency toward the empirical relationship [64-67]

$$g' = g^{1.2\pm0.2} \tag{5.33}$$

As mentioned above, the problem of the precise characterization of highly branched polymers is really open, but SEC has made great strides in that field, and should define its scope if progress is made at a theoretical level. We will illustrate by some examples the powerful use of SEC in branching studies when coupled with either off-line or on-line viscometry or light-scattering determinations.

5.3.2. Examples of Long-Chain Branching Studies by SEC

Experimentally, SEC studies of long-chain branching do require a universal calibration established from well-characterized fractions of the corresponding linear polymer. Hence for each fraction of a branched sample chromatogram,

$$Q_i = [\eta]_{i,b} M_{i,b} = [\eta]^*_{i,l} M^*_{i,l} \equiv K M^{*(a+1)}_{i,l} \tag{5.34}$$

where $[\eta]^*_{i,l}$ and $M^*_{i,l}$ are related to the linear polymer fraction that would have the same retention volume. From equation (5.34), the expression of the branching parameter [equation (5.27)] can be written as

$$g' = \left(\frac{M^*_{i,l}}{M_{i,b}}\right)^{a+1} \tag{5.35}$$

or

$$g' = \left(\frac{[\eta]_{i,b}}{[\eta]^*_{i,l}}\right)^{a+1} \tag{5.36}$$

Off-Line Viscometry or Light-Scattering Measurements

The preceding considerations show that g' can be determined by SEC if the viscosity law of the linear polymer is known and if each $[\eta]_{i,b}$ value is measured. In this case, either of the relationships above can be used, since $M_{i,b}$ is deduced from $[\eta]_{i,b}$ through equation (5.34).

A first, but time-consuming way is to prepare fractions by coacervation [68] or preparative SEC [69,70] from the whole sample, and to analyze them simultaneously by SEC. Table 5.6 gives g' values obtained from both equations (5.35) and (5.36), with fractions of a LDPE industrial sample. The small differences observed between $\langle g' \rangle_\eta$ and $\langle g' \rangle_{M_w}$ values show that prepared fractions are, in general, monodisperse.

Drott and Mendelson [71] proposed an iterative method for determining branching parameters from SEC data and intrinsic viscosity $[\eta]$

Table 5.6. Branching Parameter Values of High-Pressure Polyethylene Fractions[a]

Fractions	$\bar{M}_w \times 10^{-3}$	$\langle g' \rangle_\eta$	$\langle g' \rangle_{M_w}$	Fractions	$\bar{M}_w \times 10^{-3}$	$\langle g' \rangle_\eta$	$\langle g' \rangle_{M_w}$
1	721	0.472	0.286	7	36.8	0.776	0.784
2	124	--	--	8	28.7	0.882	0.825
3	97	--	--	9	24.2	0.846	0.746
4	68.2	0.764	0.726	10	18.5	0.892	0.788
5	53.8	--	--	11	14.9	0.957	0.792
6	41.7	0.811	0.802	12	12	0.753	0.749

[a] $\langle g' \rangle_{M_w}$ and $\langle g' \rangle_\eta$ are determined from equations (5.35) and (5.36), respectively; $\bar{M}_w$, from light scattering. Solvent: trichlorobenzene at 135°C.

Source: Ref. 68, by permission of the publishers, Hüthig & Wepf Verlag, Basel.

of the whole sample. From equations (5.23), (5.31), and (5.35), $M^*_{i,l}$ can be written

$$(M^*_{i,l})^{a+1} = (M_{i,b})^{a+1}\left[\left(1 + \frac{\lambda_b M_{i,b}}{7}\right)^{1/2} + \frac{4\lambda_b M_{i,b}}{9\pi}\right]^{-x/2} \tag{5.37}$$

with a given value for x and where λ_b is supposed to be constant throughout the range of molecular weights. From chromatographic heights h_i, a value of the intrinsic viscosity of the branched sample can be computed on the basis of the viscosity law of the linear polymer

$$[\eta]_{calc} = K \frac{\sum_i h_i (M^*_{i,l})^a}{\sum_i h_i} \tag{5.38}$$

As the sample is branched, $[\eta]$ is lower than $[\eta]_{calc}$ and calculations are iterated with various λ_b values until $[\eta] = [\eta]_{calc}$. At this point, average molecular weights can be deduced. This method was principally applied to polyethylene branching (e.g., Refs. 72 to 75) and gave rather good results in long-chain branching characterization. However, the fact that λ_b is considered to be constant (necessary to achieve iterations) is not always true [76], and this is an important limitation of this method. Among the other methods based on sample intrinsic viscosity and SEC data, the one proposed recently by Foster et al. [77] seems the most reliable so far.

On-Line Viscometry or LALLSP Measurements

As we have seen in Section 5.2.2, DRI/viscometer and DRI/LALLSP detector couplings have brought remarkable improvements in SEC data treatment particularly in long-chain branching studies. In a first step, the use of on-line discontinuous viscometers have promoted easier determinations, by measuring directly the $[\eta]_{i,b}$ values corresponding to the fractions delivered by a siphon counter; g' values are then deduced from equation (5.27). Figure 5.8 gives the branching parameter distribution of an industrial polybutadiene sample, determined by on-line discontinuous viscometry. As illustrated in

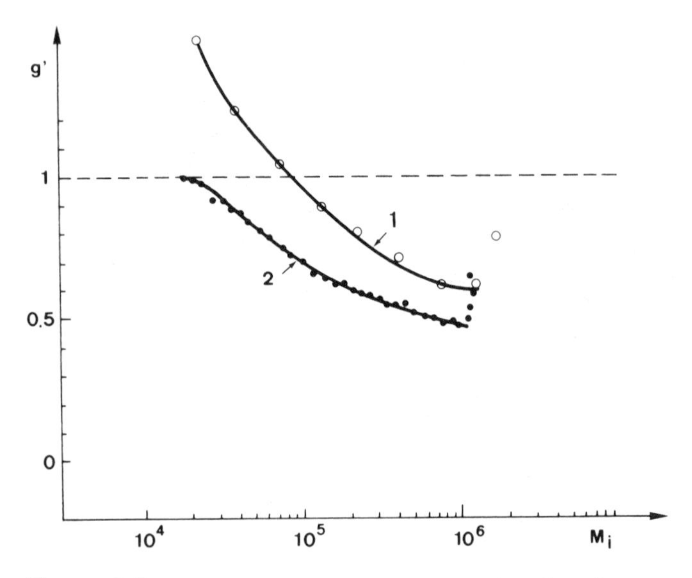

Figure 5.8. Branching parameter distribution [uncorrected (1) or corrected (2) for axial dispersion] of a Buna CB 10 Bayer polybutadiene sample, determined by using refractometer-discontinuous viscometer coupling. (From Ref. 78.)

Figure 5.8, Marais et al. [78] showed that axial dispersion corrections are absolutely necessary to get rational values for g'.

In the special case of branched polyethylene studies, the on-line automatic viscometer must be conditioned at a high temperature (typically with trichlorobenzene at 135°C) with the SEC apparatus [79-81]. Table 5.7 presents the dependence on molecular weight of long-chain branching frequency λ_b for two LDPE samples. Numerous values are obtained according to the assumptions made for the branch point functionality (3 or 4) and for the exponent x [see equation (5.31)]. The choice is rather arbitrary, which demonstrates the complexity of the branching investigations.

Continuous LALLSP detection has been recently used for branching studies [43,49-51]. For example, Axelson and Knapp [51] tested the reliability of the SEC/LALLSP coupling in the determination of g and g' values from equations (5.31) and (5.35). Figure 5.9 shows the good agreement between the results obtained by this coupling and by other techniques in the case of a LDPE NBS standard, and leads us to look forward to promising developments in the near future.

More recently, the use of the on-line continuous viscometer [25,35-38] described in Section 5.2.2 has been successfully extended by Lecacheux et al. [82,83] to high temperatures (i.e., 135-140°C), allowing precise and fast characterization of chain branching of polyethylene and complex polyolefins such as ethylene-vinyl acetate copolymers. Preliminary results of intrinsic viscosity and average molecular weights determined for various linear and branched polyethylene samples with this refractometer/viscometer coupling are in quite good agreement with those obtained by the classical method (i.e., SEC/flow time viscometer coupling), as illustrated in Table 5.8 [83].

As an example, Figure 5.10 shows a complete analysis of a LDPE sample by SEC/continuous viscometry detection. The experimental viscosity law deduced from the viscometric signal is smoothed by a third-degree polynomial regression and quantitatively compared to the Mark-Houwink relationship for linear polyethylene (see Figure (5.10). g'

Table 5.7. Long-Chain Branching Frequency Distribution of Two LDPE Samples, Determined by Using Refractometer-Discontinuous Viscometer Coupling[a]

		Long-chain frequency values $\lambda_b \times 10^{+4}$							
		Tetrafunctional hypothesis				Trifunctional hypothesis			
Sample	M	x = 1/2	x = 1	x = 3/2	x = 1/4 x (log M - 1)	x = 1/2	x = 1	x = 3/2	x = 1/4 x (log M - 1)
β	10^4	3.2	1.2	0.8	1.8	8.1	3.0	1.8	4.4
	10^5	1.9	0.4	0.2	0.4	5.0	1.1	0.6	1.1
	10^6	7.7	0.4	0.1	0.2	23.0	1.0	0.3	0.5
α	10^4	3.0	1.1	0.7	1.7	7.5	2.8	1.7	4.1
	10^5	3.0	0.6	0.3	0.6	8.0	1.5	0.7	1.5
	10^6	13.5	0.5	0.2	0.2	38.0	1.4	0.4	0.7
	10^7	255.0	0.7	0.1	0.1	755.0	2.2	0.3	0.3

[a] $\bar{M}_w$ values from light scattering are 800,000 and 195,000 for α and β, respectively.
Source: Ref. 81.

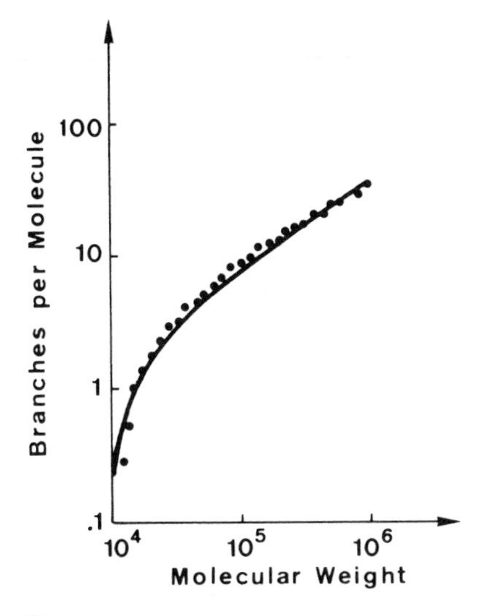

Figure 5.9. Number of branches per macromolecule (weight-average) versus molecular weight for a LDPE NBS 1476 sample. Solid curve represents data from literature and solid circles, data from SEC/LALLSP coupling. Exponent x value in equation (5.31) is taken as 0.5. (From Ref. 51.)

Table 5.8. Comparison of the Intrinsic Viscosity and Average Molecular Weight Values Obtained by SEC Coupled with Either a Continuous or Discontinuous Viscometer for Several Linear (HDPE) and Branched (LDPE) Polyethylene Samples

	[η] (ml/g)		Continuous viscometer		Classical coupling	
Sample	Continuous viscometer	Classical coupling	$\bar{M}_n \times 10^{-3}$	$\bar{M}_w \times 10^{-3}$	$\bar{M}_n \times 10^{-3}$	$\bar{M}_w \times 10^{-3}$
HDPE SRM 1475	100	99	23	55	18	54
HDPE A	163	170	15	141	13	148
HDPE B	121	127	19	84	12	92
LDPE 1	107	108	31	271	26	293
LDPE 2	107	104	28	189	27	209
LDPE 3	95	99	25	183	22	167
LDPE 4	86	84	23	145	23	131
LDPE 5	83	80	22	88	20	78
LDPE 6	79	81	21	123	18	103

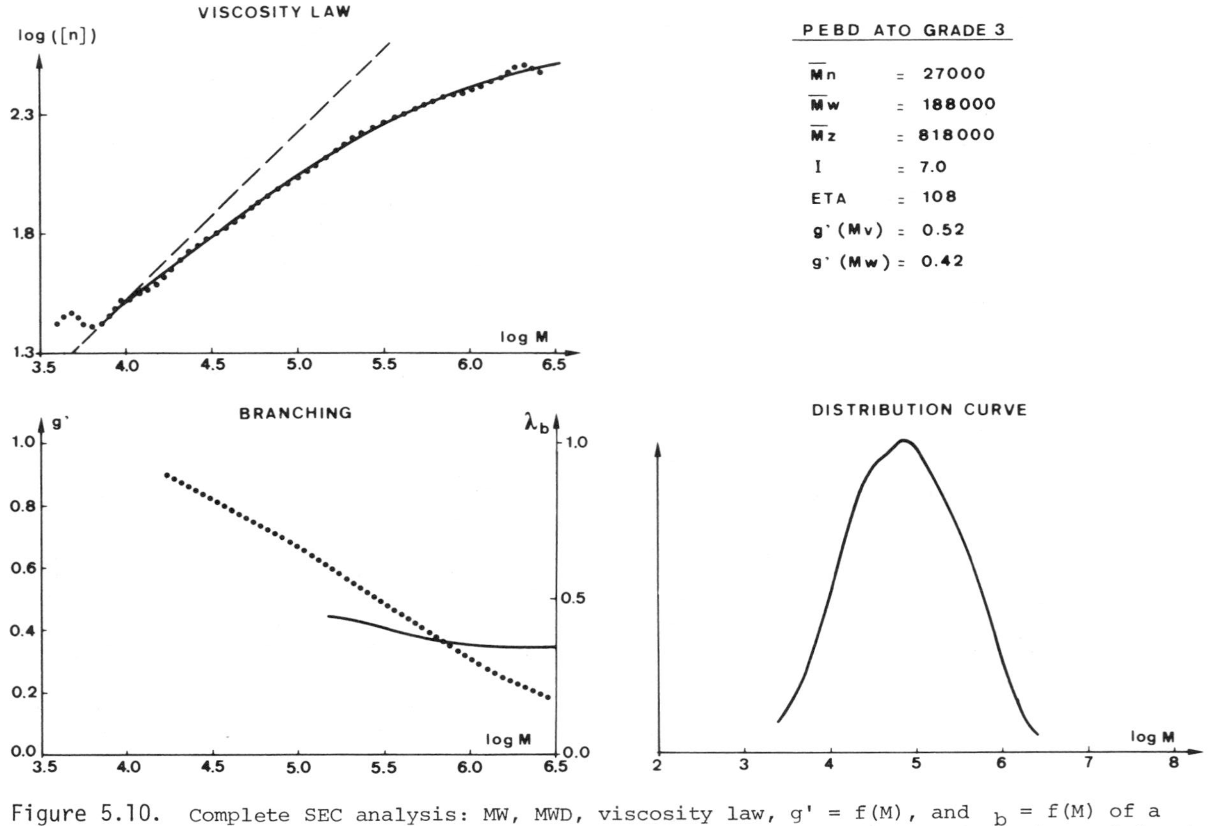

Figure 5.10. Complete SEC analysis: MW, MWD, viscosity law, g' = f(M), and $_b$ = f(M) of a branched polyethylene sample, using a refractometer-viscometer continuous detection. Solvent: trichlorobenzene at 135°C. (From Ref. 82.)

and λ_b dependence on molecular weight can then be calculated by taking the exponent x value in equation (5.31) as 1.2 and by using the simple relationship [58]

$$g = \frac{3}{2}\left(\frac{\pi}{m}\right)^{1/2} - \frac{5}{2m} \tag{5.39}$$

which leads to results similar to those obtained with equation (5.23). For different LDPE samples studied, g' varies linearly with molecular weight and, consequently, there is no significant change for the long-chain branching frequency value ($\lambda_b \sim 0.5 \times 10^{-4}$). The results are in close agreement with the ^{13}C nuclear magnetic resonance (NMR) results of Bovey et al. [84], who obtained the same order of magnitude (0.4 and 0.6×10^{-4}) and no marked dependence on molecular weight.

In conclusion, SEC coupled with molecular-weight-sensitive detectors, especially continuous viscometers, is now an excellent method for characterizing long-chain branching at both ambient and high temperatures. As mentioned above, the comparison with other techniques--flow time viscometry, ^{13}C NMR, and also ultracentrifugation [85]--demonstrates the reliability of SEC. Besides, the SEC/LALLSP/continuous viscometer coupling, which has not been published so far, should increase the powerful potential of SEC in the field of polymer branching.

5.4. CHEMICAL COMPOSITION OF COPOLYMERS

Average MW and MWD of copolymers can be determined by the procedures described in Sections 5.2 and 5.3. For this purpose, universal calibration has been found to be valid for many copolymers over a broad range of molecular weights and chemical compositions [see, e.g., Refs. 44, 86 to 89, and 4 (review)]. However, the complete characterization of copolymers (random, block, or graft) requires the best knowledge of their chemical compositions, which highly influence their bulk properties.

In special cases, refractometric chromatograms may simply provide qualitative information on chemical composition of copolymers. Figure 5.11 gives an example of SEC analytical fractionation of

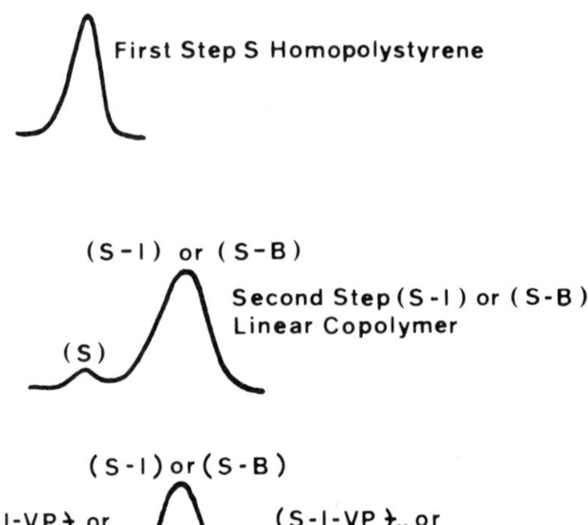

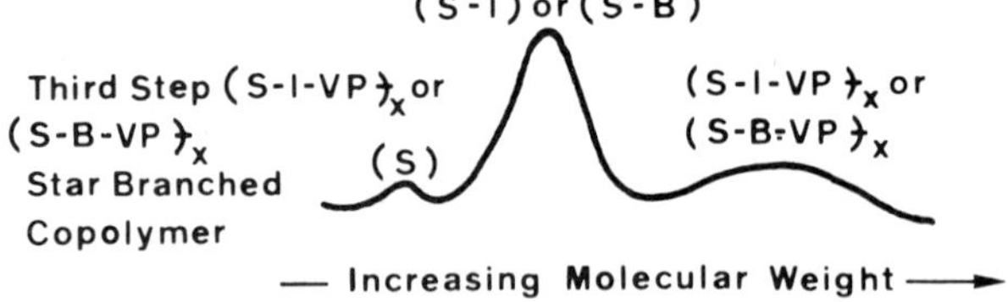

Figure 5.11. SEC chromatograms of homopolystyrene (S), styrene-isoprene (or butadiene) block copolymer (S-I) or (S-B) and star-branched copolymer $(S\text{-}I\text{-}VP)_x$ or $(S\text{-}B\text{-}VP)_x$, formed during the following steps of anionic polymerization:

$$\text{Li} \xrightarrow{\text{S}} \text{S-Li} \xrightarrow{\text{I}} \text{S-I-Li} \xrightarrow{\text{VP}} \text{S-I-VP} \xrightarrow{\text{x}} (\text{S-I-VP})_x$$

where vinyl pyridine (VP) acts as a tetrafunctional branching agent, leading to branched species of functionality x. (From Ref. 90.)

styrene/isoprene (or butadiene) block copolymer and subsequent star-branched block copolymer formed when vinyl pyridine, acting as a cross-linking agent, is added during anionic polymerization.

The best way to characterize copolymers by SEC properly is to determine chemical composition as a function of molecular weight. Nevertheless, SEC elution of copolymer samples leads to great complications since chains of different chemical compositions may have the same molecular size. Consequently, chromatographic analysis cannot be performed with a single concentration detector, such as a refractometer, but requires an additional detection system providing further information.

To solve this conceptual problem, it has recently been proposed to study copolymer composition by connecting two SEC instruments, each running with a different mobile phase [91,92]. Although this new "cross or orthogonal" chromatography is very promising not only for copolymers but also for branched ones, all the studies of copolymers have been made, up to now, by using dual detectors, generally connected on-line to the chromatograph. The nature of the additional detection obviously depends on the copolymer structures to be analyzed, but the two specific detectors widely used, in addition to the universal refractometer, are the ultraviolet (UV) and, to a lesser extent, the infrared (IR) detector. For future work, the feasibility of LALLSP detection for the quantitative characterization of chemical heterogeneities in copolymers has recently been demonstrated [93].

5.4.1. Refractometer/Ultraviolet Dual Detection

This detector coupling has been the one most investigated, due to the variety of copolymers based on styrene (or aromatic monomers), which is easily visualized by UV spectroscopy (absorption at 254 nm). Typical studies can be found, for example, in Refs. 23 and 94 to 99, and the latest are reviewed in Refs. 4 and 5. At a quantitative level, the best way to exploit refractometric and UV signals is to determine, first, the composition distribution independently of molecular weight and then the molecular weight at each composition. Figure 5.12 shows the calculated chemical composition curve versus retention volume for a styrene-butadiene-styrene triblock copolymer sample. This conversion can be made through the ratio

$$R_i = \frac{h_{UV}^i}{h_{RI}^i} \tag{5.40}$$

of the UV and refractometer responses (peak heights or areas) at each retention volume, with

$$h_{UV}^i = K_S^{UV} W_S^i \tag{5.41}$$

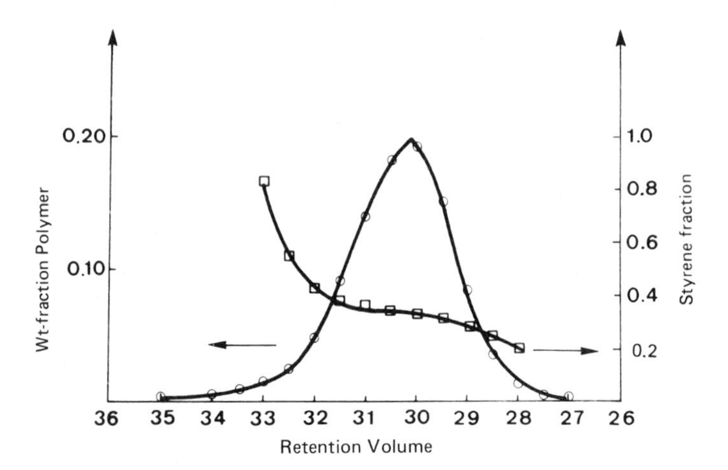

Figure 5.12. Chemical composition as a function of retention volume for a styrene-butadiene-styrene block copolymer. ○, Refractometric chromatogram; □, UV-detected styrene concentration curve. (From Ref. 100.)

and

$$h_{RI}^{i} = K_S^{RI} W_S^{i} + K_C^{RI}(1 - W_S^{i}) \tag{5.42}$$

where K_S^{UV}, K_S^{RI}, and K_C^{RI} are the response factors of the corresponding detectors for styrene (S) and comonomer (C), determined from homopolymer samples, and W_S^i the styrene weight fraction, which can be deduced from the consequential relationship [98]

$$W_S^i = \frac{K_C^{RI} R_i}{K_S^{UV} + (K_C^{RI} - K_S^{RI}) R_i} \tag{5.43}$$

In all cases, the dead volume between the two detector cells must be taken into account [40,41].

Ultimate calculation of copolymer molecular weight at each retention volume requires a correction of the refractometer response, according to W_S^i values. The best way is to apply universal calibration, but it requires, in the modern approach, an on-line additional viscometer [23]. Otherwise, classical calibration curves based on

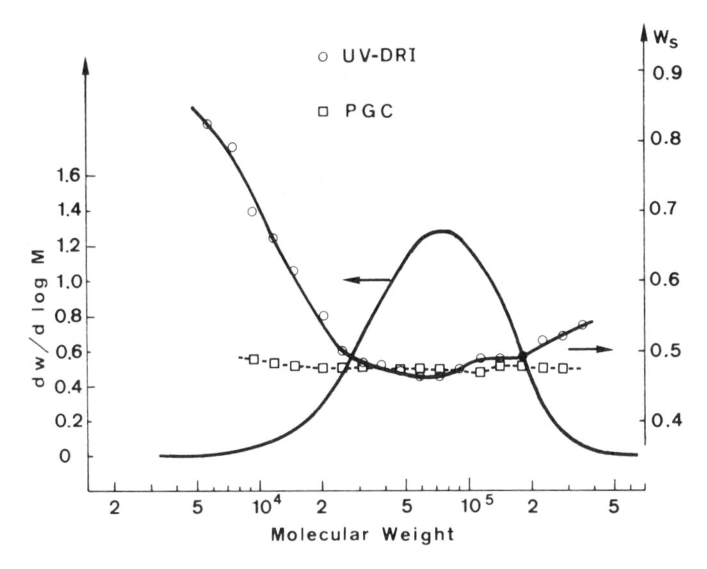

Figure 5.13. Molecular weight distribution and chemical composition of a styrene-methyl methacrylate copolymer sample, determined by refractometer-ultraviolet detector coupling (○). Chemical composition curve from pyrolysis gas chromatography (PGC, □) is also plotted. (From Ref. 98.)

the two homopolymer standards can be used [94,98] for each W_S^i value to convert the comonomer content into equivalent styrene units. Final chemical heterogeneity and MWD curves are given in Figure 5.13 for a random styrene-methyl methacrylate copolymer sample.

A particular problem involved in such determinations and recently pointed out [98] involves abnormal variations of W_S^i at the extreme parts of MWD curves, observed even for homopolymers and for copolymers for which uniform chemical composition were checked by pyrolysis gas chromatography [101] (see Figure 5.13). It seems that response factors, especially the UV response, are not always constant in the concentration range studied, and some interpretations and correction procedures have been proposed [98-102].

5.4.2. Refractometer/Infrared Dual Detection

This type of coupling has been performed simultaneously for chemical composition studies, particularly when comonomer units are insensitive to common UV detection (i.e., 254 nm). We must distinguish here between on-line IR detection [94,103-104], and off-line spectroscopic measurements of eluted solutes, sampled either in films [105] or in pressed pellets [106].

The on-line detector requires an appropriate solvent with regard to its infrared transparency; thus its use in series with a refractometer is not always possible in practice. Figure 5.14 illustrates on-line dual detection for an acrylate-methyl carbamate copolymer, where the IR detector is set at 1735 cm^{-1} to monitor the carbonyl band. As for the UV detection, chromatograms can be exploited to obtain both chemical compositions and MWDs, as shown in Figure 5.15 for a vinyl chloride-vinyl acetate copolymer sample.

In addition to on-line single-wavelength IR detection, another technique, developed by Mirabella et al. [107,108], is the *rapid stop-and-go SEC/IR method*, where the mobile-phase flow is stopped

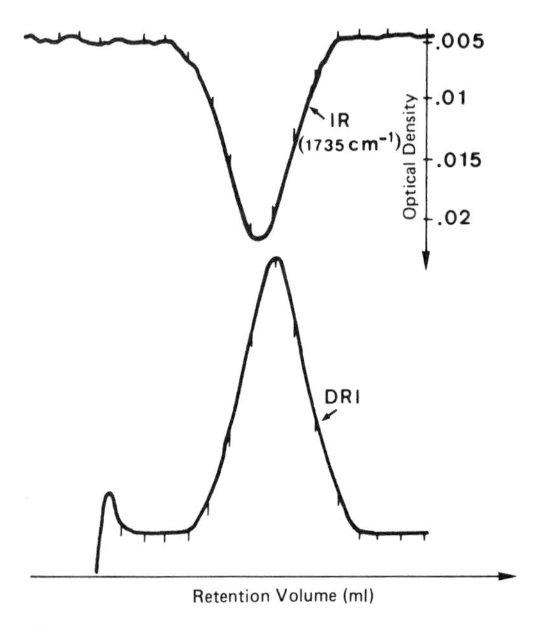

Figure 5.14. Infrared (IR) and refractometer (DRI) chromatograms of a methyl acrylate/methyl N-vinylcarbamate copolymer sample in THF. The dead volume between the two cells is indicated by the later appearance of the IR peak. (From Ref. 94.)

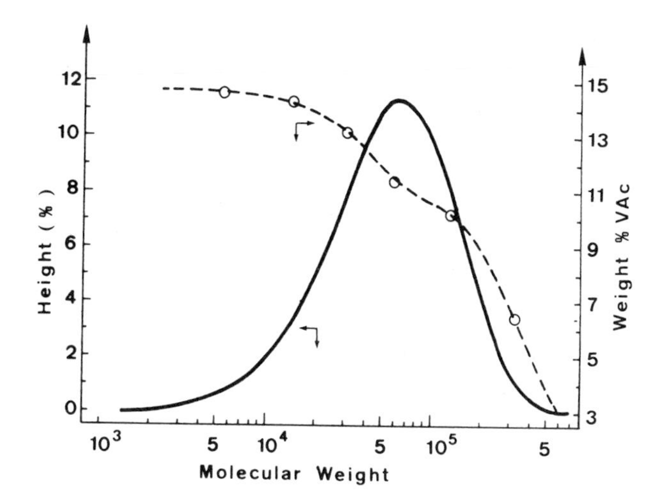

Figure 5.15. Molecular weight distribution and chemical composition of a vinyl chloride-vinyl acetate copolymer sample determined by refractometer-infrared detector coupling. (From Ref. 105.)

to record an IR spectrum over a given frequency range. Chemical composition data obtained for a styrene-vinyl stearate copolymer are found to be in good agreement with those determined from samples fractionated by preparative SEC [108]. We should also mention the use of a stop-flow, multiple internal reflectance infrared method for the SEC analysis of styrene-acrylonitrile copolymers [109].

5.5. CHAIN GROWTH AND DEGRADATION MECHANISMS

Because of its wide versatility, SEC has also been used as a tool to deduce, from soluble mixtures, information about polymerization, polycondensation, and degradation mechanisms. In this field, SEC is often combined with other techniques of liquid chromatography, such as liquid-liquid or reverse-phase chromatography, for which retention data may be complementary. An exhaustive review of polymer analysis by SEC was made by Janča [4], and many references dealing with chain growth and degradation investigations can be found. More recently, Hagnauer [5] reviewed the latest SEC applications in these areas. The study, by SEC, of chemical changes occurring during polymerization or polycondensation processes, storage, or degradation will be illustrated below by some examples.

5.5.1. Polymerization Kinetics

As SEC with modern packings can rapidly provide MW and MWD values, it is an irreplaceable method for following polymerization processes versus time. A first example is given by the recent SEC study of the solution polymerization of hexachlorocyclotriphosphazene [110].

$$m\ (\mathrm{NPCl_2})_3 \longrightarrow \left[\mathrm{N{=}PCl_2}\right]_n$$

The effect of polymerization conditions on polymer yield has been investigated, but little attention has been given to the characterization of the resulting polydichlorophosphazene. Figure 5.16 shows chromatograms obtained after 4 and 10 hr of polymerization. The SEC chromatogram at 10 hr has a similar retention time, but is somewhat broader than the one at 4 hr. By using polydichlorophosphazene samples as broad MWD standards for calibration, MW and polydispersity values can be deduced at different polymerization times. Even for

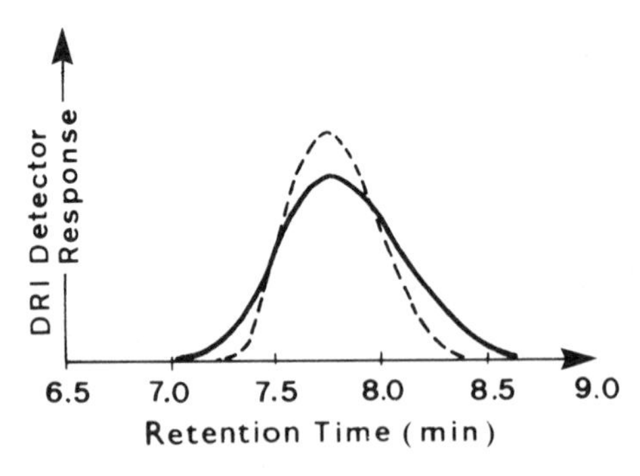

Figure 5.16. SEC chromatograms of solution polymerized polychlorophosphazene. ———, At 4 hr; -----, at 10 hr. Solvent: trichlorobenzene at 135°C. Columns: DuPont PSM Bimodal 60S + 1000S. (From Ref. 110.)

the sample at 10 hr, MWD was quite narrow, which led authors to suggest that polydichlorophosphazene appears to undergo a living polymerization [110].

The determination of chemical heterogeneity changes during polymerization processes can also be carried out by SEC. Figure 5.17 gives examples of refractometer and ultraviolet chromatograms for samples of α-methylstyrene/butadiene copolymer, obtained at various conversion rates. The use of 254-nm absorption detection

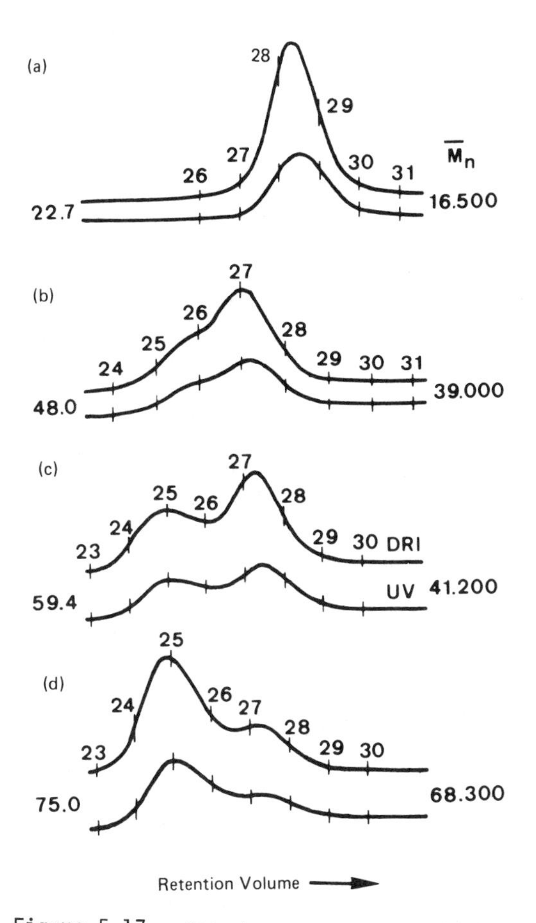

Figure 5.17. SEC chromatograms of α-methylstyrene/butadiene samples prepared from the same starting monomer mixture (75 mol % α-methylstyrene) at various conversions. Detection: refractometer and 254 nm UV. (From Ref. 96.)

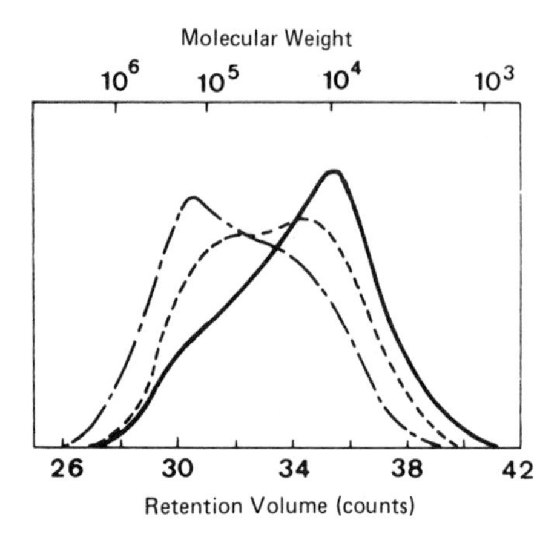

Figure 5.18. SEC chromatograms of three polyethylene samples obtained at various pressures (kg/cm^2): ———, 105; ----, 300; — - —, 395. Reaction conditions: γ-ray irradiation from ^{60}Co; temperature, 30°C. SEC solvent: trichlorobenzene at 130°C. (From Ref. 111.)

allows confirmation of chemical heterogeneity, particularly the occurrence of copolymer chains with long styrene sequences and high molecular weights (count 25), whereas the distribution is almost symmetrical before the consumption of butadiene (see curve a). Chemical heterogeneities can be calculated from SEC chromatograms.

SEC is advantageously used to detect multimodal MWDs and their dependence on experimental conditions. For example, Figure 5.18 shows chromatograms of polyethylene samples obtained by radiation-induced polymerization at various pressures, and illustrates the reliability of SEC in the detection of bimodal MWD that changes drastically. SEC data have been explained by the assumption that polymerization proceeds through the propagating chains in two different physical states: a loose state and a rigid one, where the chain mobility is different [111].

Additionally, methods for determining kinetic rate constants of polymerization from MWDs have been proposed (see, e.g., Refs. 112 and 113) and SEC can be useful to bring about agreement with theoretical MWD curves.

5.5.2. Polycondensation Kinetics

In the same way, SEC allows us to study the course of polycondensation reactions leading to linear structures. However, the most important contribution of SEC in this field is to solve partially problems involved in tridimensional polycondensation, particularly by identifying soluble species existing prior to the gelation of thermosets, and also their evolution versus time. Figure 5.19 shows, for example, the chemical change of an industrial epoxy resin, after 1 year of storage at ambient temperature, and the resulting growth of higher molecular weight species.

Phenol-formaldehyde and (melamine or urea)-formaldehyde systems have been particularly studied by SEC. The initial stage of condensation of formaldehyde (i.e., formation of methylolated phenol, melamine, or urea) and dimers is well known, since these components can be isolated as chemical individuals. The knowledge of the further course of polycondensation requires, for characterizing reaction mixtures, either the determination by SEC of average MW values versus time [115] or the resolution of higher oligomers with very efficient packings [116].

Figure 5.20 illustrates the identification, by SEC, of polynuclear methylol melamines occurring at the beginning of a melamine-formaldehyde polycondensation. Peaks A and B are mononuclear and dinuclear (methylene ether or methylene bridge) methylol melamines, respectively. Peaks C, D, and E are estimated to be tri-, tetra-, and pentanuclear methylol melamines. If SEC may be useful for the process or quality control of resins and for studying the thermal stability of oligomers, quantitative analysis requires precise calibrations which are not easy to perform in the oligomer field.

5.5.3. Chain Degradation

Accurate determinations of average MWs and MWDs by SEC can also be utilized for the elucidation of chain scission kinetics under various conditions of degradation: photo-oxidation [117], thermal treatment

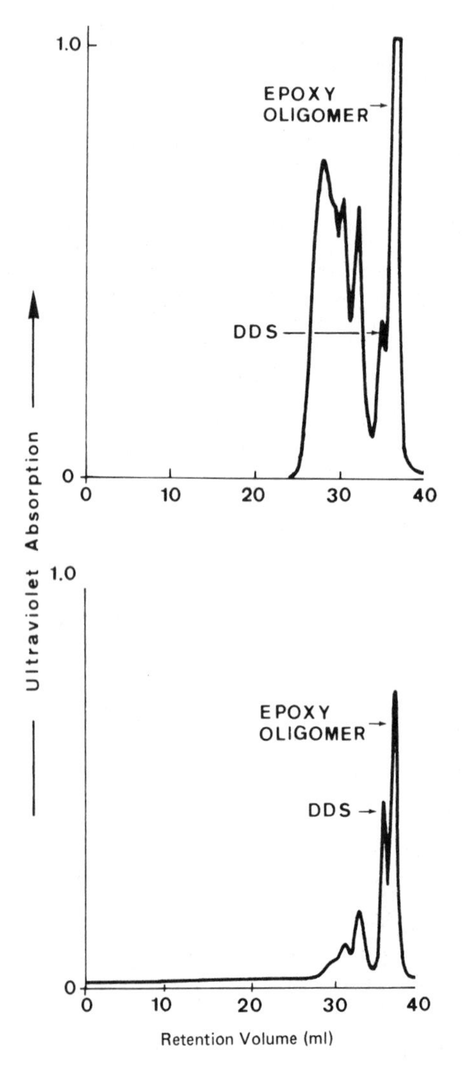

Figure 5.19. SEC chromatograms of an epoxy resin formulation based on tetraglycidyl-4,4'-methylenedianiline and diaminodiphenyl sulfone (DDS) as curing agent. Upper curve: analysis after 1 year of storage at 22°C. (From Ref. 114.)

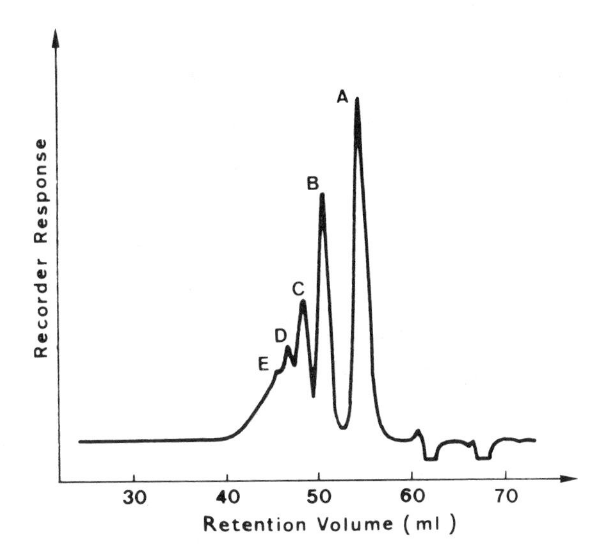

Figure 5.20. SEC chromatogram of a melamine-formaldehyde condensate sample in N,N-dimethylformamide. Sample concentration: 1%. Column: 25 cm x 8 mm inside diameter, packed with Shodex A 802. (From Ref. 116.)

[118-120], hydrolysis [121], mechanical shear [122-124], repetitive extrusion [125], freezing and thawing [125], and cryogenic crushing [126], for example. Furthermore, mechanochemical degradation of polymeric solutes may occur during the SEC process itself, and this has recently been investigated [127,128]. We will illustrate these chain degradation studies by SEC with some examples.

Thermal degradation of polymer chains has been analyzed from SEC data of the resulting samples. For example, Abbas [119] has thermally degraded bisphenol A polycarbonate in a rheometer barrel at various temperatures, and samples collected versus time have been characterized by SEC (see Table 5.9). Molecular weight decrease was observed, particularly at the highest degradation temperature (330°C). A random scission mechanism, often claimed in literature, would require that the heterogeneity index approaches 2, but this is not the case at 330°C. The comparison of the scissions per molecule values deduced from SEC data (see Table 5.9) with those calculated from the

Table 5.9. MW and Heterogeneity Index Values of a Bisphenol A Polycarbonate Sample (Makrolon 2805: $\overline{M}_n$ = 15,100, $\overline{M}_w$ = 33,000, and $\overline{M}_z$ = 53,100) After Thermal Degradation in a Rheometer Barrel[a]

Temperature (°C)	Degradation time (min)	$\overline{M}_n$	$\overline{M}_w$	$\overline{M}_z$	$\overline{M}_w / \overline{M}_n$	Scissions per mole: $\overline{DP}_n^0 / \overline{DP}_n - 1$
--	--	15,100	33,000	53,100	2.19	--
300	20	15,100	32,900	53,500	2.19	~0
	40	14,200	31,300	50,200	2.21	0.064
	60	14,000	30,700	50,100	2.18	0.075
	80	13,400	29,300	46,700	2.19	0.130
308	20	14,800	31,200	49,600	2.11	0.020
	40	13,600	29,500	46,800	2.16	0.106
	60	13,200	28,900	46,300	2.18	0.140
	80	12,100	27,100	44,200	2.24	0.247
315	20	13,800	30,100	49,000	2.19	0.095
	40	12,400	29,100	47,300	2.35	0.216
	60	11,900	26,400	42,400	2.22	0.269
	80	11,300	25,200	40,900	2.24	0.337
330	20	12,800	30,100	50,000	2.35	0.177
	40	11,100	26,700	44,500	2.40	0.357
	60	10,500	24,500	40,400	2.34	0.440
	80	10,200	23,500	39,400	2.31	0.486

[a]SEC solvent: methylene chloride at 25°C; SEC data treatment: universal calibration.
Source: Ref. 119, by permission of the publishers, Butterworth & Co. Ltd. ©.

well-known Scott's relationship corroborates that the chain fracture did not take place randomly.

Since mechanical shear degradation is often encountered in plastics conditioning processes, SEC has significantly contributed to the study of chain scissions in laminar or turbulent flows. Figure 5.21 shows the change of MWD of a polystyrene sample after its solution was subjected to a high extensional flow field. To find the location of breakage on polymer chains, the experimental MWD is compared with those calculated according to several scission assumptions: random, midpoint, or 66% midpoint/33% quarter-point breakages. In the cases

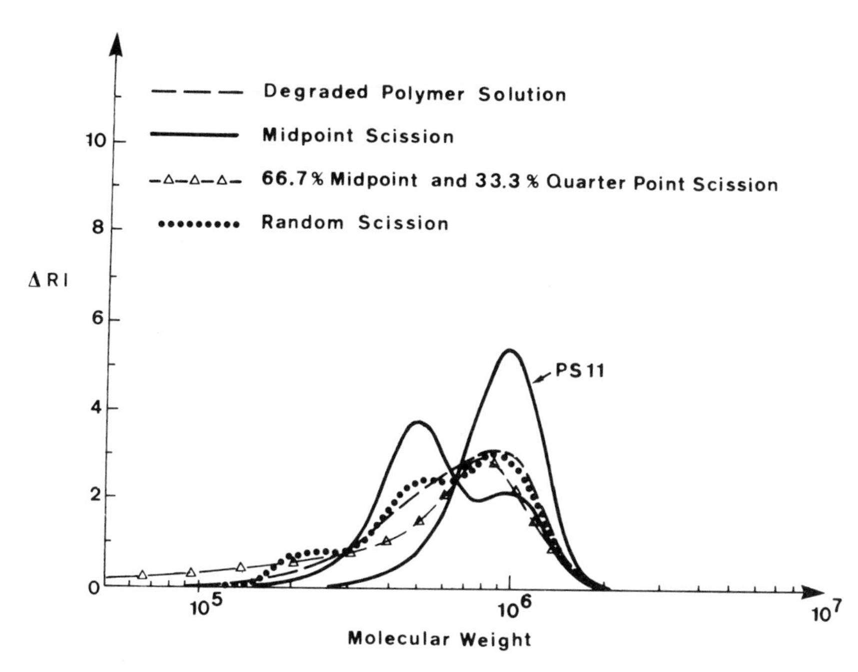

Figure 5.21. SEC chromatograms of a solution of a polystyrene sample ($\overline{M}_w$ = 800,000; $\overline{M}_w/\overline{M}_n$ = 1.14) before (PS11) and after degradation in laminar flow (average flow velocity, 11,850 cm/sec) compared with some calculated MWDs. Solvent: toluene. (From Ref. 122.)

studied, the chromatograms corresponding to the latter hypothesis fit those of the degraded solution best. The authors have concluded that as scission does not take place exclusively at midpoint, polymer chains are probably not extended fully when they are degraded [122].

It is more and more accepted now that shear degradation can also occur during SEC experiments [127-130], depending on both flow rate and sample molecular weight. Asserting such a degradation phenomenon in SEC is not very obvious, since sample dissolution may bring chain breakage, for instance when a ultrasonic bath is used. The most credible results concerning this effect have recently been obtained by Rooney and Ver Strate [128], who used a LALLSP detector, allowing absolute $\overline{M}_w$ determinations across the chromatogram. Chain degradation is absent without columns, but with typical microparticulate packings, above a given $\overline{M}_w$ varying with the polymer type, $\overline{M}_w$ becomes flow rate dependent, down to at least 0.3 cc/min. For example, a typical $\overline{M}_w$ at which degradation (at 0.5 cc/min) becomes severe is about 700,000 for polyethylene [128]. Consequently, researchers should make efforts to demonstrate that degradation does not take place in the SEC results they present.

5.6. POLYMER SOLUTION PROPERTIES

We present in this section other applications of SEC, concerning, specifically, the study of dimensional or interactional properties of polymer chains in solution: molecular dimensions, aggregates or gel-like microparticles, and polymer-(solvent or ligand) and polymer-polymer interactions. We will not discuss interactions involving stationary phases, which are developed in Chapter 4.

5.6.1. Unperturbed Molecular Dimensions

As SEC is now a very accurate method for determining MW and intrinsic viscosity of any chromatographic fraction through universal calibration, it has been used to measure unperturbed molecular dimensions (i.e., in theta conditions) of polymer solutes from SEC experiments (see, e.g., Refs. 45 and 131). It is not always possible to find a theta solvent for a given polymer, and several classical procedures have been proposed for evaluating unperturbed molecular dimensions from intrinsic viscosity and MW measurements in good solvents [132]. The best known are those of Flory and Fox [133], and of Burchard [134], Stockmayer and Fixman [135], respectively:

$$\frac{[\eta]^{2/3}}{M^{1/3}} = K_\theta^{2/3} + 2C_M\psi^*\left(1 - \frac{\theta}{T}\right)K_\theta^{5/3}\,\frac{M}{[\eta]} \tag{5.44}$$

$$\frac{[\eta]}{M^{1/2}} = K_\theta + 0.51\Phi_0\beta_c M^{1/2} \tag{5.45}$$

with K_θ defined by equation (2.55), where Φ_0 is the Flory's universal constant, ψ^* the classical entropic parameter, and β_c the binary cluster integral. $\langle r_0^2\rangle$ is therefore determined by plotting the quantities above versus either $M/[\eta]$ [Equation (5.44)] or $M^{1/2}$ [equation

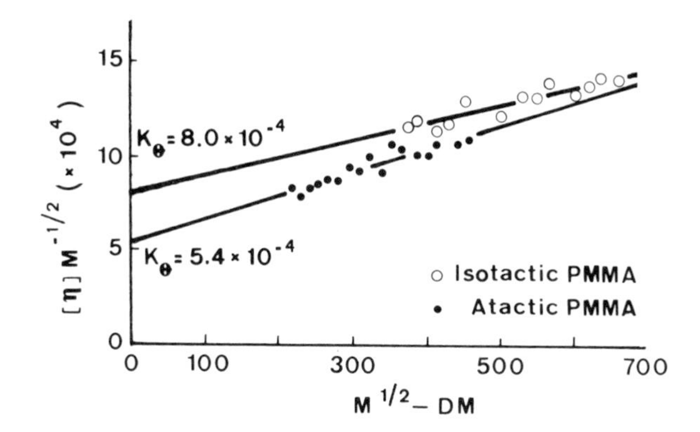

Figure 5.22. Graphical determination of K_θ values of isotactic and atactic PMMA samples, from SEC data obtained through on-line refractometer/LALLSP detection. Solvent: THF. (From Ref. 45.)

(5.45)] and by measuring intercepts related to K_θ. This general method requires several narrow molecular weight fractions of the polymer to be analyzed.

Jenkins and Porter have recently reported [45] a very interesting procedure to determine, by SEC, root-mean-square end-to-end distance from a single broad MWD sample. By using an on-line DRI/LALLSP detection and universal calibration, a series of $([\eta]_i, \bar{M}_{w_i})$ pairs can be obtained for the chromatograms, allowing the plotting of any graphical procedure leading to K_θ and, consequently, to $\langle r_0^2 \rangle$. Accurate measurements have been performed by taking into account the real universal calibration parameter $[\eta]\bar{M}_n$ [136] and axial dispersion band broadening.

Figure 5.22 illustrates determinations of K_θ for isotactic and atactic poly(methyl methacrylate) samples, through the relationship proposed by Dondos and Benoit [137]:

$$\frac{[\eta]}{M^{1/2}} = K_\theta + 0.51\Phi_0\beta_c(M^{1/2} - DM) \tag{5.46}$$

D being related to the Mark-Houwink exponent by $D = 12 \times 10^{-4}$ $(a - 0.5)$. The K_θ values deduced for these two samples and given in Table 5.10 are in excellent agreement with those obtained

Table 5.10. Comparison of K_θ Values Obtained by Direct Measurements in Theta Solvent by Classical Procedures in Good Solvent and by SEC/LALLSP Coupling

PMMA	$K_\theta \times 10^4$				
	Direct Measurement		Fox-Flory	Stockmayer-Fixman	SEC/LALLSP
77% syndiotactic	5.7	5.0	4.9	4.6	5.4
98% isotactic	8.7	7.7	7.0	8.2	8.0

Source: Ref. 45.

classically either by direct measurement in theta solvent or by graphical procedures in a good solvent [45].

5.6.2. Aggregates or Gel-Like Microparticles

Because of its fractionation due to differences in hydrodynamic volumes, SEC may detect large but soluble species present in the injected solution, particularly when a LALLSP detector is used. Such particles can be aggregates resulting from incomplete dissolution of samples, or gel-like microparticles (microgels) arising from slight cross-linking of chains in treelike structures.

Figure 5.23 gives a good illustration of the contribution of the SEC/LALLSP coupling to the detection of aggregates or microgels. Chromatograms (a) are related to an EPM rubber sample free of microgel, whereas chromatograms (b), corresponding to an EPDM sample, drastically exhibit different results at smaller retention volumes, due to the occurrence of microgels. Generally, microgels give a strong LALLSP response at about the exclusion limit and their effect can be reduced by filtration through finer filters [46].

Aggregation of poly(vinyl chloride) in THF solutions has been extensively studied by SEC [105,138-145], since it causes anomalies during precipitation fractionation with water as the nonsolvent. In SEC chromatograms, the occurrence of aggregates in PVC solutions appears sometimes as a tail or a shoulder at small retention volumes, sometimes as a second, high molecular weight peak, or gives the

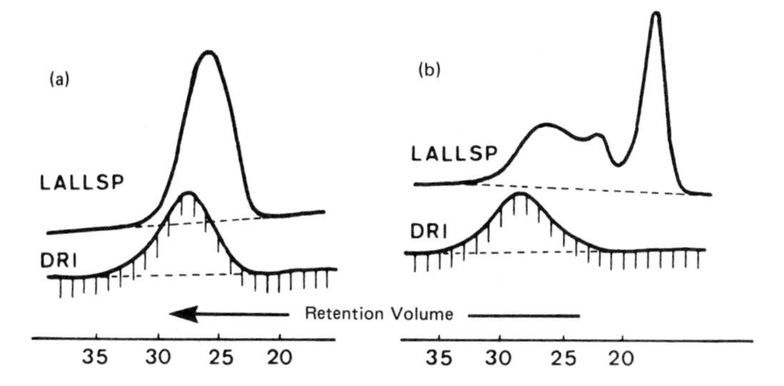

Figure 5.23. Refractometer (DRI) and LALLSP chromatograms of (a) *EPM*: ethylene-propylene (41 wt %) copolymer; (b) *EPDM*: ethylene/propylene (34 wt %)/5-ethylidene-2-norbornene (5.1 wt %) terpolymer. Solvent: trichlorobenzene at 140°C. (From Ref. 47, by permission of the publishers, Hüthig & Wepf Verlag, Basel.)

appearance of a bimodal distribution [144]. Aggregates generally disintegrate into single molecules when solutions are heated (typically, 85°C under N_2, for 4 hr). With optimum experimental conditions, the ratio of aggregation can be evaluated quantitatively [140]. In addition, as it has been shown that aggregation in PVC solutions is directly related to stereoregularity, SEC has been used successfully to determine percentages of syndiotacticity [105,141, 142,145], also measured by NMR or IR spectroscopy. This method leads to satisfactory agreement and provides a nonconventional application of SEC.

5.6.3. Preferential Interactions

Preferential solvation generally occurs when a polymer sample is dissolved in a binary solvent mixture. At a molecular level, better solvation by one of the solvents renders the solvent composition within the chain coils, different from that outside (bulk solvent). Phase separation or particular viscometric properties in ternary solvent 1/solvent 2/polymer 3 systems may be explained by a preferential solvation phenomenon [146]. Quantitative measurements of preferential solvation are classically carried out mainly by light scattering [147] or dialysis equilibrium [148-150]. In such ternary systems, preferential solvation is characterized by a preferential adsorption coefficient

$$\lambda_1' = \frac{\phi_1 - \phi_{b1}}{c_3} = \frac{\Delta\phi_1}{c_3} \tag{5.47}$$

defined as the excess of volume fraction of solvent 1 in ternary microphases (ϕ_1) with regard to that of bulk solvent (ϕ_{b1}) per polymer concentration unit.

SEC has been used more recently for quantitative evaluation of λ_1' [150-156] and has become a suitable and rapid technique for preferential solvation studies. In a typical SEC experiment, a ternary mixture of a given composition is injected into mobile phase consisting of the same binary solvent. If there is no preferential solvation, only the classical polymer peak is observed. In the case of preferential solvation, SEC columns separate the solvated polymer coils from the bulk solvent that remains. As this binary solvent has no longer the same composition as the eluant, a "negative" peak appears and its magnitude depends on the difference between the binary solvent compositions (i.e., on the extent of preferential solvation). Figure 5.24 gives a typical preferential solvation chromatogram, obtained for polyvinylpyrrolidone in a 1,2-dichloroethane/ethanol mixture; the so-called "ghost peak" (or "vacant peak") is located at V_R = 25 ml, while the solvated polymer is, in this experiment, totally excluded. The λ_1' value can be calculated by relating the excess volume $\Delta\phi_1$ to the solvent peak height, through a calibration obtained by injecting solvent binary mixtures with compositions close to the mobile-phase composition. Accurate determinations of preferential solvation parameter do require very dry solvents [155,157], since traces of water disturb the phenomenon and produce another vacant peak (see V_R = 20.9 ml in Figure 5.24) due to the hydration of one of the solvent.

SEC has also been applied recently to the study of the binding of ligand to biological macromolecules, such as serum albumin [158, 159]. In that field, SEC has been employed through frontal analysis

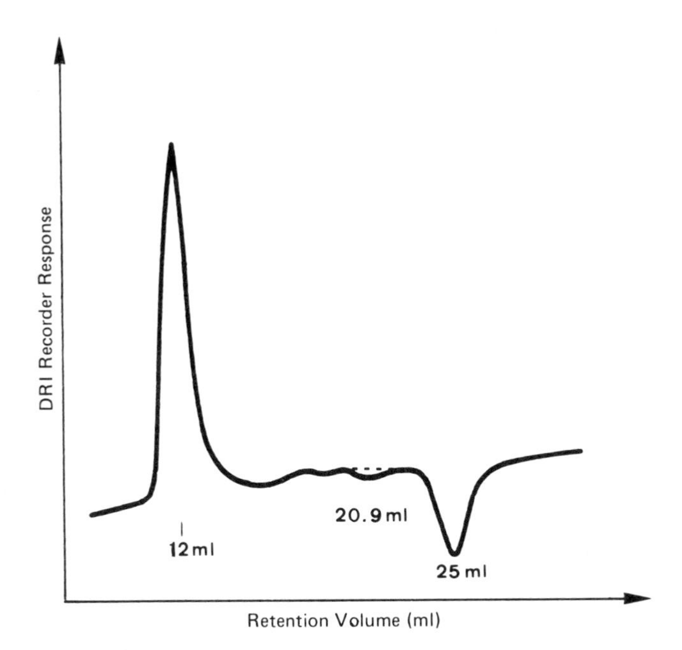

Figure 5.24. SEC chromatogram of a ternary polyvinylpyrrolidone ($\bar{M}_V$ = 700,000)/1,2-dichloroethane/ethanol (75 vol %) solution eluted by the same binary solvent mixture. Polymer concentration: 4.68 mg/ml. Column: 60 cm x 7 mm inside diameter; Styragel 10 Å; particle size: 15-25 μm. Flow rate: 2 ml/min. (From Ref. 155.)

[158,159]; a large sample of ligand-protein mixture is continuously applied to a SEC column in order to achieve a steady-state concentration on the column, and is then eluted by a buffered eluant. Figure 5.25 shows three zones in the chromatogram: the first (α) corresponds to the free protein; the second (β), to the mixture of ligand-protein complex and excess ligand; and the third (γ), to the free ligand. The third plateau height provides a direct evaluation of the free ligand concentration in the sample and leads to the calculation of the mean number of moles of ligand bound per protein molecule [158, 159].

5.6.4. Polymer Mixtures

Preferential solvation study by SEC has been logically extended to solvent 1/polymer 2/polymer 3 mixtures, by replacing one of the

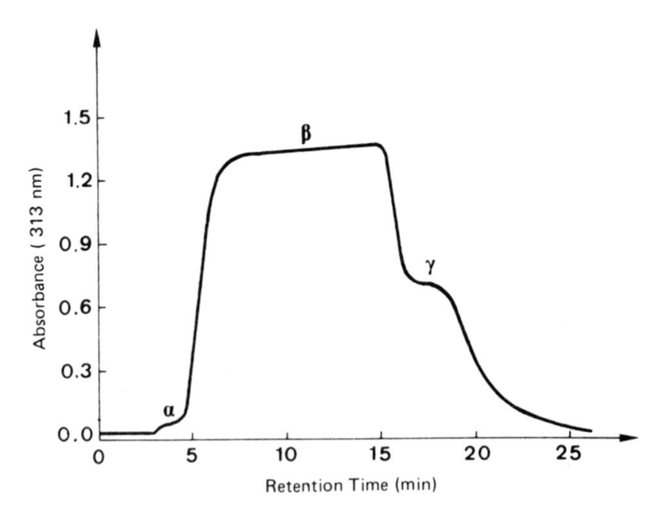

Figure 5.25. SEC chromatogram in frontal analysis of a mixture (18 ml) of warfarine (10^{-4} M) and human serum albumin (2 g/liter). Eluant: phosphate buffer, pH 7.4 (0.067 M). Flow rate: 1.5 ml/min. Temperature: 37°C. Column: 30 cm x 3.9 mm inside diameter, μ-Bondagel 125 Å. Detection: ultraviolet. (From Ref. 158.)

solvents by a polymer species [160]. No experimental complexities arise with the use of a polymer as component 2, when its molecular weight is low. Table 5.11 illustrates the good agreement between λ' values obtained by SEC and dialysis equilibrium, for benzene-low MW poly(methyl siloxane)-polystyrene mixtures.

However, anomalies occur when mixtures of polymers are analyzed in a common SEC solvent (or when a polymer is eluted by another polymer solution as mobile phase) [161-163]; peak retention volumes of both polymer solutes are shifted to higher retention volumes, with respect to their values at their nominal concentrations in the same solvent (see Figure 5.26). It can be predicted from theories concerning the behavior of macromolecules in a thermodynamically good solvent that the addition of a chemically different macromolecule to a solution of another polymer would lead to a decrease of both hydrodynamic volumes, because of the general phenomenon of polymer incompatibility. Thus SEC is a good method for the qualitative study of polymer incompatibility [161-163], particularly its dependence on

Table 5.11. Determination of Benzene Preferential Adsorption Coefficient λ' for Benzene (1)/ Poly(dimethyl siloxane) (2)/Polystyrene (3) from SEC and Dialysis Equilibrium Measurements

M_3	M_2	ϕ_1[a]	$c_2 \times 10^2$ (ml/ml)	$\Delta\phi_1 \times 10^2$[b]	h_1[c] (mm)	$c_3 \times 10^2$ (g/ml)	h_2[c] (mm)	λ'_{SEC} (ml/g)	λ'_{dia} (ml/g)
270,000	8700	0.940	0.203	0.191	52	0.190	19	0.36_6	0.33_7
						0.391	38	0.35_8	
						0.580	58	0.36_7	
135,000	1850	0.900	0.222	0.199	53	0.350	39	0.41_8	0.43_7
						0.520	59	0.42_6	
135,000	1100	0.890	0.213	0.189	52	0.200	20	0.36_3	0.32_6
						0.390	37	0.34_5	
						0.620	60	0.35_1	

[a] ϕ_1, benzene volume fraction.
[b] $\Delta\phi_1$, excess volume fraction of benzene.
[c] Peak heights, h_1 and h_2, are related to PDMS concentrations C_2 in binary PDMS-benzene mixtures and to preferential solvation in ternary mixtures, respectively.
Source: Ref. 160.

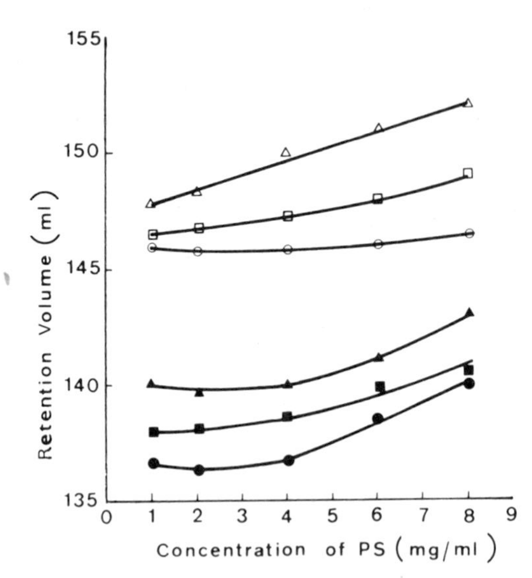

Figure 5.26. Effect of concentration of polybutadiene (PBD) on retention volumes of polystyrene (PS samples of different concentrations in tetrahydrofuran). ○, PS 110,000; □, PS 110,000 + 2 mg/ml of PBD 170,000; △, PS 110,000 + 4 mg/ml of PBD 170,000; ●, PS 390,000; ■, PS 390,000 + 2 mg/ml of PBD 170,000; ▲, PS 390,000 + 4 mg/ml of PBD 170,000. (From Ref. 162.)

system parameters such as polymer molecular weight, solvent nature, or interaction parameters. Efforts have now to be devoted to quantitative treatments (as in Ref. 163) on the shrinkage of polymer coils in such SEC experiments.

We should also mention the use of SEC for analyzing phase equilibria of incompatible polymer-polymer-solvent systems [164-166]. By sampling each of the coexisting phases in equilibrium, SEC data allow us to determine tie lines in ternary phase diagrams and, consequently, to define binodals. Nevertheless, peak shifts should be kept in mind when analyzing concentrated polymer mixtures, and dilution of phases must be made to avoid them.

5.7. OTHER APPLICATIONS

The inventory of the widespread use of SEC would not be complete if other important applications were not mentioned. They concern, mainly, the SEC study of particle size, such as that of polymer latex; of pore size of packings; and the analysis of the additives that can be found in polymers.

5.7.1. Particle Size Analysis

Chromatographic separation of colloidal particles on packed beds is based on hydrodynamic chromatography [167]. When colloidal particles are eluted on a nonporous and spherical bead packing, their rate of transport depends on the size of the colloid, the particle size of the packing, and other factors, such as composition and flow rate of the eluant. The rate of transport of a colloid or suspension is generally expressed by a R_F number, which is defined as the ratio of the retention volume of the particle to that of a low molecular weight marker and, as a result, is independent of column dimensions.

Figure 5.27 illustrates the dependence of R_F values of polystyrene latexes on particle diameter of nonporous packed beds.

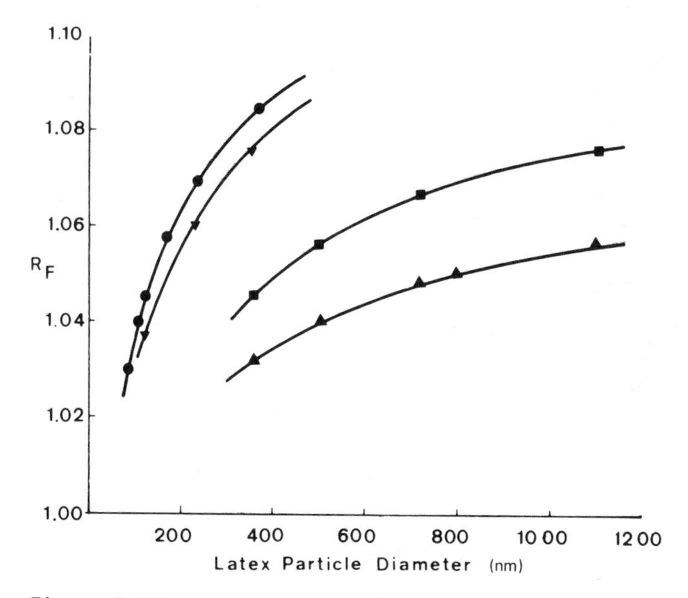

Figure 5.27. Variations of R_F of some polystyrene latexes as function of their size and of the packing bead diameter. ●, ∿18 µm; ■, ∿40 µm; ▲, ∿58 µm; ▼, ∿20 µm. (From Ref. 168.)

Typically, R_F values are always greater than unity; that is, latex particles are moving through the bed more rapidly than the eluant, and increase with increasing latex diameter. This kind of particle separation, reviewed by Small et al. [168] and more recently by Hamielec et al. [169], is explained by considering interstitial spaces as capillaries. The center of each particle is excluded from the slowest eluant streamlines closest to capillary walls, more especially as its size is large. Consequently, the particles move through interstitial spaces with a velocity that exceeds the mean velocity of the eluant by a factor that increases with increasing ratio of particle to capillary sizes (see Figure 5.27).

This fractionation mechanism obviously takes place in the pores of SEC packings, and Hamielec et al. [170,171] have demonstrated, both theoretically and experimentally, the SEC power for the analysis of dilute suspensions of submicrometer particles. These authors have established, at low ionic strength, a universal calibration between particle diameters and retention volumes of polystyrene standard latexes, as shown in Figure 5.28. Specific studies have been recently made [170-176] on the effects of eluant ionic strength, packing porosity, flow rate, axial dispersion, and on the improvement of the particle detection, generally performed by turbidimetry. Latexes of polystyrene, PMMA, poly(vinyl acetate), styrene-acrylic acid, and butadiene-acrylonitrile copolymers and suspensions of silica particles, for example, can be investigated by SEC, which is now a very effective and efficient method for particle size analysis, but also for particle growth kinetics studies.

5.7.2. Pore Size Analysis

As retention volumes of given solutes are dependent on the pore size distribution of the packing used, their measurement may bring information concerning the texture of gel beads. This is the principle of *inverse SEC*, first defined by Freeman et al. [177], and based on the experimental determination of solute distribution coefficients K_{SEC}, defined by equation (1.9). To correlate K_{SEC} and mean pore

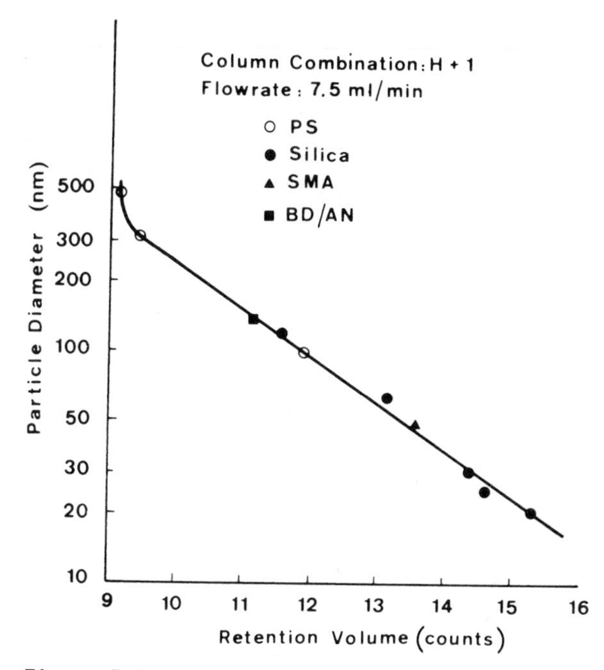

Figure 5.28. Universal particle diameter calibration for polymer latexes and silica suspensions. *PS*, Polystyrene; *SMA*, styrene-methacrylic acid copolymer; *BD/AN*, butadiene-acrylonitrile copolymer. Column combination (H + I): CPG 1500 (200-400 mesh; 1500 Å pore size) and CPG 2500 (200-400 mesh; 2500 Å pore size). Eluant: water + 1 g/liter aerosol OT as surfactant + 1 g/liter KNO_3 as electrolyte. Detector: UV-254 nm. (From Ref. 170.)

diameter $\bar{r}$ values, a theoretical expression for K_{SEC} must be used and the simplest is the Giddings expression [178] (see Chapter 1). A geometrical model for pores must then be assumed; the corresponding equations can be found in Chapter 1.

Pore size determinations have been carried out by inverse SEC on controlled-porosity glass packings [177], styrene-divinylbenzene [177,179,180], isoporous styrene [181], and dextran [182] gels, for example. Results are generally in rather good agreement with those obtained by other techniques (gas adsorption or mercury intrusion) for rigid and, to a lesser extent, semirigid packings. However, pore size data proposed for swollen gels must be considered with great suspicion, since chromatographic elutions may also be governed by nonexclusion effects, such as solute adsorption onto the gel surface and solute dissolution inside the gel (see Chapter 4). These additional mechanisms may lead to incorrect K_{SEC} values and, consequently, to apparent pore size values, independently of the assumed pore model.

5.7.3. Polymer Additive Analysis

As low-porosity packings can be extremely effective for the separation of low molecular weight compounds, analytical SEC has been successfully applied to qualitative and quantitative determinations of organic additives in polymeric materials. Analyses have been mainly devoted to plasticizers [183-189], particularly phthalates, but also to antioxidants [190-193] and other additives [193-194].

The use of high-resolution micropackings (i.e., 5-10 μm) has considerably improved the efficiency of separations. Figure 5.29 shows, for example, the separation of four phthalate esters frequently used as plasticizers of poly(vinyl chloride). Traditionally, phthalate plasticizers have been identified by gas chromatography, but with the recent advances in column packings, SEC is advantageous, since it avoids any possible ester degradation. Figure 5.30 also

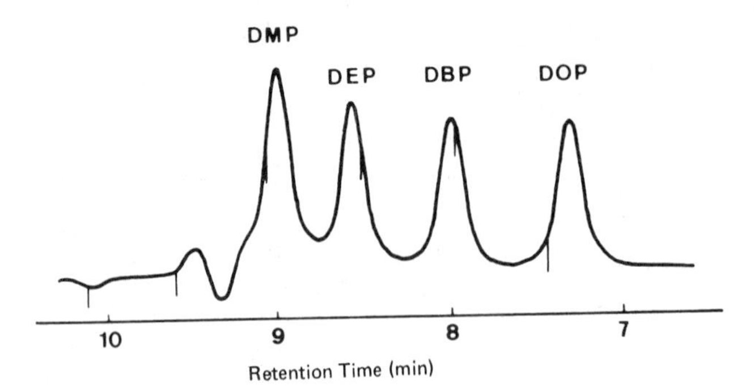

Figure 5.29. SEC chromatogram of a mixture of dioctyl (DOP), dibutyl (DBP), diethyl (DEP), and dimethyl (DMP) terephthalates. Column: TSK-GEL, type H, 66 cm x 8 mm, packed with polystyrene gel (5 μm, particle size; 100 Å, nominal porosity). Solvent: THF. Flow rate: 2 ml/mn. (From Ref. 186.)

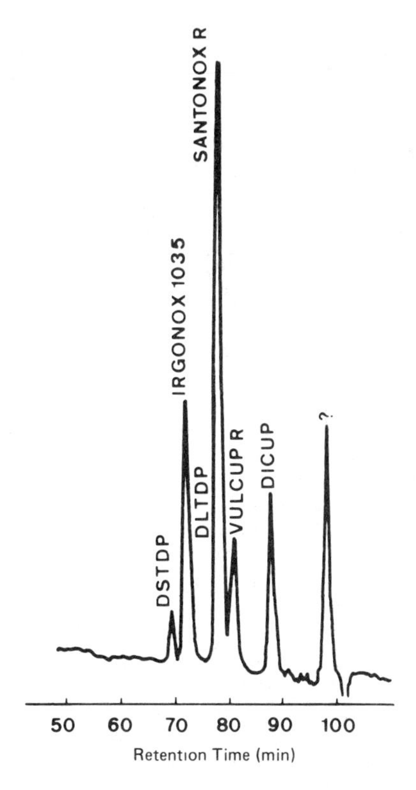

Figure 5.30. SEC chromatogram of polyethylene additives. Columns: 50 cm x 8 mm MicroPak TSK 3000H, 50 cm x 8 mm MicroPak TSK 2000H, and 80 cm x 8 mm MicroPak TSK 1000H. Solvent: THF. Flow rate: 0.5 ml/min. Detection: UV (215 nm). DSTDP, distearylthiodipropionate; DLTDP, dilaurylthiodipropionate; Irganox 1035, thiodiethylenebis(3,5-di-tert-butyl-4-hydroxy)hydrocinnamate; Santonox R, di(2-methyl-4-hydroxy-5-tert-butyl)phenyl sulfide; Vulcup R, α-α'-(bis-tert-butylperoxy)diisopropylbenzene; Dicup, Dicumylperoxide. An unknown peak is detected, probably due to a decomposition product from the peroxide initiators. (From Ref. 193.)

illustrates the SEC reliability in the additive field, with the separation of six common antioxidants (DSTDP, DLTDP, Irganox, and Santonox) and initiators (Vulcup and Dicup) found in commercial polyethylenes [193]. In addition, we can mention the use of the stop-and-go SEC/IR method (see Section 5.4), which has been extended to the quantitative analysis of both polymeric and additive portions of samples [187].

REFERENCES

1. Moore, J. C., J. Polym. Sci., A, *2*, 825, 1964.
2. Cobler, J. G., and Chow, C. D., Anal. Chem., *51*, 287R, 1979.
3. Cobler, J. G., and Chow, C. D., Anal. Chem., *53*, 273R, 1981.
4. Janča, J., J. Liq. Chromatogr., *4*(Suppl. 1), 1, 1981.
5. Hagnauer, G. L., Anal. Chem., *54*, 265R, 1982.
6. Yau, W. W., Kirkland, J. J., and Bly, K. K., *Modern Size-Exclusion Chromatography,* Wiley-Interscience, New York, 1979, Chaps. 10 and 12.
7. Gilding, D. K., Reed, A. M., and Askill, I. N., Polymer, *22*, 505, 1981.
8. Janča, J., Adv. Chromatogr., *19*, 37, 1981.
9. Yau, W. W., Kirkland, J. J., and Bly, K. K., *Modern Size-Exclusion Chromatography,* Wiley-Interscience, New York, 1979, p. 315.
10. Lesec, J., in *Liquid Chromatography of Polymers and Related Materials II* (Chromatographic Science Series, Vol. 13, J. Cazes and X. Delamare, eds.), Marcel Dekker, New York, 1980, p. 1.
11. Bly, D. D., J. Polym. Sci., A1, *6*, 2085, 1968.
12. Hellman, M. Y., in *Liquid Chromatography of Polymers and Related Materials* (Chromatographic Science Series, Vol. 8, J. Cazes, ed.), Marcel Dekker, New York, 1977, p. 29.
13. Grubisic, Z., Rempp, P., and Benoit, H., J. Polym. Sci., B, *5*, 753, 1967.
14. Pannell, J., Polymer, *13*, 277, 1972.
15. Ambler, M. R., J. Appl. Polym. Sci., *21*, 1655, 1977.
16. Lecacheux, D., Lesec, J., and Quivoron, C., J. Liq. Chromatogr., *5*, 217, 1982.
17. Weiss, A. R., and Cohn-Ginsberg, E., J. Polym. Sci., B, *7*, 379, 1969.
18. Busnel, J. P., Orvoen, J. B., and Bruneau, C. M., Analusis, *6*, 255, 1978.
19. Samay, G., Kubin, M., and Podesva, J., Angew. Makromol. Chem., *72*, 185, 1978.
20. Gallot, Z., Marais, L., and Benoit, H., J. Chromatogr., *83*, 363, 1973.
21. Kratky, O., Leopold, H., and Stalinger, H., Angew, Phys., *27*, 273, 1969.
22. Francois, J., Jacob, M., Grubisic-Gallot, Z., and Benoit, H., J. Appl. Polym. Sci., *22*, 1159, 1978.
23. Gallot, Z., in *Liquid Chromatography of Polymers and Related Materials II* (Chromatographic Science Series, Vol. 13, J. Cazes and X. Delamare, eds.), Marcel Dekker, New York, 1980, p. 113.
24. Majors, R. E., Barth, H. G., and Lochmuller, C. H., Anal. Chem., *54*, 323R, 1982.
25. Ouano, A. C., J. Polym. Sci., A1, *10*, 2169, 1972.
26. Ouano, A. C., and Kaye, W., J. Polym. Sci., Polym. Chem. Ed., *12*, 1151, 1974.
27. Meyerhoff, G., Makromol. Chem., *118*, 265, 1968.
28. Goedhart, D., and Opschoor, A., J. Polym. Sci., A2, *8*, 1227, 1970.

29. Meyerhoff, G., Sep. Sci., *6*, 239, 1971.
30. Grubisic-Gallot, Z., Picot, M., Gramain, P., and Benoit, H., J. Appl. Polym. Sci., *16*, 2931, 1972.
31. Park, W. S., and Graessley, W. W., J. Polym. Sci., Polym. Phys. Ed., *15*, 71, 1977.
32. Constantin, D., Eur. Polym. J., *13*, 907, 1977.
33. Janča, J., and Kolinsky, M., J. Chromatogr., *132*, 187, 1977.
34. Janča, J., and Pokorný, S., J. Chromatogr., *134*, 273, 1977.
35. Ouano, A. C., Horne, D. L., and Gregges, A. R., J. Polym. Sci., Polym. Phys. Ed., *12*, 307, 1974.
36. Peyrouset, A., and Prechner, R., French Patent 2,268,262 (1974).
37. Lesec, J., and Quivoron, C., Analusis, *4*, 456, 1976.
38. Letot, L., Lesec, J., and Quivoron, C., J. Liq. Chromatogr., *3*, 427, 1980.
39. Ouano, A. C., Rubber Chem. Technol., *54*, 535, 1981.
40. Bressau, R., in *Liquid Chromatography of Polymers and Related Materials II* (Chromatographic Science Series, Vol. 13, J. Cazes and X. Delamare, eds.), Marcel Dekker, New York, 1980, p. 73.
41. Lecacheux, D., and Lesec, J., J. Liq. Chromatogr., *5*, 2227, 1982.
42. Ouano, A. C., J. Chromatogr., *118*, 303, 1976.
43. Jordan, R. C., J. Liq. Chromatogr., *3*, 439, 1980.
44. Ouano, A. C., Dawson, B. L., and Jonhson, D. E., in *Liquid Chromatography of Polymers and Related Materials* (Chromatographic Science Series, Vol. 8, J. Cazes, ed.), Marcel Dekker, New York, 1977, p. 1.
45. Jenkins, R., and Porter, R. S., J. Polym. Sci., Polym. Lett. Ed., *18*, 743, 1980.
46. Cunningham, A. F., Heathcote, C., Hillman, D. E., and Paul, J. I., in *Liquid Chromatography of Polymers and Related Materials II* (Chromatographic Science Series, Vol. 13, J. Cazes and X. Delamare, eds.), Marcel Dekker, New York, 1980, p. 173.
47. Scholtens, B. J. R., and Welzen, T. L., Makromol. Chem., *182*, 269, 1981.
48. Kim, C. J., Hamielec, A. E., and Benedek, A., J. Liq. Chromatogr., *5*, 425, 1982.
49. Hamielec, A. E., Ouano, A. C., and Nebenzahl, L. L., J. Liq. Chromatogr., *1*, 527, 1978.
50. McRury, T. B., and McConnell, M. L., J. Appl. Polym. Sci., *24*, 651, 1979.
51. Axelson, D. E., and Knapp, W. C., J. Appl. Polym. Sci., *25*, 119, 1980.
52. Quivoron, C., in *Chimie Macromoleculaire II* (G. Champetier, ed.), Hermann, Paris, 1972, p. 163.
53. Letot, L., Lesec, J., and Quivoron, C., J. Liq. Chromatogr., *3*, 1637, 1980.
54. Small, P. A., Adv. Polym. Sci., *18*, 1, 1975.
55. Drott, E. E., in *Liquid Chromatography of Polymers and Related Materials* (Chromatographic Science Series, Vol. 8, J. Cazes, ed.), Marcel Dekker, New York, 1977, p. 161.

56. Miller, M. L., in *The Structure of Polymers*, Reinhold, New York, 1966, p. 124.
57. Stockmayer, W. H., and Fixman, M., Ann. N.Y. Acad. Sci., *57*, 334, 1953.
58. Zimm, B. H., and Stockmayer, W. H., J. Chem. Phys., *17*, 1301, 1949.
59. Zimm, B. H., and Kilb, R. W., J. Polym. Sci., *37*, 19, 1959.
60. Thurmond, C. D., and Zimm, B. H., J. Polym. Sci., *8*, 477, 1952.
61. Morton, M., Helminiak, T. E., Galkary, S. D., and Bueche, F., J. Polym. Sci., *57*, 471, 1962.
62. Altares, T., Wyman, D. P., Allen, V. R., and Meyersen, K., J. Polym. Sci., A, *3*, 4131, 1965.
63. Berry, G. C., and Casassa, E. F., J. Polym. Sci., D, *4*, 1, 1970.
64. Moore, W. R. A. D., and Millns, W., Br. Polym. J., *1*, 81, 1969.
65. Volker, H., and Luig, F. J., Angew. Makromol. Chem., *12*, 43, 1970.
66. Hama, T., Yamagushi, K., and Suzuki, T., Makromol. Chem., *155*, 283, 1972.
67. Casper, R., Biskup, V., Lange, H., and Pohl, U., Makromol. Chem., *177*, 1111, 1976.
68. Prechner, R., Panaris, R., and Benoit, H., Makromol. Chem., *156*, 39, 1972.
69. Peyrouset, A., and Panaris, R., J. Appl. Polym. Sci., *16*, 135, 1972.
70. Peyrouset, A., Prechner, R., Panaris, R., and Benoit, H., J. Appl. Polym. Sci., *19*, 1363, 1975.
71. Drott, E. E., and Mendelson, R. A., J. Polym. Sci., A2, *8*, 1361, 1970.
72. Drott, E. E., and Mendelson, R. A., J. Polym. Sci., A2, *8*, 1373, 1970.
73. Westerman, L., and Clark, J. C., J. Polym. Sci., Polym. Phys. Ed., *11*, 559, 1973.
74. Williamson, G. R., and Cervenka, A., Eur. Polym. J., *10*, 295, 1974.
75. Coleman, M. M., and Fuller, R. E., J. Macromol. Sci., B, *11*, 419, 1975.
76. Otocka, E. P., Roe, R. J., Hellman, M. Y., and Muglia, P. M., Macromolecules, *4*, 507, 1971.
77. Foster, G. N., McRury, T. B., and Hamielec, A. E., in *Liquid Chromatography of Polymers and Related Materials II* (Chromatographic Science Series, Vol. 13, J. Cazes and X. Delamare, eds.), Marcel Dekker, New York, 1980, p. 143.
78. Marais, L., Gallot, Z., and Benoit, H., Analusis, *10*, 443, 1976.
79. Servotte, A., and De Bruille, R., Makromol. Chem., *176*, 203, 1975.
80. Park, W. S., and Graessley, W. W., J. Polym. Sci., Polym. Phys. Ed., *15*, 85, 1977.
81. Constantin, D., Eur. Polym. J., *13*, 907, 1977.
82. Lecacheux, D., Lesec, J., and Quivoron, C., 184th Am. Chem. Soc. Meet., Kansas City, Kans., 1982.
83. Lecacheux, D., Lesec, J., and Quivoron, C., J. Appl. Polym. Sci., *27*, 4867, 1982.

84. Bovey, F. A., Schilling, F. C., McCrackin, F. L., and Wagner, H. L., Macromolecules, *9*, 76, 1976.
85. Dietz, R., J. Appl. Polym. Sci., *25*, 951, 1980.
86. Dondos, A., Rempp, P., and Benoit, H., Makromol. Chem., *175*, 1659, 1974.
87. Janča, J., and Kolinsky, M., J. Appl. Polym. Sci., *21*, 83, 1977.
88. Janča, J., Mrkvickova, L., Kolinsky, M., and Brar, A. S., J. Appl. Polym. Sci., *22*, 2661, 1978.
89. Janča, J., Pokorný, S., and Kolinsky, M., J. Appl. Polym. Sci., *23*, 1811, 1979.
90. Ambler, M. R., in *Liquid Chromatography of Polymers and Related Materials III* (Chromatographic Science Series, Vol. 19, J. Cazes, ed.), Marcel Dekker, New York, 1981, p. 29.
91. Balke, S. T., and Patel, R. D., J. Polym. Sci., Polym. Lett. Ed., *18*, 453, 1980.
92. Balke, S. T., and Patel, R. D., Polym. Prep., *21*, 290, 1981.
93. Tanaka, T., Omoto, M., and Inagaki, H., Makromol. Chem., *182*, 2889, 1981.
94. Runyon, J. R., Barnes, D. E., Rudd, J. F., and Tung, L. H., J. Appl. Polym. Sci., *13*, 2359, 1969.
95. Adams, H. E., Sep. Sci., *6*, 279, 1971.
96. Stojanov, C., Shirazi, Z. H., and Audu, T. O. K., Chromatographia, *11*, 63, 1978.
97. Revillon, A., J. Liq. Chromatogr., *3*, 1137, 1980.
98. Mori, S., and Suzuki, T., J. Liq. Chromatogr., *4*, 1685, 1981.
99. De Chirico, A., Arrighetti, S., and Bruzzone, M., Polymer, *22*, 529, 1981.
100. Adams, H. E., in *Gel Permeation Chromatography* (K. H. Algelt and L. Segal, eds.), Marcel Dekker, New York, 1971, p. 402.
101. Mori, S., J. Chromatogr., *194*, 163, 1980.
102. Ogawa, T., J. Appl. Polym. Sci., *23*, 3515, 1979.
103. Terry, S. L., and Rodriguez, F., J. Polym. Sci., C, *21*, 191, 1968.
104. Dawkins, J. V., and Hemming, M., J. Appl. Polym. Sci., *19*, 3107, 1975.
105. Mori, S., J. Chromatogr., *157*, 75, 1978.
106. Anderson, D. G., and Isakson, K. E., Prepr. Am. Chem. Soc., Div. Org. Coating Plast. Chem., *30*, 123, 1970.
107. Mirabella, F. M., Jr., Barrall, E. M., II, and Johnson, J. F., J. Appl. Polym. Sci., *19*, 2131, 1975.
108. Mirabella, F. M., Jr., Barrall, E. M., II, and Johnson, J. F., J. Appl. Polym. Sci., *20*, 959, 1976.
109. Bartick, E. G., J. Chromatogr. Sci., *17*, 336, 1979.
110. Hagnauer, G. L., and Koulouris, T. N., in *Liquid Chromatography of Polymers and Related Materials III* (Chromatographic Science Series, Vol. 19, J. Cazes, ed.), Marcel Dekker, New York, 1981, p. 99.
111. Yamaguchi, K., Watanabe, H., Sugo, T., Watanabe, T., Takehisa, M., and Machi, S., J. Appl. Polym. Sci., *25*, 1633, 1980.
112. Braks, J. G., and Huang, R. Y. M., J. Polym. Sci., Polym. Phys. Ed., *13*, 1063, 1975.

113. Braks, J. G., Mayer, G., and Huang, R. Y. M., J. Appl. Polym. Sci., *25*, 449, 1980.
114. Crabtree, D. J., and Hewitt, D. B., in *Liquid Chromatography of Polymers and Related Materials* (Chromatographic Science Series, Vol. 8, J. Cazes, ed.), Marcel Dekker, New York, 1977, p. 63.
115. Katuscak, S., Tomas, M., and Schiessl, O., J. Appl. Polym. Sci., *26*, 381, 1981.
116. Matsuzaki, T., Inoue, Y., Ookubo, T., and Mori, S., J. Liq. Chromatogr., *3*, 353, 1980.
117. Wandelt, B., Brzezinski, J., and Kryszewski, M., Eur. Polym. J., *16*, 583, 1980.
118. Goldfarb, L., Hann, N. D., Dieck, R. L., and Messersmith, D. C., J. Polym. Sci., Polym. Chem. Ed., *16*, 1505, 1978.
119. Abbas, K. B., Polymer, *21*, 936, 1980.
120. Malhotra, S. L., Lessard, P., and Blanchard, L. P., J. Macromol. Sci., A, *15*, 121, 1981.
121. Hellman, M. Y., and Johnson, G. E., in *Liquid Chromatography of Polymers and Related Materials III* (Chromatographic Science Series, Vol. 19, J. Cazes, ed.), Marcel Dekker, New York, 1981, p. 115.
122. Leopairat, P., and Merrill, E. W., J. Liq. Chromatogr., *1*, 21, 1978.
123. Yu, J. F. S., Zakin, J. L., and Patterson, G. K., J. Appl. Polym. Sci., *23*, 2493, 1979.
124. Muller, H. G., and Klein, J., Makromol. Chem., Rapid Commun., *1*, 27, 1980.
125. Abbas, K. B., in *Liquid Chromatography of Polymers and Related Materials II* (Chromatographic Science Series, Vol. 13, J. Cazes and X. Delamare, eds.), Marcel Dekker, New York, 1980, p. 123.
126. Vivatpanachart, S., Nomira, H., and Miyahara, Y., J. Appl. Polym. Sci., *26*, 1485, 1981.
127. Huber, C., and Lederer, K. H., J. Polym. Sci., Polym. Lett. Ed., *18*, 535, 1980.
128. Rooney, J. G., and Ver Strate, G., in *Liquid Chromatography of Polymers and Related Materials III* (Chromatographic Science Series, Vol. 19, J. Cazes, ed.), Marcel Dekker, New York, 1981, p. 207.
129. Slagowski, E. L., Fetters, L. J., and McIntyre, D., Macromolecules, *7*, 394, 1974.
130. Kirkland, J. J., J. Chromatogr., *125*, 231, 1976.
131. Atkinson, C. M., and Dietz, R., Makromol. Chem., *177*, 213, 1976.
132. Quivoron, C., in *Chimie macromoléculaire II* (G. Champetier, ed.), Hermann, Paris, 1972, p. 242.
133. Flory, P. J., and Fox, T. G., Jr., J. Am. Chem. Soc., *73*, 1904, 1951.
134. Burchard, W., Makromol. Chem., *50*, 20, 1961.
135. Stockmayer, W. H., and Fixman, H., J. Polym. Sci., C, *1*, 137, 1963.
136. Hamielec, A. E., and Ouano, A. C., J. Liq. Chromatogr., *1*, 111, 1977.

137. Dondos, A., and Benoit, H., Polymer, *19*, 523, 1978.
138. Rudin, A., and Benschop-Hendrychova, I., J. Appl. Polym. Sci., *15*, 2881, 1981.
139. Chan, R. K. S., and Worman, C., Polym. Eng. Sci., *12*, 437, 1972.
140. Lingaae-Joergensen, J., Makromol. Chem., *167*, 311, 1973.
141. Abdel-Alim, A., and Hamielec, A. E., J. Appl. Polym. Sci., *16*, 1093, 1972.
142. Abdel-Alim, A., and Hamielec, A. E., J. Appl. Polym. Sci., *17*, 3033, 1973.
143. Jisova, V., Janča, J., and Kolinsky, M., J. Polym. Sci., Polym. Chem. Ed., *15*, 533, 1977.
144. Chartoff, R. P., and Lo, S. K. T., in *Liquid Chromatography of Polymers and Related Materials* (Chromatographic Science Series, Vol. 8, J. Cazes, ed.), Marcel Dekker, New York, 1977, p. 135.
145. Sorvick, E. M., J. Appl. Polym. Sci., *21*, 2769, 1977.
146. Lety-Sistel, C., Sebille, B., and Quivoron, C., Eur. Polym. J., *9*, 1297, 1973.
147. Strazielle, C., in *Light Scattering from Polymer Solutions* (M. Huglin, ed.), Academic Press, London, 1972, Chap. 15.
148. Zivny, A., Pouchly, J., and Solc, K., Coll. Czech. Chem. Commun., *32*, 2753, 1967.
149. Tuzar, Z., and Kratochvil, P., Coll. Czech. Chem. Commun., *32*, 3358, 1967.
150. Berek, D., Bleha, T., and Pevna, Z., J. Polym. Sci., Polym. Lett. Ed., *14*, 323, 1976.
151. Berek, D., Bleha, T., and Pevna, Z., J. Chromatogr. Sci., *14*, 560, 1976.
152. Chaufer, B., Lesec, J., and Quivoron, C., C.R. Acad. Sci. Paris, Ser. C, *284*, 881, 1977.
153. Campos, A., Borque, L., and Figueruelo, J. E., J. Chromatogr., *140*, 219, 1977.
154. Soria, V., Campos, A., and Figueruelo, J. E., An. Quim., *74*, 1026, 1978.
155. Chaufer, B., Lesec, J., and Quivoron, C., J. Liq. Chromatogr., *2*, 633, 1979.
156. Straub, P. R., and Brant, D. A., Biopolymers, *19*, 639, 1980.
157. Spychaj, T., and Berek, D., Polymer, *20*, 1108, 1979.
158. Sebille, B., Thuaud, N., and Tillement, J. P., J. Chromatogr., *167*, 159, 1978.
159. Sebille, B., Thuaud, N., and Tillement, J. P., J. Chromatogr., *180*, 103, 1979.
160. Campos, A., and Figueruelo, J. E., Polymer, *18*, 1296, 1977.
161. Bakos, D., Berek, D., and Bleha, T., Eur. Polym. J., *12*, 801, 1976.
162. Narasimhan, V., Huang, R. Y. M., and Burns, C. M., J. Appl. Polym. Sci., *26*, 1295, 1981.
163. Kok, C. M., and Rudin, A., Makromol. Chem., *182*, 2801, 1981.
164. Narasimhan, V., Lloyd, D. R., and Burns, C. M., J. Appl. Polym. Sci., *23*, 749, 1979.
165. Lloyd, D. R., Narasimhan, V., and Burns, C. M., J. Liq. Chromatogr., *3*, 1111, 1980.

166. Hashizume, J., Teramoto, A., and Fujita, H., J. Polym. Sci., Polym. Phys. Ed., *19*, 1405, 1981.
167. Small, H., J. Coll. Int. Sci., *48*, 147, 1974.
168. Small, H., Saunders, F. L., and Solc, J., Adv. Coll. Int. Sci., *6*, 237, 1976.
169. Husain, A., Hamielec, A. E., and Vlachopoulos, J., J. Liq. Chromatogr., *4* (Suppl. 2), 295, 1981.
170. Hamielec, A. E., and Singh, S., J. Liq. Chromatogr., *1*, 187, 1978.
171. Singh, S., and Hamielec, A. E., J. Appl. Polym. Sci., *22*, 577, 1978.
172. Silebi, C. A., and McHugh, A. J., J. Appl. Polym. Sci., *23*, 1699, 1979.
173. Nagy, D. J., Silebi, C. A., and McHugh, A. J., J. Appl. Polym. Sci., *26*, 1555, 1981.
174. Nagy, D. J., Silebi, C. A., and McHugh, A. J., J. Appl. Polym. Sci., *26*, 1567, 1981.
175. Husain, A., Vlachopoulos, J., and Hamielec, A. E., J. Liq. Chromatogr., *2*, 193, 1979.
176. Hamielec, A. E., J. Liq. Chromatogr., *1*, 555, 1978.
177. Freeman, D. H., and Poinescu, I. C., Anal. Chem., *49*, 1183, 1977.
178. Giddings, J. C., Kucera, E., Russel, C. P., and Myers, M. N., J. Phys. Chem., *72*, 4397, 1968.
179. Schram, S. B., and Freeman, D. H., J. Liq. Chromatogr., *3*, 403, 1980.
180. Halasz, I., and Vogtel, P., Angew. Chem. Int. Ed. Engl., *19*, 24, 1980.
181. Tsyurupa, M. P., and Davankov, V. A., J. Polym. Sci., Polym. Chem. Ed., *18*, 1399, 1980.
182. Kuga, S., J. Chromatogr., *206*, 449, 1981.
183. Alliet, D. F., and Pacco, J. M., in *Gel Permeation Chromatography* (K. H. Algelt and L. Segal, eds.), Marcel Dekker, New York, 1971, p. 417.
184. Hallwachs, M. R., Hanson, H. E., Link, W. E., Salomons, N. S., and Widder, C. R., J. Chromatogr., *55*, 5, 1971.
185. Gladen, R., Chromatographia, *5*, 396, 1972.
186. Kato, Y., Kido, S., Watanabe, H., Yamamoto, M., and Hashimoto, T., J. Appl. Polym. Sci., *19*, 629, 1975.
187. Mirabella, F. M., Jr., Barral, E. M., II, and Johnson, J. F., Polymer, *17*, 17, 1976.
188. Pacco, J. M., and Mukherji, A. K., J. Chromatogr., *144*, 113, 1977.
189. Hellman, M. Y., J. Liq. Chromatogr., *1*, 491, 1978.
190. Protivova, J., and Pospisil, J., J. Chromatogr., *88*, 99, 1974.
191. Protivova, J., Pospisil, J., and Holcik, J., J. Chromatogr., *92*, 361, 1974.
192. Walter, R. B., and Johnson, J. F., J. Polym. Sci., Macromol. Rev., *15*, 29, 1980.
193. Majors, R. E., and Johnson, E. L., J. Chromatogr., *167*, 17, 1978.
194. Shepherd, M. J., and Gilbert, J., J. Chromatogr., *178*, 435, 1979.

—6—

AUTOMATIC DATA TREATMENT

BENGT STENLUND and CARL-JOHAN WIKMAN / *University of Åbo Akademi, Turku/Åbo, Finland*

6.1. INTRODUCTION

The use of computers in research and industrial laboratories is rapidly increasing. There are many reasons for this development. One of the most important is the need for the automation of analytical procedures. This can sometimes lead to a substantial reduction in the time used per analysis and, furthermore, relieves trained personnel of tedious and time-consuming routines. The possibility of human errors in lengthy calculations is also eliminated. Another primary reason for using computers is that calculations may be too complex to handle manually.

In connection with SEC, computers are used for a variety of purposes, including:

Data acquisition
Correction of chromatograms for axial dispersion
Calculation of calibration functions
Calculation of average molecular weights, the degree of branching, and other parameters
Simulation of chromatograms
Automatic control of laboratory equipment, such as pumps, injectors, and detectors

The computing equipment can vary considerably depending on purpose and availability and comprises:

Mainframe computers [1,2]
Multipurpose laboratory data systems [3-6]
Programmable integrators for chromatography [7-9]
Dedicated micro- or minicomputers [10-12]

The authors will restrict themselves primarily to the acquisition and handling of SEC data using low-cost microcomputers, since they feel that these will play an increasing role in laboratory automation in the future. The study of systems of this kind, furthermore, provides an excellent means of acquiring an understanding of how data collection and handling are performed.

Microcomputers

The central component of microcomputers is an 8- or 16-bit microprocessor. There are usually between 4 and 64 kilobytes of random access memory (RAM), which can be used to store programs and data. A minimum of 16 kilobytes is recommended.

The microcomputer also has a keyboard, a monitor, and possibly a printer for communication with the operator. It has an operating system including a high-level programming language such as BASIC or Pascal. It is also possible to program a microcomputer in machine language using octal or hexadecimal code, either directly or using a machine language compilator (assembler). The assembler uses three-letter mnemonic codes for the processor instructions, and an octal or hexadecimal code for operands.

For the long-term storage of programs and data a cassette recorder or disk drive is used. Data can be transferred to and from the computer via input-output (I/O) ports.

6.2. DATA ACQUISITION

Real-time or on-line data acquisition can be defined as the automatic recording and saving of data at known intervals during the run of a sample. Data can be stored on a punched tape, cassette tape, diskette, or the like, and are later transferred to the computing unit. Data can also be stored in the memory of the computer. Storage of data and parameters for a run on tape or diskette makes the raw data accessible at any time for new calculations.

Real-time calculations are defined as calculations performed during data collection procedure. In extreme cases all calculations can be performed on a real-time basis, but normally most of them are performed after data collection is complete.

6.2.1. Conversion of Analog to Digital Information

The signal from a SEC detector is always in analog form, and somewhere in the course of data collection it has to be converted to digital information. Chromatograms can be evaluated by manual measurement of the heights over the baseline at different retention volumes. The usual way to automate this procedure is to use an analog-to-digital (A/D) converter. This is a device which samples the voltage from the detector and converts it into digital information.

The voltage range covered by an A/D converter is normally 0-10 V. The detector signal range is often considerably smaller, perhaps 0-10 mV. To take full advantage of the resolution of the A/D converter, the detector signal has to be amplified. There should be no problem in amplifying the signal up to a thousand times.

The resolution of the A/D converter is determined by the number of bits. An 8-bit A/D converter has a resolution of 0-255, and a 12-

bit converter has a resolution of 0-4095. The conversion time is on the order of 20-50 μsec.

The amplifiers of most detectors are equipped with lag filters. The main use for filters is in eliminating the common-mode ac signal and in reducing the signal-to-noise ratio. Extra filters may sometimes be necessary, but since they always distort the signal to some extent, they have to be used with care [13].

Integrators for chromatography and different laboratory data systems are equipped with A/D interfaces. Some detectors permit reading of digitalized data via serial or parallel standard connectors. In other cases a digital voltmeter, data logger, or separate A/D converter can be used. Some microcomputer dealers offer A/D boards which can be directly connected to the computer and handled as part of the addressing space of the computer. The principle of conversion is the same in all cases.

6.2.2. A/D Boards for Microcomputers

A block diagram of the circuitry of an A/D board is shown in Figure 6.1. The hardware may include:

Adjustable gain amplifier for the detector signal
Multiplexer for choice of channel (e.g., 16 channels)
Sample-and-hold circuit holding the voltage constant during conversion
A/D converter to perform the actual conversion
Data and control lines to the computer

There is usually some software included, enabling control and reading of the A/D board using the high-level language. Among the programmable characteristics are:

Degree of amplification of the detector signal.
Selection of the channels to be read (one, some, or all).
Summation of a present number of readings (bunching).
Setting of thresholds. The difference between two readings has to exceed a threshold before the latter are stored; the procedure is used to save memory capacity.

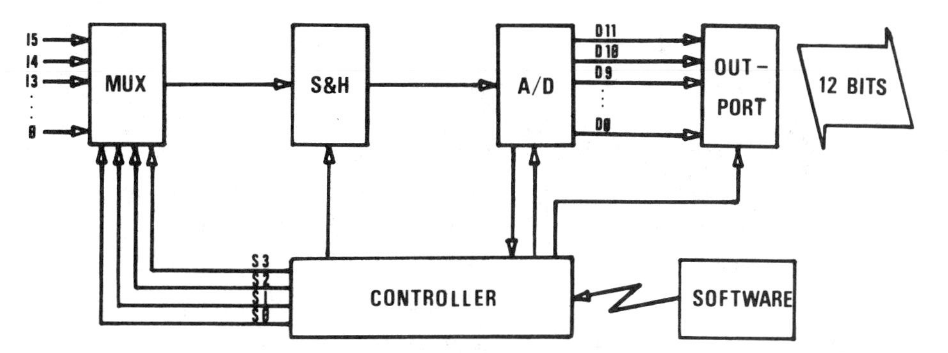

Figure 6.1. Block diagram of an A/D board for a microcomputer. MUX, Multiplexer; S & H, sample-and-hold circuit. The multiplexer is used to choose the channel to be read (0-15).

6.2.3. Number of Data Points per Peak

It is necessary to describe a peak with a minimum number of data points in order to achieve the desired precision in the calculations. This can be exemplified by calculating the total area covered by a Gaussian distribution G(x), using an increasing number of data points [14]. For a more extensive discussion, see Chapter 7.

If σ is the standard deviation, it is necessary to vary x from -5σ to $+5\sigma$ in order to achieve an accuracy of 10^{-6} in the area measurement. Using an increasing number of points in this interval, and calculating the area using triangulation, the relative precision is

$\pm 10^{-4}$ for 10 points per peak
$\pm 10^{-5}$ for 15 points per peak
$\pm 10^{-6}$ for 20 points per peak

Pickett et al. [1] have investigated the influence of the number of data points on the calculation of average molecular weights and found that as few as five points placed properly gave usable results (see Chapter 7).

If a peak of a SEC chromatogram has a duration of 1 min, a sampling frequency of 1 Hz gives 60 data points for this peak. We therefore conclude that a sufficient sampling frequency is easily achieved. As we shall see, however, there are reasons for using a sampling frequency higher than the required minimum.

6.2.4. Software for Data Acquisition with Microcomputers

It is possible to write programs for data acquisition using a high-level programming language, especially if a fast language such as Pascal is used, but machine programming is recommended. The main advantage is speed; machine language programs can be more than 100 times faster. The speed allows for:

- Summation of a large number of readings to attenuate random noise
- Other types of digital filtering to reduce the signal-to-noise ratio, such as moving-average filters [15]
- Simultaneous reading of more than one channel, enabling the use of dual detector systems, or data acquisition from more than one chromatography system
- Plotting of real-time chromatograms on a monitor screen or printer
- Control and recording of external events

Programs for Continuous Data Collection

One way of compiling a machine language data acquisition program is to collect data continuously and bunch them during the run. Start and end times are read from an internal clock. The structure of a program of this kind is shown in Figure 6.2.

This program can be backed up by a high-level program, which may perform as follows:

- Draws the coordinate system on the monitor screen
- Shows the baseline, which can then be adjusted to the desired level
- Calculates the number of readings to be bunched as a function of the run time given by the operator
- Asks for the number of data points desired
- Stores data and parameters on diskette after the run for later analysis

Interrupt Programs

An interrupt occurs when a device, often the internal clock, signals the processor that a program routine with a higher priority has to be executed. The processor then halts the execution of the program on

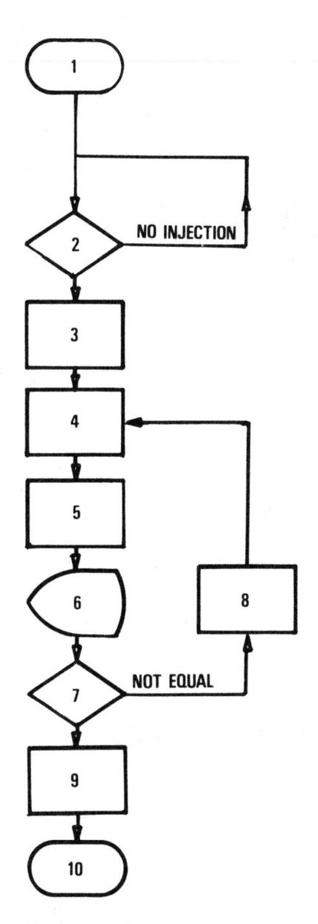

Figure 6.2. Flowchart for a machine language program for continuous data collection. (1) Jump from the backup program. (2) Start loop waiting for an injection to occur. (3) Read and store the start time. (4) Take and bunch a predetermined number of readings. (5) Store the data point in the memory. (6) Plot the data point on the monitor screen. (7) Test whether the number of data points is equal to a predetermined number. (8) Increase the data point counter by one. (9) Read the end time and calculate the run time. (10) Jump to the backup program.

hand and jumps to the interrupt routine. The registers necessary when the main program is later resumed are saved, and the interrupt program is executed. The registers are then restored and the processor allowed to continue the earlier program. The interrupt frequency can be in the range 1-100 Hz in connection with SEC.

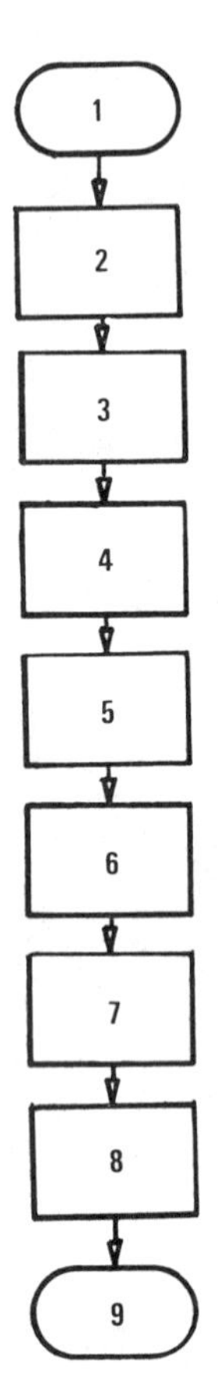

Figure 6.3. Flowchart for a machine language interrupt routine for data collection. (1) Start. (2) Disable other interrupts. (3) Store registers. (4) Take a reading from the A/D board. (5) Store the data in the memory. (6) Increase the data pointer by one. (7) Restore registers. (8) Enable other interrupts. (9) Return from interrupt routine.

The advantage of using interrupts is the possibility of running two or more programs seemingly simultaneously, making better use of the computer's abilities. The structure of an interrupt routine is shown in Figure 6.3.

The main program, either in the high-level language or in machine language, handles such functions as:

Bunching and filtering of data
Plotting of chromatogram on monitor screen or printer
Transferring of data to diskette
Making real-time calculations

6.3. DATA HANDLING

Data-handling programs are often written in a high-level language, which is easier than writing programs in machine code. The reason is that time is not critical in off-line calculations, and programs are easier to modify if necessary. There are two main types of data-handling programs: programs for calibration and programs for calculation of average molecular weights and other parameters.

6.3.1. Calibration Programs

The calibration is either a straightforward computation of a calibration function, or an iterative procedure. The calibration function is generally assumed to be a polynomial given, for example, by equation (2.8), as discussed in Chapter 2.

Universal calibration is performed in a similar way. An example of a third-degree polynomial adopted to data from runs of 12 narrow polystyrene standards is shown in Figure 6.4.

It is also possible to use standards with varying dispersities and one known molecular weight average [16]. In this case the coefficients of the polynomial that best fits the known molecular weight averages, corresponding to average retention volumes, are obtained through an iterative procedure.

One way to calibrate a SEC system when narrow standards are unavailable is to use a polydisperse standard. The corresponding methods are described in Chapter 2. Malawer et al. [17,18] have developed an algorithm for this purpose. Knowledge of two different molecular weight averages, $\overline{M}_w$ and $\overline{M}_n$, for example, is required. The coefficients of the calibration function given by equation (2.3) are obtained using the iterative procedure described in Figure 6.5.

First, the slope C_2 [see equation (2.3)] is varied until the lines are parallel (i.e., w - n = 0). Then C_2 is held constant and

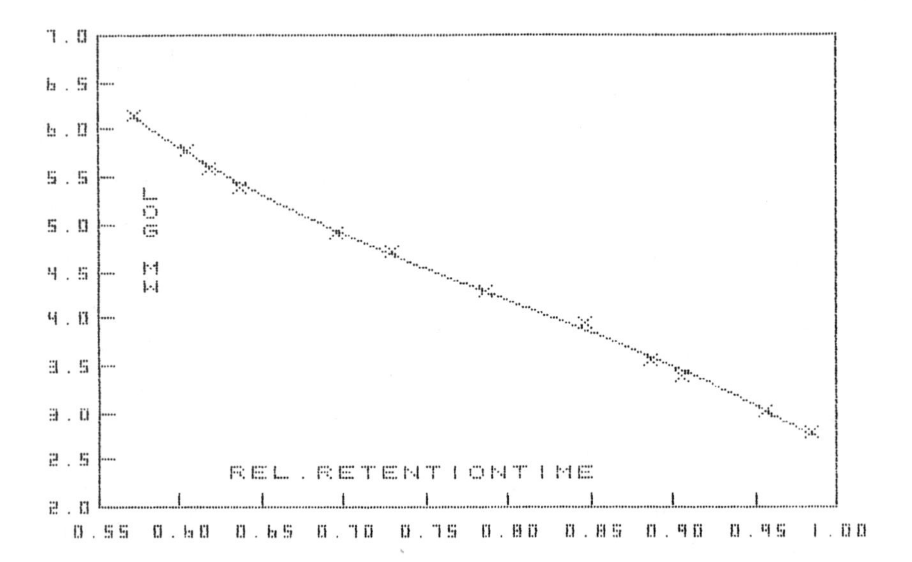

Figure 6.4. Third-degree polynomial adopted to data from runs of 12 narrow polystyrene standards.

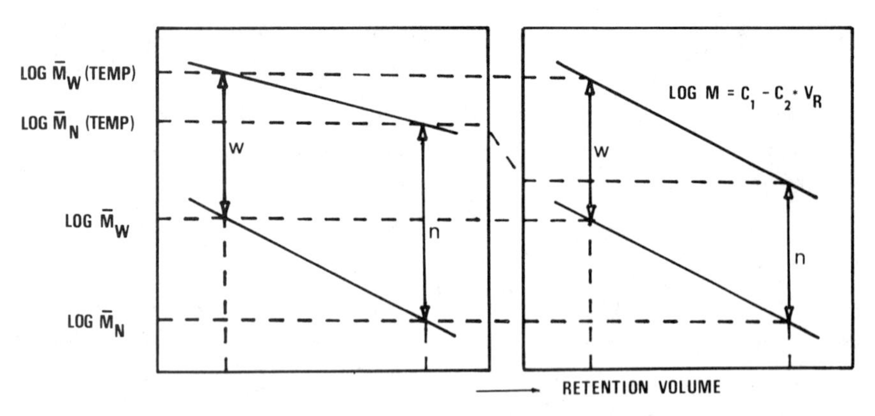

Figure 6.5. Visualization of method for calibration with a polydisperse standard with two known average molecular weights ($\bar{M}_w$ and $\bar{M}_n$). First the slope of the temporary calibration line is varied until w - n = 0 (left). Then the intercept is varied until w + n = 0 (right).

the intercept C_1 is varied until the lines coincide (i.e., w + n = 0). The frame of a program for this calibration is shown in Figure 6.6. In using this calibration system caution has to be taken for samples

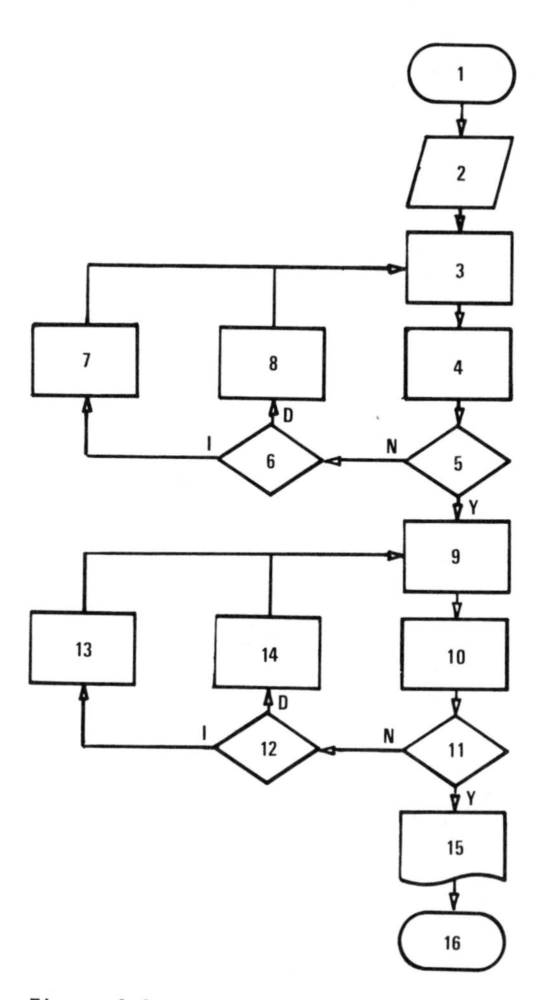

Figure 6.6. Simplified flowchart of a program for calibration with one polydisperse standard. (1) Start. (2) Input first estimates of C_1 and C_2. (3) Calculate $\overline{M}_w$(temp.) and $\overline{M}_n$(temp.). (4) Calculate w - n. (5) Is (w - n) smaller than limit? (6) Increase or decrease C_2? (7) Increase C_2. (8) Decrease C_2. (9) Calculate $\overline{M}_w$(temp.) and $\overline{M}_n$(temp.). (10) Calculate w + n. (11) Is (w + n) smaller than limit? (12) Increase or decrease C_1? (13) Increase C_1. (14) Decrease C_1. (15) Print results. (16) End.

that differ much in dispersity and average molecular weights from the standard, and for standards and samples approaching the exclusion limits.

6.3.2. Baselines

Before calculations of molecular weights can be attempted, a baseline under the peak has to be drawn. This can be done between two predetermined points, such as the start and end of the chromatogram, or as a continuation of the baseline before the first peak. Considering the possibility of variations in the slope of the baseline during a run, the danger in using methods of this kind are obvious. A better way is to draw the baseline from the start of the peak to the end of the peak, as determined by the first or second difference ratio of the experimental distribution.

First-derivative methods use the difference ratio $\Delta x/\Delta y$. Δx may, for example, cover five data points. A peak start is accepted when the ratio is larger than a limit. The limit is preferably a function of the signal-to-noise ratio. The disadvantage of this method is that the start sensitivity is dependent on the slope of the baseline. A rising baseline increases this sensitivity.

Second-derivative methods are insensitive to the slope of the baseline. One variant of a second-derivative method uses the criteria shown in Figure 6.7. All criteria have to be fulfilled before a start is accepted. Usually, the end of a peak is more difficult to detect than the start, due to tailing. The peak can also end in a valley before the next peak, in which case the minimum can be taken as the endpoint. A very instructive peak-detection algorithm has been published by Woerlee and Mol [19].

A printout of the chromatogram for a polystyrene standard is shown in Figure 6.8. The baseline has been plotted together with tick marks to indicate the start and the end of the peak. From a graphic representation like this, the validity of the automatic baseline drawing can be established.

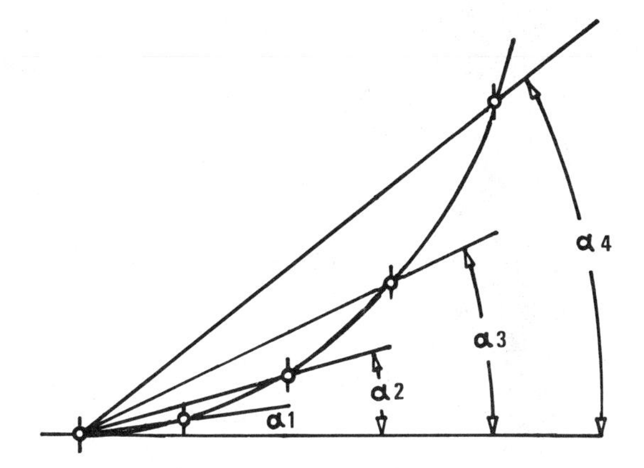

Figure 6.7. Peak start detection with a second-derivative method. The criteria for start are: (1) $\tan \alpha_2 > \tan \alpha_1$; (2) $\tan \alpha_3 > \tan \alpha_2$; (3) $\tan \alpha_4 > \tan \alpha_3$. (Redrawn from Ref. 19.)

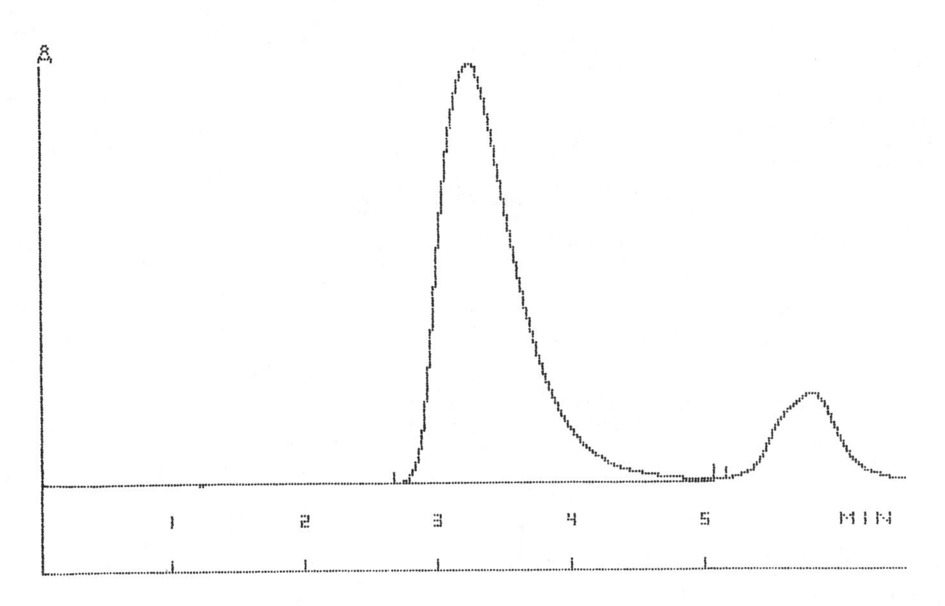

Figure 6.8. Printout of a run of a narrow polystyrene standard. The baseline has been drawn automatically using a peak-detection algorithm. The tick marks indicate the start and the end of the peak.

6.3.3. Calculation of Average Molecular Weights

Once the baseline has been established, the height under the baseline is subtracted from the total height for each point of the peak. Next, a correction for axial dispersion can be made.

Different average molecular weights can then be calculated. For each point of the chromatogram the molecular weight corresponding to that retention volume is calculated from the calibration function. The formulas for calculating $\overline{M}_n$, $\overline{M}_w$, and $\overline{M}_v$ from an experimental chromatogram are given in Section 5.2.1.

6.4. AUTOMATIC CONTROL OF SEC EQUIPMENT

Control of external events with a microcomputer is possible in a number of ways. The functions of many laboratory devices can be controlled externally via serial (e.g., RS 232) or parallel (e.g., IEEE 488) standard connectors. This sometimes permits full control from a microcomputer.

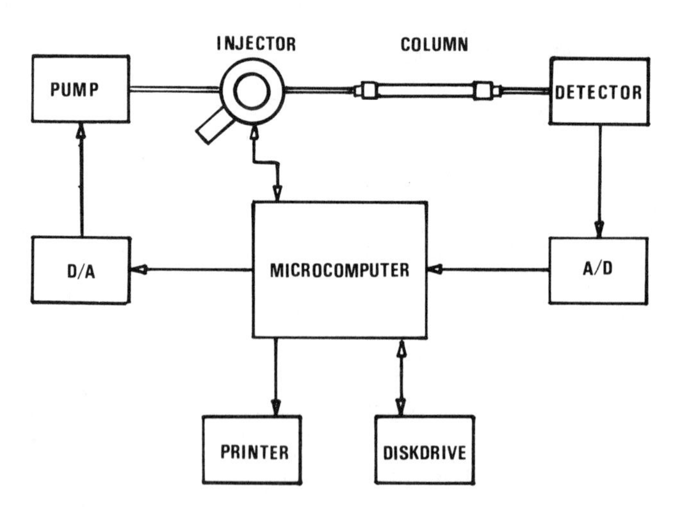

Figure 6.9. Block diagram of an automated SEC system. The microcomputer controls the pump via an D/A converter and the injector via the I/O port. The detector signal is recorded using an A/D converter. Data are stored on a diskette.

For control of off/on events the parallel I/O port of the microcomputer can be used. Using special parallel connecting boards, some 50-100 on/off events can be controlled or registered.

The microcomputer can also generate an analog signal via an digital-to-analog converter. In this way the flow rate of pumps, for example, can be directly controlled. These possibilities together enable the assembly of a fully automated SEC system such as the one outlined in Figure 6.9.

ACKNOWLEDGMENT

The authors express their appreciation to Jouko Aavikko, M.Sc., for cooperation and stimulating discussions.

REFERENCES

1. Pickett, H. E., Cantow, M. J. R., and Johnson, J. F., J. Appl. Polym. Sci., *10*, 917, 1966.
2. Bly, D. D., DuPont Innovation, *5*, 16, 1974.
3. Girard, J. E., Goehner, R. P., Hatfield, W. T., and Gundlach, P. D., J. High Resolut. Chromatogr. Chromatogr. Commun., *1*, 206, 1978.
4. Jordan, R. C., and Christ, P. J., Am. Lab., *11*, 71, 1979.
5. Baker, D. R., and George, S. A., Am. Lab., *12*, 41, 1980.
6. George, S. A., and Baker, D. R., Ind. Res. Dev., *22*, 113, 1980.
7. MacLean, W. W., J. Chromatogr., *99*, 425, 1974.
8. MacLean, W. W., Am. Lab., *6*, 63, 1974.
9. Ouchi, G., and Flarity, C., Am. Lab., *13*, 71, 1981.
10. Hamielec, A. E., Walther, G., and Wright, J. D., Adv. Chem. Ser., *125*, 138, 1973.
11. Mukherji, A. K., and Ishler, J. M., J. Liq. Chromatogr., *4*, 71, 1981.
12. Chan, C. Y., Irwin, R., and O'Brien, R. J., Am. Lab., *13*, 54, 1981.
13. Haddad, P. R., Keating, R. W., and Low, G. K. C., J. Liq. Chromatogr., *5*, 853, 1982.
14. Sutre, P., and Malengé, J. P., Chromatographia, *5*, 141, 1972.
15. Cram, S. P., Chesler, S. N., and Brown, A. C., III, J. Chromatogr., *126*, 279, 1976.
16. Szewczyk, P., J. Polym. Sci., Polym. Symp., *68*, 191, 1980.
17. Malawer, E. G., and Montana, A. J., J. Polym. Sci., Polym. Phys. Ed., *18*, 2303, 1980.
18. Malawer, E. G., Montana, A. J., Cheng, H. N., and Smith, T. E., Chromatogr. Newsl., *9*, 30, 1981.
19. Woerlee, E. F. G., and Mol, J. C., J. Chromatogr. Sci., *18*, 258, 1980.

— 7 —

PRECISION AND ACCURACY OF RESULTS

SVATOPLUK POKORNÝ / *Institute of Macromolecular Chemistry, Czechoslovak Academy of Sciences, Prague, Czechosolvakia*

7.1. INTRODUCTION

This chapter reviews potential sources of errors, their effect on the accuracy and precision of SEC results, and possibilities for minimizing the errors. Instrumentation is dealt with first; it represents sources of errors caused either by inadequate construction or by poor functioning of some of the parts. Errors due to an inadequate choice of experimental conditions are dealt with next. This is followed by a survey of basic errors in the interpretation of experimental data, and the chapter is concluded by an outline of limitations of the method based on an analysis of short-term and long-term reproducibility.

7.2. INSTRUMENTAL ERRORS

In apparatus used for SEC, similarly to those used for liquid chromatography in general, the key role in obtaining correct experimental data is played by those parts that either affect the chromatographic process directly or are connected with the recording of the experimental chromatogram. These are primarily the pumping system, the sampling system, the separation system of the columns, detectors, and the device used for measurement of the retention volume (V_R). One should also mention the necessity of using a capillary line leading from the injection site of the sample to the outlet from the detector system with the smallest volume as possible, including the detector cells, to avoid an excessive broadening of the chromatographic band outside the columns.

7.2.1. Effect of Pumping System Instability

Due to subsequent pressure and temperature changes in the detector cells, the unstable flow of the eluant causes increased noise and instability of the baseline on the chromatographic record. This gives rise to problems in the interpretation of the chromatogram, particularly in polymers with a larger width of molecular weight distribution (MWD) (see Section 7.4.2).

Large errors in the determination of average molecular weights are produced by the instability of the flow of the mobile phase in high-speed SEC, where V_R values are usually very small and are derived directly from the retention time by multiplying with an adjusted flow rate value. The instability evidenced by the pumping system may be of various types and may variously affect the accuracy of results [1]:

Variation in flow rate about the adjusted value (Figure 7.1a). This is a short-term variation of the flow rate compensated by the feedback of up-to-date pumping systems. The long-term average flow rate value remains constant during the analysis, and the effect on the accuracy of results is not important. If the flow rate varies about the adjusted constant value in the range ±1%, the error in the determination of molecular weights is about 0.5% relative; if the variation is ±5%, the respective error would be about 5% relative.

A constant change in flow rate during the elution of the SEC curve (Figure 7.1b). This type of flow instability is usually due to change in the hydrodynamic resistance of the system related to the passage of the viscous zone of the polymer sample through the column. A 5% change of this type in the flow rate has as a result an 8% error in the determination of $\overline{M}_n$ and a 3% error in the determination of $\overline{M}_w$. A similar change in the flow rate amounting to only 1%, which is more realistic, leads to only a 1-2% change in the calculated average molecular weight values.

A long-term drift of the flow rate (Figure 7.1c). This type of instability of the flow rate causes more significant errors. If the flow rate changes by 5% from the injection of the sample to the end of the analysis, the resulting error of determination of $\overline{M}_n$ is 37% or 54%, and the error of determination of $\overline{M}_w$ is 31% or 43%, depending on a positive or negative change in the flow rate during the experiment. A mere 1% drift in the flow causes errors of about 9% in the determination of $\overline{M}_n$ and 7% in the determination of $\overline{M}_w$.

A constant deviation of the flow rate from the specified value (Figure 7.1d) affects the results of measurement in a very important way. A 1% deviation of the actual flow rate from the required value shifts the resulting average molecular weight values by about 20% compared with the real ones.

Similar types of errors due to the instability of the pumping system also affect the calibration measurement. If changes in the flow rate during the calibration and analysis of the sample occur in opposite directions, that is, the deviation from the required value

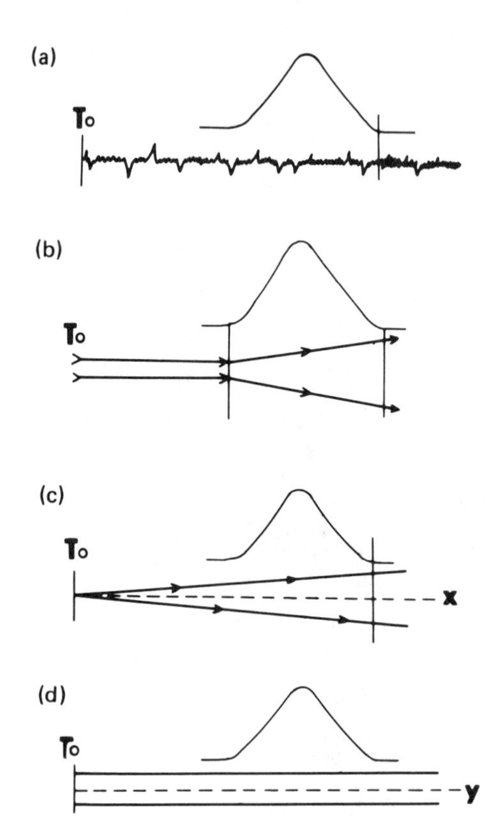

Figure 7.1. Various types of instability of the pumping system: (a) variation in flow rate about the adjusted value; (b) constant change in flow rate during the elution of the SEC curve; (c) Constant change in flow rate throughout the analysis (x, constant flow); (d) constant flow rate deviation from specified value (y, specified value). (From Ref. 1.)

is positive during the calibration and negative during the analysis of the sample, and vice versa, the errors increase accordingly.

Up-to-date pumping systems provided with a continuous checkup of the flow rate and with feedback show no perceptible drift in the flow rate, but a short-term variation about the constant value (Figure 7.1a) must be borne in mind, together with a constant deviation of the flow rate from the specified value after the establishment of stationary conditions (Figure 7.1d). The constancy of the flow rate is given in the range 0.6-1% relative [2]. As pointed out above, the short-term variation of the flow rate in this range does not affect the accuracy of measurement to any important extent, but a constant deviation of the flow rate from the specified value in the given range becomes significant with respect to the accuracy of measurement.

It is therefore recommended that the retention time to V_R be recalculated by multiplying the retention time, not by the required values, but by the actual flow rate value measured after the establishment of stationary conditions in the pumping system with feedback by employing an independent method (e.g., by measuring time needed to fill a certain volume with the mobile phase). After that, the pumping system maintains the flow rate of the mobile phase constant while stationary conditions last, with only an insignificant variation about the stationary value. This simple method gives good results and prevents considerable errors in the molecular weight determinations of polymers by means of high-speed SEC with high-performance columns. The values reported for a practical checkup of the precision of the molecular weight determination of polymers by the high-performance SEC method using a precise pump with feedback and a time base for the measurement of V_R are 1% and 1.2% (relative standard deviation) [3], for $\overline{M}_n$ and $\overline{M}_w$, respectively.

A more precise method involves the use of an internal standard for the elimination of the irreproducibility of determination of V_R. The relative standard deviation in the determination of $\overline{M}_n$ is thus reduced to 0.42% and that in the determination of $\overline{M}_w$, to 0.58% [3]. The problem still remains of how to choose the internal standard to prevent it from interfering with peaks of the compounds undergoing separation. An interesting situation arises in the use of elemental sulfur as the internal standard in the high-speed SEC of polymers in tetrahydrofuran (THF) on columns packed with µ-Styragel [4]. Under such conditions, sulfur applied as a 0.1% solution gives a sharp symmetrical elution curve, well separated from all compounds undergoing separation ($K_{SEC} > 1$), and the long-term reproducibility then reaches values of better than 0.3%.

7.2.2. Sample Injection Systems

The precision of sampling systems usually applied for the sample injection in SEC (i.e., of the loop injector or of automatic injectors used in up-to-date liquid chromatographs) is given as a relative standard deviation of the injected volume of better than 1% [2,5], which is fully satisfactory. One should, however, avoid poor functioning of such devices (e.g., gradual plugging, leakage, etc.), which may lead to broadening of the chromatographic band or to tailing of the elution curves, and consequently to negative effects, mainly on the determination of $\bar{M}_n$ and the polydispersity index [6].

7.2.3. Separation Column System

The separation columns are the core of the apparatus, and the success of the entire separation process depends on their properties. The head of the column must be designed to distribute the eluant evenly throughout the cross section, thus ensuring that the deformation of the chromatographic band will be minimal. On the other hand, it has been reported that an undesirable effect on the chromatographic band appears near the column wall (the *wall effect*); from this standpoint, the best column would be one with an infinite diameter, so that during the passage through the column the chromatographic band would never come into contact with the region near the wall [7]. Obviously, no such columns are used in practice, but we try to minimize the wall effect by using columns with the smoothest possible inner surface.

Regarding the choice of columns with respect to the separation range, the rule is that retention volumes in the whole extent of the elution curve must lie in the range of retention volumes for which the calibration dependence has been approximated by some function on the basis of calibration using suitable standards (see Figure 7.2). Extrapolation of the calibration function to a range not covered by suitable standards (dashed line in Figure 7.2b) may serve only as an emergency solution, and a negative effect on the results should be anticipated in this case. A satisfactory solution is described in Section 2.2.3 and is demonstrated in Figure 2.2.

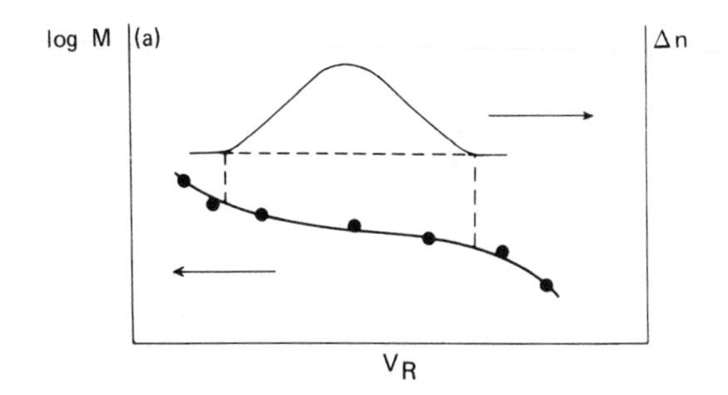

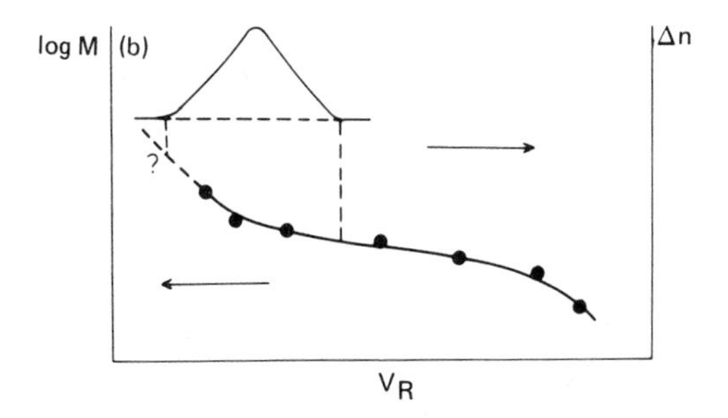

Figure 7.2. Example of a correct (a) and an incorrect (b) choice of column system with respect to the calibrated separation range. ●, Experimental points due to calibration with the standards; ——, approximated calibration plot; ---, region of uncertain course of the calibration plot.

To reduce the effects on the resulting molecular weights of inaccuracies in the determination of the retention volume, columns with maximal selectivity are used to advantage. Also, resolution on columns with higher selectivity is better compared with less selective, *universal columns*. Examples of the calibration plots for both types of columns are given in Figure 7.3. Obviously, *selective columns* are able to carry out separation within a much narrower range of molecular weights than are universal columns, and they are therefore applied in cases of precise measurements of samples in a known range of molecular weights.

Also, the use of columns with the highest possible efficiency, given, for example, by the number of theoretical plates per unit

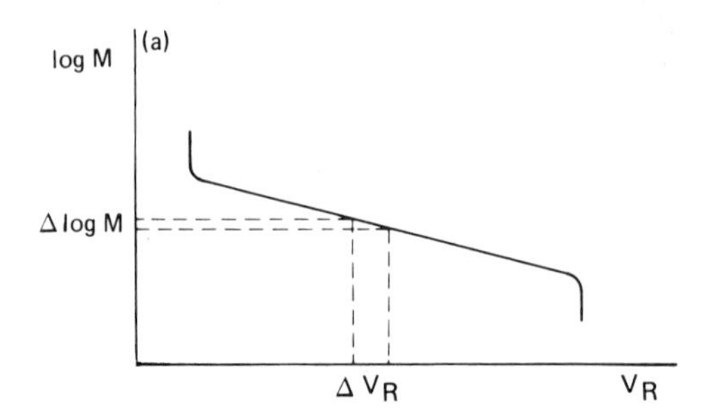

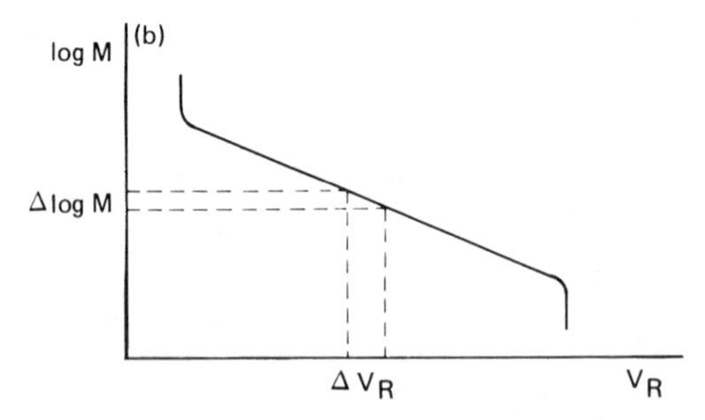

Figure 7.3. Schematic view of the calibration plot of a selective (a) and a universal (b) column system.

length, renders the results more precise and simplifies the calculation and correction procedures. Recently, high-performance columns packed with 5-μm particles have been used successfully, allowing the polydispersity index to be determined without corrections due to spreading, and also with high reliability for relatively narrow standards ($M_w/M_n > 1.01$) [8].

7.2.4. Detectors

With the exception of detectors monitoring quantities directly proportional to molecular weights [such as the low-angle laser light-scattering photometer (LALLSP) and an automatic viscometer], whose principles and possible applications are described in Chapter 5, the response of the detector is proportional either to the concentration or to the mass of the eluting polymer in the measuring cell. The basic requirement regarding the detectors is a feasibly linear dependence of the response on concentration or mass. This requirement is met with most commercial detectors within the usual concentration range. Another assumption to be fulfilled in the treatment of the elution curve is that the response be independent of molecular weight. This assumption is usually more or less satisfied in most structurally uniform homopolymers with a sufficiently high degree of polymerization ($P > 10$). Major errors may arise, however, by neglecting a change in the response of the detector with molecular weight in analyses of oligomers, structurally nonuniform polymers, or copolymers.

In the analyses of oligomers, one should know the dependence of the response of the differential refractive index (DRI) detector on molecular weight and use appropriate correction factors in calculations [9,10]. With an increasing number of monomeric units in the oligomer, the response of the DRI detector increases up to a certain degree of polymerization, and if this fact is not respected in the subsequent treatment, calculations give higher than real values of the average molecular weights.

The determination of the MWD of copolymers is also complicated by the fact that due to different copolymerization parameters of comonomers and to the polymerization proceeding to high conversion, fractions with different molecular weight usually have different composition, and hence a different detector response [11] (for details, see Section 5.4.1).

If there are doubts regarding the uniform composition of the polymer undergoing analysis, or if the polymer used is of a relatively low molecular weight, it is better, by injecting samples of various degrees of polymerization, to make sure that the detector's response is independent of molecular weight. Neglecting this fact will produce more or less distorted results.

7.2.5. Measurement of the Retention Volume

A correctly measured V_R is the primary problem in the evaluation of SEC chromatograms, because in the calibration dependence a certain

V_R has a just corresponding value of molecular weight. In conventional devices for SEC of an older type, where major volumes are used, the type of V_R measurement most widely in use is the siphon counter. The accuracy of measurement of V_R with a siphon is sufficiently high in short-term measurements (better than 0.1%), but in long-term applications the error may increase above 1%; larger disproportions appear most often after the siphon has been cleaned and readjusted [6]. Table 7.1 presents results of three series of measurements, carried out in major time intervals using a siphon counter in constant arrangement.

The dependence of V_R measured with a siphon on the flow rate of the eluant must also be borne in mind. With increasing flow rate, the apparent siphon volume increases, because during the constant time of its discharge it is filled with a further volume proportional to the flow rate. Hence the apparent siphon volume is given by

$$V_{app} = V_S + qt \tag{7.1}$$

where V_{app} is the apparent siphon volume, V_S the real siphon volume, q the flow rate of the mobile phase, and t the discharge time of the siphon. Hence in precise measurements V_R should be calculated from the number of siphons by using V_{app} and not V_S. On the contrary, at very low flow rates the experimental error due to the siphon is caused by evaporation of the mobile phase during the slow filling of the

Table 7.1. Example of Errors Involved in Three Series of Retention Volume Measurements Achieved at Different Times with a 1-ml Siphon Counter and a Tetrahydrofuran Flow Rate of 2 ml/min

Mean time (sec)	Deviation (%)	Retention volume of PS[a] 111,000 (ml)	Peak molecular weight	Error (%)
25.62	<0.1	34.90	122,000	10
25.43	<0.1	35.15	111,000	0
25.26	<0.1	35.40	102,000	-10

[a]Polystyrene standard.
Source: Ref. 6.

siphon. Some siphon types have been designed which protect the eluant from evaporation [12].

With polar eluants that do not wet the glass surface of the siphon, this procedure cannot be used. Sometimes the glass surface of the siphon is hydrophilized, but this precaution is only a temporary one and must be renewed. In this case, V_R can be measured by using a drop counter. The measurement also depends on the flow rate of the mobile phase, the temperature, and the types of compounds present in the eluant, because the latter affect the surface tension, which determines the drop size at given conditions. By maintaining constant conditions, such a procedure may be employed in those cases when a siphon cannot be used, for the reasons outlined above.

The highest accuracy of measurement of V_R may be achieved by direct derivation from the retention time in connection with exact pumps provided with feedback. At present, this procedure is used primarily in high-performance SEC, where relatively small volumes are applied. Problems of accuracy of this procedure, including the use of internal standards, are discussed in greater detail in Section 7.2.1.

Now it should be pointed out to what extent inaccuracy in the determination of V_R affects the final results of SEC measurements (i.e., the molecular weight values and the polydispersity index). It is possible to derive theoretically the magnitude of the error in the determination of M ensuing from an error in the determination of V_R at the maximal attainable selectivity when d ln M/d K_{SEC} is minimal. Then for the error in the determination of M expressed through a relative standard deviation, the following relation is valid [13]:

$$\% M = \left| 1 - \exp \left(\frac{6\Delta K_{SEC}}{a + 1} \right) \right| \times 100 \tag{7.2}$$

where K_{SEC} is the distribution coefficient and a is the exponent of the Mark-Houwink relation for the polymer and mobile phase used. Similarly, the magnitude of the error in the determination of the polydispersity index may be calculated for various widths of distribution. Table 7.2 presents the calculated errors in the deter-

Table 7.2. Calculated Errors in the Molecular Weight (% M) and Polydispersity Index (% I) Determination in Dependence on the Precision of Measurement of the Retention Volume at Maximal Attainable Selectivity

Precision of measurement of retention volume (% rel.)	Percent M	Percent I (I = 1.1)	Percent I (I = 2.0)
1.0	7.30	0.095	0.70
0.5	3.60	0.048	0.35
0.1	0.71	0.009	0.07
0.01	0.35	0.005	0.03

Source: Ref. 13.

mination of M and of the polydispersity index I due to inaccuracy in the determination of V_R at the maximal attainable selectivity. Although the error involved in the determination of M may be quite significant, that appearing in the determination of the polydispersity index, especially with narrow standards, is negligible.

7.3. ERRORS DUE TO EXPERIMENTAL CONDITIONS AND VARIABLES

Due care should be paid to the choice and maintenance of experimental conditions in SEC analyses, because they have a considerable effect on the separation process. They may also affect the functioning of parts of the chromatograph; for example, temperature variations may affect the functioning of the pumping system and of the detectors, and a high or irregular flow rate of the mobile phase may affect the magnitude of noise of the detectors [5].

The effect of experimental conditions on the chromatographic process is analyzed in detail in Chapter 4. Here we describe only briefly the effect of the individual variables and conditions on the accuracy of the results of measurements and suggest some procedures that permit minimizing the errors as much as possible.

7.3.1. Effect of the Mobile Phase Flow Rate

Within rather a broad extent, the flow rate of the mobile phase does not influence conclusively the experimentally determined V_R values. Changes in V_R at very low and very high flow rates are probably caused by the siphon counter [14,15] (see also Section 7.2.5).

The flow rate of the mobile phase, of course, affects column efficiency in a decisive manner, in agreement with theory [15]. It is therefore recommended that a compromise be made between the speed of the analysis and the degree of efficiency. If we choose such an optimal flow rate of the mobile phase and also maintain it constant for a long time (i.e., during calibration measurements and analyses of the samples), the results obtained are only minimally affected.

7.3.2. Effect of Temperature

In contrast with adsorption chromatography, for example, where even small temperature changes (related to the mechanism of separation) may affect retention to a considerable extent [16], the influence of temperature on the SEC separation is much less important. Experimental measurements and an elucidation of factors causing a change in V_R depending on temperature [17] showed, however, that in exact measurements the effect of temperature on the results of SEC measurements cannot be neglected. Changes in V_R produced by a change in the column temperature arise for the following reasons:

1. Change in the volume of the eluant entering the columns with temperature
2. Change in the viscosity of the eluant and solute with temperature
3. Change in the hydrodynamic volume of the solute with temperature
4. Change in the gel volume with temperature
5. Effect of temperature on the extent of gel-solute interaction (adsorption, partition)
6. Change in the diffusion coefficient of the solute with temperature

These factors are operative to varying degrees, the most important being causes 1 and 5 in the list above. The extent of the error in the result of a SEC measurement due to a change in temperature can be seen in an example where calibration measurement and the measurement of samples were carried out at temperatures differing by 10°C [17]. The resulting relative error in the determination of average $\bar{M}_n$ and $\bar{M}_w$ values amounted to 12-18%. Hence precise measurements require that both calibration and the measurement itself be carried out at the same temperature, maintained within ±1°C of the adjusted value.

With increasing temperature the efficiency of the column system increases due to a decrease in the viscosity of the mobile phase or to an increase in the diffusion coefficient of the solute. From this standpoint, elevated temperatures seem to be called for, but with respect to the poorer reproducibility of measurements and a lower durability of columns at higher working temperatures, the temperatures used are nearer to room temperature. Only with polymers insoluble at such temperatures (e.g., polyolefins), elevated temperatures should be used.

Precise thermostating of temperature-dependent detectors (e.g., of the differential refractometers) operating with high sensitivity is of great importance, because even small temperature variations have a negative effect on the stability of the baseline, which leads to errors in the evaluation of the chromatogram (see Section 7.4.2).

7.3.3. Effect of the Choice of Combination: Mobile Phase-Solute-Gel

The principle of universal calibration (see Chapter 2) usually employed in the interpretation of SEC chromatograms is valid only assuming separation based on the hydrodynamic volume of molecules in solution. This assumption is often not fulfilled and depends on a suitably chosen mobile phase-solute-gel combination [18] (see also Chapter 4). The measure of solute-gel interaction (i.e., K_p value) is also often a function of molecular weight [19].

The following procedures are suggested to minimize negative effects on the results due to the interactions outlined above:

1. Mobile phase-solute-gel combinations reported in the literature as suitable (i.e., without side interactions) should be used.
2. If no such system has been described, or it cannot be used for some reasons, a system should be employed in which the absence of side interactions may be assumed (polarities of the mobile phase, solute, and gel are approximately the same). Results obtained from SEC using the principle of universal calibration are compared with molecular weight values measured by absolute methods. If the results coincide with the whole range of molecular weights within the limits of the usual experimental error, the system is adequate. If this is not the case, another system must be used [20].
3. The last possibility consists in the construction of a calibration dependence based on fractions of the polymer to be analyzed (without applying the principle of universal calibration). In this case, interactions (if any) are directly reflected in the calibration dependence, and the latter may be used without complication in the evaluation of elution curves of unknown samples of polymers of the same type. This method, which seems to be simple, is complicated by the fact that fractions of the respective polymer suited for calibration of the system are usually not readily available.

7.3.4. Effect of the Sample Concentration

It has been proven experimentally that with increasing concentration of the injected sample, V_R also increases, the more so the higher the molecular weight [14]. It has been derived theoretically and confirmed experimentally that the main role in the concentration dependence of V_R is played by the viscosity of the injected polymer solution [21-26]. In particular, on high-performance columns packed with small particles, too high a concentration of the high molecular weight polymer may even lead to splitting of the elution curve into several maxima (called "*viscous fingering*") [8]. The concentration effects are treated in greater detail in Chapter 4. Not respecting these effects in practical measurements introduces considerable errors into the results (up to 50% relative), especially in the case of high molecular weight samples [6].

If errors due to the concentration effects are to be minimized, the injected samples must be used in minimal concentrations still acceptable with respect to the required detector signal-to-noise ratio (see Section 7.4.2). Also, if the measurement is performed at the same concentrations as calibration, then assuming the same MWD, the influence of the concentration effects on the results is minimal. The influence of polymer concentration can be removed by using the method of data treatment according to the calibration polynomial of the third degree, the coefficients of which are a function of concentration [6]:

$$\log M = A_0(C) + A_1(C)V_R + A_2(C)V_R^2 + A_3(C)V_R^3 \qquad (7.3)$$

where the $A_i(C)$ are concentration-dependent coefficients. The concentration dependence of the coefficients may be expressed by a similar polynomial:

$$A(C) = B_0 + B_1C + B_2C^2 + B_3C^3 \qquad (7.4)$$

where C is the concentration of the solute. Equation (7.3) represents an infinite number of calibration curves. In the treatment of the experimental elution curve, the correct calibration curve is determined for each point according to the actual concentration read off from detector response. This method, although complicated with respect to calculation, gives excellent results. A more detailed discussion of the elimination or quantitative evaluation of the concentration effects is given in Chapter 4.

7.3.5. Effect of the Injected Volume

The volume of the sample injected into the column contributes to the total broadening of the chromatographic band. The injected volume should be chosen so as to make band broadening due to injection negligible within the limits of experimental error. By taking the experimental error of determination of the retention volume as ±0.1%, it is possible to maintain the maximal band broadening due to injection negligible within these limits. Its value is reported as 4.5% of the total width of the elution curve [27]. From the practical standpoint, the volume of the injected sample should be maximally one-third of the volume of the outflowing low molecular weight standard (e.g., benzene) used in the determination of column efficiency [28].

V_R is also affected by the magnitude of the injected volume, in the sense that it is increased by one-half of the injected volume [14]. For this reason, a constant sample volume should be injected both during the calibration measurements and during the analysis.

For practical reasons, it is very important to know if it is better to use a smaller injected volume and a higher polymer concentration, or vice versa. Using theoretically derived relations regarding concentration effects [21,24,26], it is possible to calculate changes in V_R due to the individual contributing processes (e.g., changes in the effective size of macromolecules, the viscosity phenomenon) and the total change in V_R at various ratios of the injected volume and concentration, but at a constant injected amount. The results of calculations showed, and the experiments confirmed, that if band broadening due to injection can be neglected within the limits of experimental error, and the total amount of the injected polymer is constant, the ratio between concentration and the injected volume may within broad limits be varied without any pronounced effect on the result [27,29].

7.3.6. Effect of Sample Preparation

In some cases, the procedure of sample preparation may considerably affect the result of the SEC analysis. Often, the sample is not perfectly dissolved, filtration prior to injection removes the undissolved fractions, and the elution curve thus obtained does not correspond to the original polymer. Figure 7.4 shows chromatograms and apparent molecular weight values of a sample of cellulose nitrate obtained in THF after various times of dissolution of the sample. Quite obviously, the results depend very strongly on the time of dissolution of the sample. This means that sample preparation requires great care. Dissolution of samples in an ultrasonic bath has been found useful, but is not recommended for polymers of higher molecular

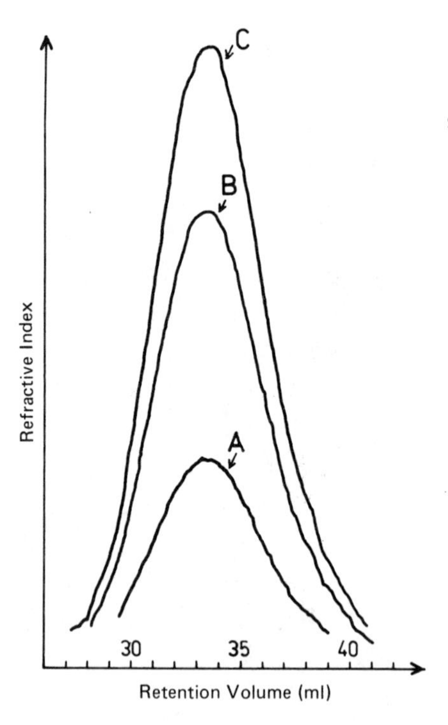

solvent THF 30°

	dissolution	$\overline{M_n}$	$\overline{M_w}$
A	apparent	97,000	220,000
B	1 hour	65,000	130,000
C	3 hours	40,000	100,000

Figure 7.4. Differences in SEC chromatograms and apparent molecular weights in dependence on the time of dissolution of the same cellulose nitrate sample in tetrahydrofuran. (From Ref. 6.)

weight (over 300,000), where the energy of the ultrasound leads to their degradation [6].

7.4. ERRORS INTRODUCED BY INTERPRETATION OF EXPERIMENTAL DATA

By obeying principles outlined in the preceding parts of this chapter, we obtain a correct elution curve as a result of SEC measurements. The curve must be subjected to treatment in order to obtain the required values (i.e., molecular weight averages or a suitable form of the distribution curve). Also, the data treatment may be subjected to a number of errors which more or less impair the results of a correctly performed measurement.

The whole procedure of interpretation of experimental data may be divided into two mutually closely connected stages. The first stage consists of the construction of a calibration plot and approximation of the calibration function. The second stage is the evaluation of a SEC chromatogram of the analyzed sample and calculation of the required quantities.

7.4.1. Construction of the Calibration Dependence and Approximation of the Calibration Function

The calibrations of a separation system are dealt with in greater detail in Chapter 2. Below, only those main principles are given that affect the accuracy of the results.

Calibration Standards

Molecular weights of standards as given by the manufacturer are not always correct, and in some cases the error may be as high as 20% [6,7,30]. For calibration measurements, sufficiently characterized and verified standards should be chosen, with the molecular weight values covering the separation range of the column system with sufficient density within the whole range. Usually, polystyrene standards are employed for universal calibration; they are the easiest to prepare with a sufficiently low polydispersity index within a wide range of molecular weights and they are also rather easy to characterize. If the calibration measurements are carried out with standards characterized by absolute methods, the SEC error in the case of a correctly performed measurement is much lower than the error of absolute methods (usually given as about 5%), and the calibration accuracy is determined by the accuracy of the absolute methods rather than by that of SEC [3].

If very narrow standards with respect to polydispersity are not available, a question arises as to what molecular weight value should be used for M_{peak} in the approximation of the calibration dependence. This problem is dealt with theoretically in Chapter 2. With respect to the minimization of errors in the most frequently applied principle of universal calibration, it was found that with the Wesslau distribution the best parameters for universal calibration are $[\eta]\sqrt{\bar{M}_n\bar{M}_w}$ (for a < 0.7) and $[\eta]\bar{M}_n$ (for a > 0.7). With the Schulz distribution, $[\eta]\bar{M}_w$ appears to be the best parameter [31,32].

In the calibration of the column system by means of polydisperse standards (see Chapter 2), errors that appear in the results of the final analyses of unknown samples ensue from inaccuracy in the characterization of these standards by employing absolute methods. Another error is caused by a possible difference between the MWD of polydisperse standards used in the calibration and that of the analyzed sample (e.g., one part of the distribution curve of the analyzed polymer lies beyond the calibrated range). If the inaccuracies of the chromatographic process itself are added, the total errors of determination of the final molecular weight averages and of the polydispersity index are given as 17-19% for the determination of $\bar{M}_n$, 14-16% for the determination of $\bar{M}_w$, and 8-10% for the determination of the polydispersity index [33].

Approximation of the Calibration Plot

Experimental points obtained in calibration measurements must be transformed into an analytical form of the function defined by equations (2.2) and (2.3) used in further calculations. An analysis of the problems shows that this function is usually adequately represented by a polynomial of third degree [30,34]. Here one should point out the danger involved in an unjustified linearization of the calibration function, which, although simplifying the calculation, nevertheless has as a consequence a distortion of the results, especially in marginal regions of the calibration curve. The same is also true for the "pseudolinear" columns [31,34,35].

When constructing a universal calibration plot, a suitable constant of the Mark-Houwink equation must be used in order to calculate the $[\eta]$ value of the calibration polymer. The problems involved are dealt with below.

7.4.2. Evaluation of the Elution Curve and Calculation of Molecular Weights

The usual procedure for evaluating the elution curve is described in Chapter 5. In this stage the accuracy and precision of SEC results may be negatively affected by a wrong choice of the number of segments of the chromatogram, by an uncertainty in the baseline due to the noise or instability of the detector, by a neglected or wrongly performed spreading correction, and by a wrong choice of the constants of the Mark-Houwink equation in the application of universal calibration.

Effect of the Number of Chromatogram Segments

The number of segments into which the elution curve is divided in the evaluation has a completely unambiguous effect on the accuracy of the results. With a single exception [36], model calculations concurrently indicate that with an increasing number of segments the relative error in the determination of various types of molecular weight averages decreases. Figure 7.5 shows deviations in percent in a simulated calculation of $\overline{M}_n$, $\overline{M}_w$, and $\overline{M}_z$ as functions of the number of segments. The figure shows a rise in accuracy with an increasing number of segments. At the same time, it can be seen

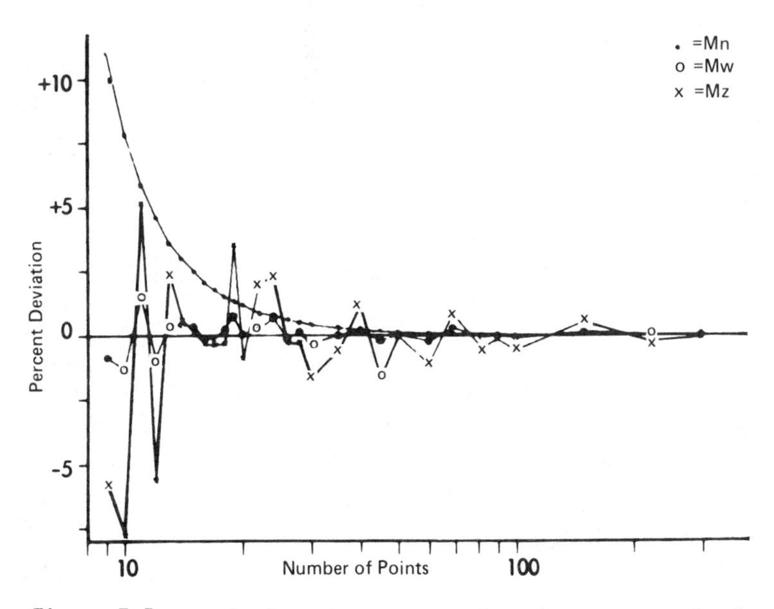

Figure 7.5. Deviations in percent of various types of molecular weight averages from the theoretical value in dependence on the choice of the number of segments of the elution curve. (From Ref. 37.)

that about 20 segments are sufficient to achieve satisfactory accuracy. A further increase in the number of segments, which raises the laboriousness of the calculation, does not lead to any further marked rise in accuracy [38].

At a constant number of segments, the error involved in the calculation of molecular weight averages is proportional to the magnitude of the polydispersity index, which probably is due to the increased possibility of neglecting marginal parts of the chromatogram. This is related to the stability of the baseline and to the detector noise, and is dealt with below. The error in the evaluation of the elution curve given for 20 segments and a polydispersity index of 2 is 2-3% relative [38].

Effect of Instability of the Baseline and of Detector Noise

Exact recording of the detector signal and distinction between marginal parts of the elution curve and the baseline (in very low concentration ranges of the solute in the mobile phase) are very important with respect to the accuracy of the results. Here an important role is played by the stability of the baseline and by the detector signal-to-noise ratio. Figure 7.6 shows a schematic view of a SEC chromatogram at various detector signal-to-noise ratios. At low values of this ratio, problems arise in evaluation of marginal parts of the chromatogram and in the determination of the heights or areas of the individual segments. At a constant noise value the inaccuracy in the determination of areas in the evaluation of the elution curve is inversely proportional to the amount injected (i.e., to the height of the elution curve) and is given at a signal-to-noise ratio of 20 (Figure 7.6b) as 3.8% relative, and at a signal-to-noise ratio of 100 (Figure 7.6a) as 1.2% relative [5]. The noise level is raised by poor functioning of the SEC apparatus, including detectors, and by not maintaining constant conditions of analysis, as discussed in Sections 7.2 and 7.3 (e.g., instability of the flow, temperature variations).

The error arising by neglecting marginal parts of the chromatogram and due to the imperfect stability of the baseline is particu-

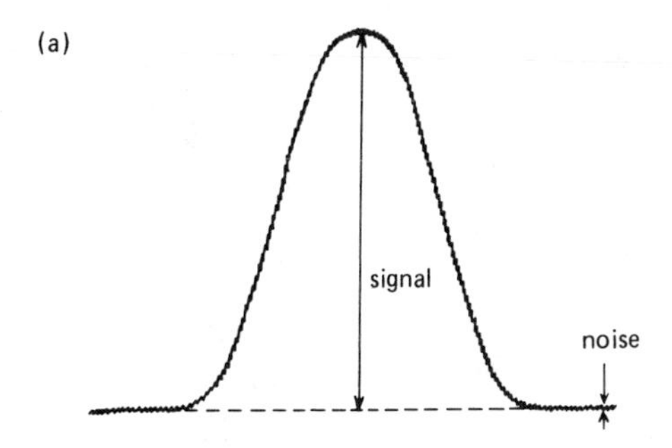

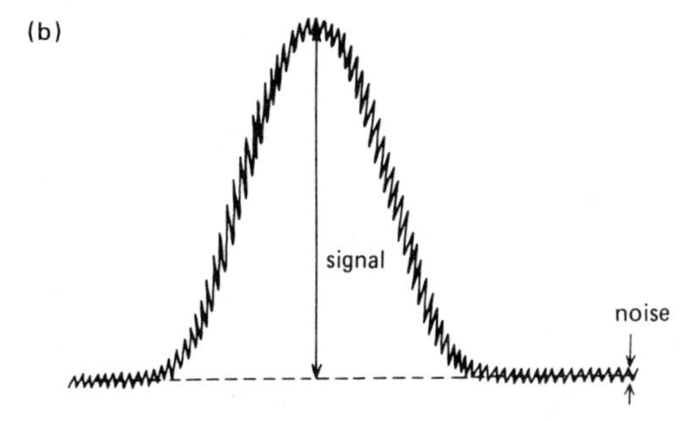

Figure 7.6. Schematic view of SEC chromatograms with various signal-to-noise ratios: (a) signal-to-noise ratio = 100; (b) signal-to-noise ratio = 20.

larly important with samples having a high polydispersity index and in the calculation of molecular weight averages of higher orders (e.g., $\overline{M}_z$) [31,37,38]. On the contrary, the results of model calculations for the log-normal and Poisson distribution indicate that for samples with low polydispersity indices ($I < 1.10$), this negligence does not lead to any important error [13].

Correction for Axial Dispersion

In exact calculations, especially with samples having lower polydispersity indices, the correction methods described in Chapter 3 for correction of axial dispersion in columns must often be used with respect to the total efficiency of the separation system [6]. A simple method allowing us to estimate if correction methods should be used with the column system in question is based on the determination of values of the symmetrical correction factor for spreading Λ and of the asymmetrical correction factor for spreading sk [39]. These correction factors are defined as

$$\Lambda = \frac{1}{2}\left[\frac{(\bar{M}_n)_{true}}{(\bar{M}_n)_{exp}} + \frac{(\bar{M}_w)_{exp}}{(\bar{M}_n)_{true}}\right] \tag{7.5}$$

$$sk = \frac{\Phi^* - 1}{\Phi^* + 1} \tag{7.6}$$

$$\Phi^* = \frac{(\bar{M}_n)_{true}(\bar{M}_w)_{true}}{(\bar{M}_n)_{exp}(\bar{M}_w)_{exp}} \tag{7.7}$$

molecular weight average values with the subscript exp are uncorrected values calculated directly from the chromatogram, while values with the subscript true are correct values obtained by, for example, applying absolute methods. If $\Lambda < 1.05$ and $sk < 0.05$, the correction procedures need not be applied, because corrected molecular weight average values differ from uncorrected ones by less than 5%. If one of the values of the correction factors is higher, a suitable correction procedure i.. used as described in detail in Chapter 3.

With high-performance columns having high efficiency, correction need not be used in some cases [8]. It seems of interest that also with a conventional column system (five columns 1200 x 9.5 mm, particle size of the packing 35 μm), a sufficient efficiency may be reached by reducing the flow rate of the mobile phase to 0.1 ml/min, and corrections may thus be avoided [40]. The time of analysis becomes inadequately long, however.

Choice of Constants of the Mark-Houwink Equation

When applying the principle of universal calibration to calculation, one should know [η] both for the calibration polymer and for the polymer undergoing the analysis. [η] is calculated using suitable forms of the Mark-Houwink equation (2.29) which hold for a certain polymer, temperature, and mobile phase.

If the constants K and a of this equation are taken from the literature for a certain system, there is a danger of error appearing in the final results, because the values of these constants often differ considerably for the same system. For this reason, only those constants K and a should be employed that have been properly checked by their use in SEC, and only in the range of molecular weights for which they have been determined (see Section 2.4). A table of values of constants successfully applied in SEC for various common polymer types has been compiled by the American Society for Testing and Materials (ASTM) [41].

If there are doubts regarding the suitability of constants K and a, for example because of a different structure of the polymer, different range of molecular weights, copolymer composition, and the like, it is better to check the reliability of results obtained by universal calibration by comparing them with values determined by absolute methods of measurement of molecular weights.

It was found during an investigation of the effect of the exponent a of the Mark-Houwink equation on the accuracy of determination of the polydispersity index of the polymer that the effect of the a value in polymers with $I < 1.1$ can be neglected. In the range $1.1 < I < 2.5$ an error in exponent a equal to ± 0.05 leads to an error in the determination of the polydispersity index in the limits of the usual experimental error. Only with an extremely high index value, $I > 6$, is the error due to an inaccurate a value of importance [42].

7.5. REPRODUCIBILITY OF MEASUREMENTS

In order to estimate possibilities offered by the SEC method for the characterization of polymers, one should know the reproducibility of measurements which may be achieved by carefully respecting the principles outlined above. From the practical standpoint, there exists *short-term reproducibility*, ranging between several hours and several days and affected only by correct functioning of the apparatus and by maintenance of constant conditions of the separation process, and *long-term reproducibility*, in an interval between several months and several years, where a gradual deterioration of columns becomes increasingly important.

7.5.1. Short-Term Reproducibility

A scatter of results obtained under constant conditions in the determination of short-term reproducibility determines the limit of precision of the method under the given conditions and may be regarded as the usual experimental error. The short-term reproducibility of V_R is given in the range 0.1-0.3% relative [5,14]. Short-term reproducibility in the determination of $\bar{M}_n$ and $\bar{M}_w$ has been studied in sufficient detail. As reported in the literature, the relative standard deviations in the determination of $\bar{M}_n$ and $\bar{M}_w$ vary in the range 3.1-6.8% and 1.2-5.0%, respectively, depending on the precision of the conditions used [43-47]. Also in an interlaboratory test carried out in eight laboratories, the scatter of the determined molecular weight averages varied within the same limits [48].

7.5.2. Long-Term Reproducibility

The long-term reproducibility of the results of measurements is strongly influenced, predominantly by the durability of the column packing. The latter depends on the quality of the gel used, the packing procedure, and the care taken in the operation. Columns may be impaired by, for example, gradual fouling of the packing with undissolved fractions in the injected samples, by plugging of the columns with reacting components of the sample, and by exceeding the maximal permitted pressure when the packing is compressed and mixing spaces arise. The packing has also been found to be impaired more quickly if elevated temperatures are used in the analysis of polyolefins [45,49].

The relative standard deviation observed in the investigation of the reproducibility of V_R of polystyrene standards within two years was higher than for short-term reproducibility and varied in the range 0.56-1.30% [50]. If some essential changes take place in the gel bed, the reproducibility of V_R is still poorer [14].

Even though with carefully used stable packings no major change in the reproducibility of measurement could be detected for several years, an occasional recalibration of the system is still recommended to obtain current information on its state [45]. Minor changes in V_R may be corrected by using an internal standard, and decreased efficiency may be amended by correction for axial dispersion with an altered dispersion factor. In the case of a major loss of efficiency, which suggests more important changes in the gel bed, it is better to repack or to exchange the column system.

REFERENCES

1. Bly, D. D., Stoklosa, H. J., Kirkland, J. J., and Yau, W. W., Anal. Chem., *47*, 1810, 1975.
2. Halász, I., and Vogtel, P., J. Chromatogr., *142*, 241, 1977.
3. Anderson, L., J. Chromatogr., *216*, 35, 1981.
4. Schulz, W. W., J. Liq. Chromatogr., *3*, 941, 1980.
5. Barth, H., Dallmeier, E., Courtois, G., Keller, H., and Karger, B. L., J. Chromatogr., *83*, 289, 1973.
6. Letot, L., Lesec, J., and Quivoron, C., J. Liq. Chromatogr., *3*, 1637, 1980.
7. Knox, J. H., Laird, G. R., and Raven, P. A., J. Chromatogr., *122*, 129, 1976.
8. Kato, Y., Kido, S., and Hashimoto, T., J. Polym. Sci., A2, *11*, 2329, 1973.
9. Mori, S., J. Chromatogr., *156*, 111, 1978.
10. Barrall, E. M., Cantow, M. J. R., and Johnson, J. F., J. Appl. Polym. Sci., *12*, 1373, 1968.
11. Mori, S., and Suzuki, T., J. Liq. Chromatogr., *4*, 1685, 1981.
12. Yau, W. W., Suchan, H. L., and Malone, C. P., J. Polym. Sci., A2, *6*, 1349, 1968.
13. Janča, J., and Klepárník, K., J. Liq. Chromatogr., *5*, 193, 1982.
14. Boni, K. A., Sliemers, F. A., and Stickney, P. B., J. Polym. Sci., A2, *6*, 1567, 1968.
15. Cooper, A. R., Johnson, J. F., and Bruzzone, A. R., Eur. Polym. J., *9*, 1381, 1973.
16. Gilpin, R. K., and Sisco, W. R., J. Chromatogr., *194*, 285, 1980.
17. Mori, S., and Suzuki, T., Anal. Chem., *52*, 1625, 1980.
18. Dawkins, J. V., J. Liq. Chromatogr., *1*, 279, 1978.
19. Mencer, H. J., and Grubisic-Gallot, Z., J. Liq. Chromatogr., *2*, 649, 1979.
20. Ho-Duc, N., Daoust, H., and Gourdenne, A., Polym. Prepr., *12*, 639, 1971.
21. Janča, J., J. Chromatogr., *134*, 263, 1977.
22. Janča, J., and Pokorný, S., J. Chromatogr., *148*, 31, 1978.
23. Janča, J., and Pokorný, S., J. Chromatogr., *156*, 27, 1978.

24. Janča, J., J. Chromatogr., *170*, 309, 1979.
25. Janča, J., and Pokorný, S., J. Chromatogr., *170*, 319, 1979.
26. Janča, J., Anal. Chem., *51*, 637, 1979.
27. Janča, J., J. Liq. Chromatogr., *4*, 181, 1981.
28. Yau, W. W., Kirkland, J. J., and Bly, D. D., *Modern Size-Exclusion Liquid Chromatography*, Wiley-Interscience, New York, 1979, p. 240.
29. Moore, J. C., Sep. Sci., *5*, 723, 1970.
30. Janča, J., Kolínský, M., and Mrkvičková, L., J. Chromatogr., *121*, 23, 1976.
31. Ogawa, T., and Inaba, T., J. Appl. Polym. Sci., *20*, 2101, 1976.
32. Lecacheux, D., Lesec, L., and Quivoron, C., J. Liq. Chromatogr., *5*, 217, 1982.
33. Pollock, M. J., MacGregor, J. F., and Hamielec, A. E., J. Liq. Chromatogr., *2*, 895, 1979.
34. Mori, S., and Suzuki, T., J. Liq. Chromatogr., *3*, 343, 1980.
35. Yau, W. W., and Fleming, S. W., J. Appl. Polym. Sci., *12*, 2111, 1968.
36. Cooper, A. R., and Matzinger, D. P., J. Liq. Chromatogr., *2*, 67, 1979.
37. Goedhart, D. J., J. Liq. Chromatogr., *2*, 1255, 1979.
38. Füzes, L., J. Liq. Chromatogr., *3*, 615, 1980.
39. Yau, W. W., Kirkland, J. J., and Bly, D. D., *Modern Size-Exclusion Liquid Chromatography*, Wiley-Interscience, New York, 1979, p. 324.
40. Tymczynski, R., and Turska, E., J. Liq. Chromatogr., *4*, 1491, 1981.
41. ASTM Method D 3593-77, Standard Method of Test for the Determination of Molecular Weight Averages and the Molecular Weight Distribution of Certain Polymers by Liquid Exclusion Chromatography Using Universal Calibration.
42. Janča, J., and Mrkvičková, L., Polymer, *20*, 388, 1979.
43. Duerksen, J. H., and Hamielec, A. E., Symp. Anal. GPC, Am. Chem. Soc., Chicago, 1967.
44. Nakajima, T., J. Appl. Polym. Sci., *15*, 3089, 1971.
45. Samay, G., and Füzes, L., J. Polym. Sci., Polym. Symp., *68*, 185, 1980.
46. Hill, J. A., Int. Semin. GPC, Boston, 1965.
47. Harmon, D. J., J. Appl. Polym. Sci., *11*, 1333, 1967.
48. Adams, H. E., et al., J. Appl. Polym. Sci., *17*, 269, 1973.
49. Hazell, J. E., Prince, L. A., and Stapelfeldt, H. E., J. Polym. Sci., C, *21*, 43, 1967.
50. Mukherji, A. K., Bertrand, J. C., and Allen, D. A., Angew. Makromol. Chem., *72*, 213, 1978.

INDEX